Unified Symmetry

In the Small and in the Large

Unified Symmetry

In the Small and in the Large

Edited by

Behram N. Kursunoglu

Global Foundation, Inc.
Coral Gables, Florida

Stephen Mintz

Florida International University
Miami, Florida

and

Arnold Perlmutter

University of Miami
Coral Gables, Florida

Plenum Press • New York and London

Library of Congress Cataloging-in-Publication Data

On file

Proceedings of an International Symposium on Unified Symmetry: In the Small and in the Large, held January 27–30, 1994, in Coral Gables, Florida

This volume was taken from a series of conferences sponsored by Global Foundation, Inc., Coral Gables, Florida

ISBN 0-306-44914-5

A Division of Plenum Publishing Corporation
233 Spring Street, New York, N. Y. 10013

10 9 8 7 6 5 4 3 2 1

Printed in the United States of America

PREFACE

The twenty-second Coral Gables conference "**UNIFIED SYMMETRY: In the Small and In the Large**" continued with the efforts to unify the small and the large. The information gathered with the Huble telescope has in part, in the absence of the SCC, provided a basis for the physicists to unify cosmology and elementary particle physics. The congressional cancellation of the biggest experimental project on the frontiers of physics should not be regarded as an insurmountable obstacle to progress in theoretical physics. The physicists' rise to prominence was mostly reached through their creation of the nuclear era. The post cold war era has somewhat reduced the political, military, and, in part, the social role of the physicist. Some in the administration and the Congress would like physicists to focus on the directly utilitarian aspects of science. Thus, some people do not realize that this regimentation of science would inhibit the creativity. The contributions of solid state physics research to the advancement of technology is the result of physics freely pursued independently of its applications. Modern Physics beginning with Newton's theory of gravity has enabled us to create the space age, to contribute to various technologies, and to impact on our technological modus vivendi.

"Big physics" is evolving into a joint international effort, and it will very likely adopt this approach at an increasing rate. What may be a "*useless science*" today could become the basis to open new technological paths, and new vistas tomorrow. Even without applications good science should be part of our cultural values. Thus, finding the top quark, looking for the Higgs or Planck's bosons, detecting gravitational waves, mapping of the genome of an organism, are examples of our scientific culture. Not only the physicists but non-physicists would also find great satisfaction in the achievement of unified symmetry in the large and in the small. It is this unification that will enable us to predict the nature of the very large from our complete understanding of the very small, and conversely, to predict the nature of the very small from the total knowledge of the very large. This is the same as the unification of cosmology with elementary particle physics. This has been and will continue to be the agenda of the Coral Gables conferences. Once again we offer these proceedings of the 1994 conference as the current contribution to the field of the unified theory and presently known experimental observations.

This conference was in part supported by grants from the National Science Foundation and the U.S. Department of Energy. The conference also received support from Martin Marietta Astronautics and Grumman Aerospace Corporations. Finally, the trustees of the Global Foundation wish to acknowledge gratefully the elegant dinner reception in honor of the conference participants given by Edward and Maria Bacinich in their beautiful Palm Beach home.

Editors
Coral Gables, Florida
July 7, 1994

Table of Contents

SECTION I

UNIFIED GENERAL RELATIVITY AND QUANTUM THEORY

UNIFIED SYMMETRY IN THE SMALL AND IN THE LARGE

Behram N. Kursunoglu
Global Foundation, Inc.
Coral Gables, Florida 33146-3164

ABSTRACT

The evolution of a fundamental length r_0 from its zero value to a value as large as the size of the universe forms the basis for the unification of the elementary particle physics and cosmology.

The four fundamental interactions are mediated by the *strategic triad* of massive and massless gauge bosons with spins 0, 1, 2. All of these gauge bosons are unified in a traceless symmetric tensor $T_{\mu\nu}$ appearing in the generalized Dirac wave equations derived from the generalized theory of gravitation which includes the coupling of the massive and massless triad of spins 0, 1, and 2 with spin $\frac{1}{2}$ quarks and leptons. The four massive spin 2 bosons, which generate the short-range part (Planck range) of the generalized gravitational field, each decay (10^{-43} sec.) into two spin 1 bosons and constitute the eight gluons of the strong force. The masses of all the gauge bosons, instead of Higgs boson, are generated by the spin 2 Planck bosons. Just as the strong force is the short-range part of the generalized gravitation, the weak force arises as the short-range part of the generalized electrodynamics. The electric field of the polarized vacuum as a dielectric medium of pairs of particles and antiparticles is calculated exactly.

This paper also contains an elementary computation for the size and total mass of the universe based on the generalized theory of gravitation which predicts, inter alia, a flat universe. The primordial free monopoles are synthesized into neutral dipoles of magnetic charges, followed by a supersymmetric transition into quarks and leptons, whose structure consists of neutral distribution of magnetic charges. A Planck length-size invariant singularity arising from a transient repulsive gravity energized the explosive creation of the universe.

A SYNOPSIS OF THE GENERALIZED THEORY OF GRAVITATION

"Bruria Kaufman, Einstein's last assistant, gave a summary of his [i.e. Einstein's] work at the 1955 relativity conference in Bern. In the clear and useful report is also found a comparison with near-simultaneous work on a nonsymmetric connection by Erwin Schrödinger and Behram N. Kursunoglu. Schrödinger treats only

Unified Symmetry: In the Small and in the Large,
Edited by B.N. Kursunoglu *et al.*, Plenum Press, New York, 1995

the connection as primary and introduces the fundamental tensor via the cosmological term device of Eddington. Kursunoglu's theory is more like Einstein's but contains one additional parameter."

The above quotation is on page 348 of the book by Abraham Pais, "Subtle is the Lord... The Science and the Life of Albert Einstein" Oxford University Press, New York, 1982. The last word "parameter" in the quotation refers to the fundamental length r_0 in my version of the nonsymmetrization of general relativity which has opened new horizons for the unification of the two long-range forces of gravitation and electromagnetism. This work, performed during 1950-1952, formed the basis of my Ph.D. thesis of Cambridge University. In a long conversation on November 19, 1953 in Einstein's home in Princeton, his parting remarks to me were "Your theory because of r_0 is more general than mine, but time will show which one of us is right." Over the past forty years I have made some progress in confronting my theory with a number of experimental facts. The existence of a correspondence principle shows that in the limit $r_0 = 0$ the theory reduces to general relativity in the presence of electromagnetic fields. The absence of r_0 in Einstein's and Schrödinger's versions made it impossible to obtain physical interpretations for their theories. One main requirement to be met by all such theories was the derivation of the classical equations of motion of an electric charge in an external field. I had shown the absence of any motion in Einstein's and Schrödinger's versions of the theory. However, I was able to derive Lorentz's equations of motion for my theory. In a letter to me dated October 15, 1951, Schrödinger expressed his amazement at the dynamical contents of my theory.

The theory is based on a hermitan nonsymmetric 16 component tensor $\hat{g}_{\mu\nu} = g_{\mu\nu} + iq_0^{-1}\Phi_{\mu\nu}$ and its *supersymmetric* version, obtained by the transformations $r_0 \to ir_0$, $q_0^{-1} \to iq_0^{-1}$, where now $\hat{g}_{\mu\nu} = g_{\mu\nu} + q_0^{-1}\Phi_{\mu\nu}$ is nonhermitian. The constant q_0 has the dimensions of an electric field and therefore the antisymmetric tensor term $q_0^{-1}\Phi_{\mu\nu}$ is, like $g_{\mu\nu}$, a dimensionless quantity. The correspondence principle of the theory yields the fundamental relation

$$r_0^2 q_0^2 = \frac{2G}{c^4} \tag{1}$$

or, more explicitly, the energy density q_0^2 is related to the maximum curvature r_0^{-2} by the relation

$$q_0^2 = \rho_0 c^2 \tag{2}$$

where the mass density ρ_0 is given by

$$\rho_0 = \frac{c^2}{2G} r_0^{-2} \tag{3}$$

and where c and G represent the speed of light and the gravitational constant, respectively. The emergence of $\Phi_{\mu\nu}$ is due to the creation of electric and magnetic charges e and g, respectively, from the vacuum. The quantity q_o^2 should have had the value of 10^{114} erg, cm^{-3} when r_o was of the order of a Planck-length during the Planck time of 10^{-43} sec. When applied to the universe today the observed values $q_o^2 \sim 10^{-8}$ erg. cm^{-3} yields, as follows from (1), for the current size of the expanding universe the value $r_o \sim 10^{10}$ light years. For the total mass of the universe, volume integration of (2), yields the result

$$M_u = \frac{c^2}{2G} r_o \sim 10^{22} \text{ solar masses.} \tag{4}$$

At the primordial time the space-time curvature of the universe of Planck-length size was $r_o^{-2} \sim 10^{66}$ cm^{-2} as opposed to the curvature of the present universe $r_o^{-2} \sim 10^{-54}$ cm^{-2}. This demonstrates the extreme importance of the roles played by r_o. Thus, what started as the assumption that the fundamental length was a small ($\sim 10^{-33}$ cm) quantity has now been superseded by the fact that it can also be as large as the size of the universe.

The spherically symmetric field equations for both the hermitian and nonhermitian cases for $r_o = 0$ reduce to general relativity. However, when r_o is of the order of the size of the universe the field equations yield flat space-time solutions. Therefore, the theory predicts a flat universe consistent with observations. This result must not be confused with the flat space-time solutions of general relativity in the absence of matter. The flat space-time solutions in general relativity depend on the initial conditions.

The theory predicts monopoles bound in the form of dipoles, or stratified neutral layered distributions. There exist no free monopoles, except in the early universe when primordial matter consisted of positive and negative magnetic charges of various fractional spins, and sizes. All these monopoles under the action of very short-range forces synthesized into integral spin dipoles, and half integral spin quarks and leptons. The recently discovered (so called) top quark, is almost 200 times more massive than the proton, which does not seem consistent with a point like object. Instead it seems that, like all other quarks it should have a a structure to carry the spin $\frac{1}{2}$. All fermions obtain their spin from the sum total of their fractional spin monopole constituents where

$$\lim_{n \to \infty} g_n = 0, \quad \sum_0^\infty g_n = 0, \quad \sum_0^\infty g_n^2 = \frac{1}{2}\hbar c, \tag{5}$$

where the third equation is an Ansatz or a conjecture.

In a 1973 letter Paul Dirac advised me to look for short-range interactions in my theory. The ensuing effort resulted in the two coupling constants

$$\alpha_w = \frac{e}{\sqrt{(e^2+g^2)}}, \qquad \alpha_s = \frac{g}{\sqrt{(e^2+g^2)}}, \tag{6}$$

where e and g represent electric and magnetic charges, respectively. In view of the space dependence of the magnetic charge g for short distances, both α_w and α_s behave like *running coupling constants.* A detailed discussion of the short-range forces appeared in the book "Reminiscences About a Great Physicist: Paul Adrien Maurice Dirac" edited by Behram N. Kursunoglu and Eugene P. Wigner, Cambridge University Press, 1987, Cambridge. The letters from Einstein, Schrödinger, and Dirac are published in my paper entitled "After Einstein and Schrödinger: A New Unified Field Theory" *Journal of Physics Essays* vol. 4, No. 4, pp 439-518, 1991, printed in Canada. See also the author's paper in the proceedings of the 1993 Coral Gables Conference, Nova Science Publishers, New York. In addition many other communications, conversations with various physicists will appear in a book "*The Ascent of Gravity*: The Story of the Physics and the Human Side of a Unified Field Theory, 1950-present" in preparation for publication in 1995.

The theory yields generalized Dirac wave equations in quadruplet form describing the coupling of the spin triad of (0,1,2) massive and massless particles to matter as the carriers of all forces. There exists four spin 1 massive (weak bosons) and a massless (photon), four spin 2 massive bosons (decaying into eight spin 1 bosons), and a spin 2 massless boson (graviton). Of the four equations two contain terms violating CP symmetry. All of these properties are derived without the use of internal symmetry breaking. It appears that the breaking of internal symmetry, as a basis for unification, does not allow gravity to be part of a complete unified field theory. There appears to be no room in the standard model for the inclusion of gravity. The quest for a unified field theory has, in this approach, been based on general relativity which has enabled us to derive generalized Dirac wave equations. The reverse of this procedure is equivalent to the current efforts based on quantum gravity for a unified field theory.

Why has the above rather beautiful theory based on the generalization of general relativity been ignored by some and even received in a hostile manner? Is it the mindset of people whose long time trodden path has already become a trench deep enough to block their view of any ideas lying outside of the old path? In my 1953 visit with Einstein I told him that most physicists did not believe in his unified field theory. He replied, "After 1905, when my work began appearing on the special theory of relativity, I used to visit the University and join my physicist colleagues for afternoon tea. They used to talk about the special theory of relativity and they used to laugh, and I used to laugh with them." When he finished his story, Einstein had a broad smile on his face.

I. INTRODUCTION

Gravitation, although the weakest force of all, is coupled to all interactions. Therefore, we ask why gravitation was not adopted as the basis for the unification of all fundamental interactions? One basic reason is the fact that Einstein's theory of gravitation does not include the behavior of gravity in the Planck scale $[r_p=\sqrt{(\frac{\hbar G}{c^3})}]$ space time structure, the knowledge of which behavior is of decisive and crucial importance for the unification of gravity with the rest of physics. The Planck-scale gravitational field was, in fact, the result of the unification of gravitation and electromagnetism. The ununified gravitation and electromagnetism are described as long (infinite) range forces without any short-range components. Thus, the associated quanta are the massless graviton (spin 2) and photon (spin 1), respectively. Naturally, one wonders about the absence of short-range parts of these two long-range fundamental force fields.

The unification of gravitation and electromagnetism does reveal the existence of short-range components of the two long-range fields. The short-range gravity is manifested as a strong force mediated by a massive [i.e., $M_p = \sqrt{(\frac{\hbar c}{G})}$] spin 2 boson (or Planck boson) and it decays within Planck-time interval $\tau_p = \frac{1}{c}\sqrt{(\frac{\hbar G}{c^3})} \cong 10^{-43}$ sec. into either two spin 1 bosons (gluons) or four spin $\frac{1}{2}$ quarks. A decay into spin $\frac{3}{2}$ and spin $\frac{1}{2}$ composite systems is also possible. The spin 1 decay products may be identified as the conventional bosons binding quarks to constitute elementary particles. However, we must not miss the fact that the spin 1 bosons themselves have constituents the spin $\frac{1}{2}$ quarks and therefore the nonlinearity of the theory is a natural result.

The Standard Theory's assumed existence of the so-called Higgs boson to break the symmetry may have resulted from the absence of gravitation in the electroweak theory. The present unified theory yields four massive spin 0 bosons and four Planck-massive spin 2 bosons, along with four massive spin 1 bosons mediating weak interactions. Because of the decay of spin 2 bosons we have altogether eight spin 1 bosons (or gluons) mediating strong interactions. In this theory, unlike the Standard Theory, the weak force emerges as the short-range part of the generalized electromagnetic field and is mediated by four massive spin 1 weak bosons instead of the usual three weak bosons $W_{\pm}$, Z predicted by the Standard Theory. The implied *fourth generation* in the above statements will be discussed further when the derivation of the generalized Dirac equation is presented in the following sections.

II. UNIFIED SYMMETRY BREAKING

In contrast to internal compact gauge symmetry breaking we have based the unification of all fundamental interactions on gauge symmetry breaking which includes non-compact symmetry groups. It is in the Planck-scale physics (i.e., 10^{19} Gev or 10^{-33} cm) that all interactions are unified as, for example, prevailed during the instant of the big bang creation of the universe. According to the present theory at the instant of creation the energy density is given by,

$$q_o^2 = \frac{\pi}{3}\frac{1}{V_p}\left[\frac{GM^2(r_o)}{r_p}\right], \tag{II.1}$$

where r_p is the Planck length and $V_p = \frac{4\pi}{3} r_p^3$ is the Planck volume. The result (II.1) can also be expressed in terms of the generalized electromagnetic energy density of this theory. By using the special value of the fundamental length, r_o, $r_o^2 = \frac{2G}{c^4} N^2 (e^2+g^2)$, obtained from the solutions of the field equations, we can derive

$$q_o^2 = \frac{\pi}{3N^2}\left[\frac{c\hbar}{e^2+g^2}\right]^2 \frac{1}{V_p}\,\frac{e^2+g^2}{r_p}\,, \tag{II.2}$$

where e and g represent electric and magnetic charges, respectively, and where G is the usual gravitational constant and N ranges over the interval from 0 to 10^{60}, i.e., $0 \le N \le 10^{60}$.

The mass $M(r_o)$ in (II.1) represents a spin 2 boson mass and is related to a Planck boson mass generating the short-range gravitational field or, more precisely, the strong force between the constituents of an elementary particle. The term $GM^2(r_o)/r_p$ in (II.1) can be interpreted as an electromagnetic or short-range gravitational *self-energy* of the vacuum which, for $r_o = 0$, tends to infinity.

The mass $M(r_o)$ is defined in terms of the fundamental length r_o by the Compton wave-length expression

$$\frac{1}{2} r_o^2 = \left[\frac{\hbar}{M(r_o)c}\right]^2 ; \tag{II.3}$$

the reason for this choice will be apparent when we obtain a wave equation from the linearized field equations. The combined form of (II.1) and (II.3) is the statement

$$q_o^2 r_o^2 = \frac{c^4}{2G}\,, \tag{II.4}$$

that could have been written down, independently of (II.1) and (II.3), as the required condition to fulfill the *correspondence principle*, where in the limit $r_o = 0$ or $q_o = \infty$ the present unified field theory reduces to general relativity, and which also includes electromagnetic field without electric and magnetic charges. The constant q_o has the dimensions of an electric field. The relation (II.4) first appeared in my 1952 Phys. Rev. paper (see references in ref. 1). The relation (II.4) can also be interpreted as an *equation of state*. Furthermore, the right-hand-side of (II.4) may be representing a constant *cosmic force* to govern the expansion, steady state, (or collapse) of the universe.

Is it fully justifiable to start with quantum theory and some internal symmetry breaking schemes to unify fundamental interactions that include gravity? The conventional approach produced the electroweak theory, QCD, and GUT, its methodologies led to the concepts of supergravity, superstrings, supersymmetry, but did not accomplish the unification of gravity with the other three fundamental interactions. These theories could not sublimate general relativity to quantum theory and the quantization of gravitation.

We should begin by recognizing that general relativity is the progenitor of this theory which uses Planck-scale physics to unify all fundamental interactions. In what follows, I shall lay the foundations of this point of view, which was discussed in a conservative way in reference 1 in 1991, where *quantum theory is detivable from the generalized theory of gravitation*. The mathematical precursor of this theory was published in early 1950's when I was a graduate student in Cambridge in the P.A.M. Dirac era. A generalized Dirac equation for leptons and quarks is derived and it includes the force fields mediated by the massive and massless spin 0, 1, 2 bosons.

In the present theory, the resulting unification of all interactions includes the weak interaction as the parity and charge conjugation violating short-range part of the generalized electromagnetic interaction, unified with the generalized gravitational force. The short-range part of this force includes the strong force mediated by the four spin 2 Planck bosons decaying into eight spin 1 bosons (gluons) in a Planck-time span of 10^{-43} sec. The same spin 1 bosons mediate interactions between quarks and their own constituents. The short-range part of the generalized gravitational interaction include massive spin 0, 1, 2 bosons represented by a symmetric, traceless second rank tensor of the same form as the usual energy momentum tensor of the electromagnetic field (see reference 1).

In order to see the nature of non-compact gauge symmetry breaking we shall use the local gauge invariance properties of the gravitational field as defined by H. Weyl according to

$$g_{\mu\nu} = < e_{\mu} \, |F| \, e_{\nu} > = \eta_{ab} \, e^{a}_{\mu} \, e^{b}_{\nu} , \tag{II.5}$$

where η_{ab} represent the usual flat metric. The $g_{\mu\nu}$ are locally gauge invariant. Thus, under a Lorentz transformation of the "vierbein" $|e_{\mu}>$

$$|e_{\mu}'> = L \, |e_{\mu}> , \tag{II.6}$$

we obtain

$$\tilde{L} \, F \, L = F , \tag{II.7}$$

with F representing the flat metric η_{ab}, we see that the gauge group consists of homogeneous Lorentz transformations belonging to the non-compact group SO (3,1).

We may, by choosing complex vierbein, switch on electric and magnetic charges and obtain in place of the symmetric tensor $g_{\mu\nu}$ a hermitian tensor

$$\hat{g}_{\mu\nu} = < Z_{\mu} \, |F| \, Z_{\nu} > , \tag{II.8}$$

where

$$|Z_{\mu}> = |e_{\mu}> + i \, |f_{\mu}> , \quad <Z_{\mu} \, | = <e_{\mu}| - i < f_{\mu} \, | , \tag{II.9}$$

and where the hermitian tensor $\hat{g}_{\mu\nu}$ [= $(\hat{g}_{\mu\nu})^{\dagger}$] is a nonsymmetric complex tensor of the form

$$\hat{g}_{\mu\nu} = g_{\mu\nu} + i \, q_{o}^{-1} \, \Phi_{\mu\nu} . \tag{II.10}$$

The transition from the symmetric tensor $g_{\mu\nu}$ to a nonsymmetric tensor $\hat{g}_{\mu\nu}$ can be looked upon as a symmetry breaking process. The constant q_{o}, as defined by (II.4), is introduced to define $\hat{g}_{\mu\nu}$ as a dimensionless 16-component field variables of the generalized theory of gravitation replacing the symmetric field variables $g_{\mu\nu}$ of general relativity. The antisymmetric tensor $\Phi_{\mu\nu}$ ($= -\Phi_{\nu\mu}$) has the dimensions of an electric field and it represents a generalized electromagnetic field.

From the definition (II.8) we see that the local gauge invariance group SO(3,1) of $g_{\mu\nu}$ is now replaced by the local gauge invariance group U(3,1) = SU(3,1)xU(1). When the real and imaginary parts of (II.10) are separated, the symmetric and antisymmetric parts $g_{\mu\nu}$ and $\Phi_{\mu\nu}$, respectively, break the non-compact gauge symmetry U(3,1). Thus, based on an extended local gauge invariance requirement we have generalized the symmetric irreducible (under general coordinate transformations) field variables $g_{\mu\nu}$ into a hermitian tensor $\hat{g}_{\mu\nu}$ [$=(\hat{g}_{\mu\nu})^{\dagger}$] which is reducible under general coordinate transformations, where symmetric and antisymmetric parts transform separately. However, the $\hat{g}_{\mu\nu}$ remains irreducible under the local gauge symmetry of invariance with respect to the transformations belonging to the noncompact gauge group U(3,1). We shall shortly show that such gauge invariance is a consequence of turning on electric and magnetic charges by the transition $0 \to r_o$, to yield the result $g_{\mu\nu} \to \hat{g}_{\mu\nu}$.

The choice of a hermitian tensor (II.10) as the generalized field variables is not the only possibility. A *supersymmetry* operation (not related to the conventional supersymmetry) defined by the transitions $r_o \to ir_o$, $q_o \to iq_o$, would replace the hermitian tensor (II.10) by the non-hermitian tensor

$$\hat{g}_{\mu\nu} = g_{\mu\nu} + q_o^{-1}\Phi_{\mu\nu}, \qquad \text{(II.11)}$$

for which the corresponding generalized theory also satisfies the *correspondence principle* by yielding in the limit $r_o = 0$ or $q_o = \infty$, general theory of relativity. Thus, the *supersymmetry* operation is a transition from the hermitian (non-hermitian) to the non-hermitian (hermitian) field tensor representing *Bose-Einstein* (integral spin) and *Fermi-Dirac* (half integral spin) branches of the generalized theory of gravitation. The local gauge symmetry for the non-hermitian tensor (II.11) is the invariance with respect to the gauge transformations belonging to the group SO(3,1)xO(2), which, when the symmetric and antisymmetric parts are separated, becomes a *broken symmetry.* In particular, breaking of the non-compact gauge symmetries SU(3,1)xU(1) and SO(3,1)xO(2) with the compact parts U(1) and SO(2), yield conservation of magnetic and electric charge distributions. Furthermore, the confined magnetic monopoles of infinite varieties with special neutral distributions constitute the structure of quarks as fermions and their special combinations as bosons. The details of these gauge symmetries are discussed in reference 1.

III. THE FIELD EQUATIONS

The field equations of the generalized theory of gravitation based on the correspondence principle were derived in reference 1. The uniquely established Lagrangian is given by

$$\mathcal{L} = \hat{\mathbf{g}}^{\mu\nu}(\hat{R}_{\mu\nu} - r_o^{-2} q_o^{-1} F_{\mu\nu}) + 2r_o^{-2}[\sqrt{(-\hat{g})} - \sqrt{(-g)}], \tag{III.1}$$

which results from the nonsymmetrization of the general relativity Lagrangian, where the nonsymmetric hermitian or non-hermitian curvature tensor $\hat{R}_{\mu\nu}$ is defined, in terms of the nonsymmetric affine connection $\Gamma^{\rho}_{\mu\nu}$, by

$$\hat{R}_{\mu\nu} = -\Gamma^{\rho}_{\mu\nu,\rho} + \Gamma^{\rho}_{\mu\rho,\nu} - \Gamma^{\rho}_{\mu\nu}\Gamma^{\sigma}_{\rho\sigma} + \Gamma^{\rho}_{\mu\sigma}\Gamma^{\sigma}_{\rho\nu} \tag{III.2}$$

and where

$$\hat{\mathbf{g}}^{\mu\nu} = \sqrt{(-\hat{g})}\,\hat{g}^{\mu\nu}, \quad \mathrm{Det}\, g_{\mu\nu} = g, \tag{III.3}$$

$$\hat{g} = \mathrm{Det}\,\hat{g}_{\mu\nu} = g\,(1 \pm q_o^{-2}\,\Omega - q_o^{-4}\,\Lambda^2), \tag{III.4}$$

where the + or − is chosen for non-hermitian or hermitian $\hat{g}_{\mu\nu}$, respectively. The auxiliary field $F_{\mu\nu}$ is defined by

$$F_{\mu\nu} = \partial_\mu A_\nu - \partial_\nu A_\mu\,. \tag{III.5}$$

Now, because of the extensive use of the various quantities in the theory we must include in this paper the following definitions:

$$\Omega = \tfrac{1}{2}\Phi^{\mu\nu}\Phi_{\mu\nu}, \quad \Lambda = \tfrac{1}{4} f^{\mu\nu}\Phi_{\mu\nu}, \quad f^{\mu\nu} = \frac{1}{2\sqrt{(-g)}}\,\epsilon^{\mu\nu\rho\sigma}\Phi_{\rho\sigma}, \tag{III.6}$$

$$\hat{g}^{\nu\rho}\,\hat{g}_{\mu\rho} = \delta^{\nu}_{\mu}, \tag{III.7}$$

$$\hat{\mathbf{g}}^{\underline{\mu\nu}} = \hat{\mathbf{g}}^{\underline{\nu\mu}} = \sqrt{(-g)}\,\underline{b}^{\mu\nu}, \quad \underline{b}^{\mu\nu} = \frac{(1 \pm 1/2\, q_o^{-2}\Omega)\; g^{\mu\nu} \mp q_o^{-2}\, T^{\mu\nu}}{\sqrt{(1 \pm q_o^{-2}\Omega - q_o^{-4}\,\Lambda^2)}}, \tag{III.8}$$

$$\underline{b}^{\nu\rho} b_{\mu\rho} = \delta^{\nu}_{\mu}, \quad b_{\mu\nu} = \frac{(1 \pm 1/2\, q_o^{-2}\Omega)\, g_{\mu\nu} \mp q_o^{-2}\; T_{\mu\nu}}{\sqrt{(1 \pm q_o^{-2}\Omega - q_o^{-4}\Lambda^2)}}, \tag{III.9}$$

$$T^{\nu\rho}\, T_{\mu\rho} = \delta^{\nu}_{\mu}\,(\tfrac{1}{4}\Omega^2 + \Lambda^2),\; T_{\mu\nu} = \tfrac{1}{2}\Omega\, g_{\mu\nu} - \Phi_{\mu\rho}\Phi_{\nu}{}^{\rho},\; T^{\rho}_{\rho} = g^{\mu\nu} T_{\mu\nu} = 0\,, \tag{III.10}$$

$$\hat{g}^{\mu\nu} = \sqrt{(-g)}\,\frac{\Phi_{\mu\nu} \mp q_o^{-2}\Lambda\ f_{\mu\nu}}{\sqrt{(1 \pm q_o^{-2}\Omega - q_o^{-4}\Lambda^2)}}\,, \tag{III.11}$$

The Lagrangian (III.1) in the correspondence limit r_o=0 reduces to the Lagrangian of general relativity. The two supersymmetric branches can be expressed as

$$\hat{g}^s_{\mu\nu} = g_{\mu\nu} + i^s q_o^{-1}\Phi_{\mu\nu}\,, \tag{III.12}$$

where s = 0,1 for the non-hermitian and hermitian field variables, respectively. The nonsymmetric affine connection $\Gamma^\rho_{\mu\nu}$, as obtained from varying the action function

$$S = \int \mathcal{L}\, d^4 x\,, \tag{III.13}$$

with respect to $\Gamma^\rho_{\mu\nu}$, are to be calculated from the 64 algebraic equations

$$\hat{g}_{\mu\nu,\rho} - \hat{g}_{\sigma\nu}\Gamma^\sigma_{\mu\rho} - \hat{g}_{\mu\sigma}\Gamma^\sigma_{\rho\nu} = 0\,. \tag{III.14}$$

Furthermore, the curvature tensor $\hat{R}_{\mu\nu}$ is invariant under the gauge transformation

$$\Gamma^\rho_{\mu\nu} \longrightarrow \Gamma^\rho_{\mu\nu} + \delta^\rho_\mu\,\partial_\nu\lambda\,, \tag{III.15}$$

but this property does not hold for general relativity where one deals with the symmetric connection. For real $\hat{g}_{\mu\nu}$ the curvature tensor $\hat{R}_{\mu\nu}$ is transposition invariant viz.,

$$[\tilde{\hat{R}}_{\mu\nu}] = \hat{R}_{\nu\mu}(\tilde{\Gamma}) = \hat{R}_{\mu\nu}(\Gamma)\,,$$

provided

$$\Gamma_\rho = \Gamma^\sigma_{\underset{\vee}{\rho\sigma}} = 0\,. \tag{III.16}$$

This condition, as can be seen from the contravariant form of (III.14),

$$\hat{g}^{\mu\nu}{}_{,\rho} + \hat{g}^{\sigma\nu}\Gamma^\mu_{\sigma\rho} + \hat{g}^{\mu\sigma}\Gamma^\nu_{\rho\sigma} - \hat{g}^{\mu\nu}\Gamma^\sigma_{\underline{\rho\sigma}} = 0\,, \tag{III.17}$$

yield the four field equations

$$\hat{g}^{\mu\nu}{}_{\vee,\nu} = 0\,. \tag{III.18}$$

The field equations (III.18) also follow from varying the action function S with respect to the extra field variables A_μ. The remaining field equations are obtained , by varying S with respect to $\hat{g}^{\mu\nu}$, as

$$\hat{R}_{\mu\nu} = r_o^{-2} (\hat{b}_{\mu\nu} - \hat{g}_{\mu\nu}) , \tag{III.19}$$

where

$$\hat{b}^s_{\mu\nu} = b_{\mu\nu} + i^s q_o^{-1} F_{\mu\nu} . \tag{III.20}$$

By separating out symmetric and antisymmetric parts in (III.19) and eliminating $F_{\mu\nu}$ we obtain the final field equations as

$$\hat{R}_{\underline{\mu\nu}} = r_o^{-2} (b_{\mu\nu} - g_{\mu\nu}) , \tag{III.21}$$

$$\hat{R}_{\overset{\vee}{\mu\nu},\rho} + \hat{R}_{\overset{\vee}{\nu\rho},\mu} + \hat{R}_{\overset{\vee}{\rho\mu},\nu} + r_o^{-2} I_{\mu\nu\rho} = 0 , \tag{III.22}$$

$$\hat{g}^{\overset{\vee}{\mu\nu}}{}_{,\nu} = 0 . \tag{III.23}$$

Because two differential identities are obtainable from (III.22) and (III.23), the eighteen field equations for the sixteen field variables $\hat{g}_{\mu\nu}$ are equivalent to sixteen independent field equations. The four component fully antisymmetric conserved quantities $I_{\mu\nu\rho}$ represent magnetic current density.

$$4\pi s^\mu = \frac{1}{3!} \epsilon^{\mu\nu\rho\sigma} I_{\nu\rho\sigma} , \tag{III.24}$$

where

$$I_{\mu\nu\rho} = \Phi_{\mu\nu,\rho} + \Phi_{\nu\rho,\mu} + \Phi_{\rho\mu,\nu} , \tag{III.25}$$

and where, as follows from the third expression in (III.6) and (III.24) above, the magnetic current density can also be expressed as

$$s^\mu = \frac{1}{4\pi} \frac{\partial[\sqrt{(-g)} f^{\mu\nu}]}{\partial x^\nu} , \tag{III.26}$$

where $f^{\mu\nu}$ is the dual tensor of $\Phi_{\mu\nu}$ as defined in (III.6) above. The definition (III.26) for the magnetic charge current density s^μ does not imply the presence of free monopoles. This can be seen by using the antisymmetric part of the field equations (III.19),

$$r_o^2 \hat{R}_{\underset{\vee}{\mu\nu}} + \Phi_{\mu\nu} = F_{\mu\nu}, \tag{III.27}$$

the dual form of which is given by

$$r_o^2 \hat{R}_{*}{}^{\underset{\vee}{\mu\nu}} + f^{\mu\nu} = F_*^{\mu\nu}, \tag{III.28}$$

where

$$\hat{R}_{*}^{\underset{\vee}{\mu\nu}} = \frac{1}{2\sqrt{(-g)}} \epsilon^{\mu\nu\rho\sigma} \hat{R}_{\underset{\vee}{\rho\sigma}}, \qquad F_*^{\mu\nu} = \frac{1}{2\sqrt{(-g)}} \epsilon^{\mu\nu\rho\sigma} F_{\rho\sigma} \tag{III.29}$$

The total electric current J^μ is defined by

$$J^\mu = J_e{}^\mu + J_o{}^\mu, \tag{III.30}$$

where

$$J_e{}^\mu = \frac{1}{4\pi} \frac{\partial}{\partial x^\nu} [\sqrt{(-g)} \Phi^{\mu\nu}], \tag{III.31}$$

represents the electric current density of an electrically charged particle and where

$$J_o{}^\mu = \frac{r_o^2}{4\pi} \frac{\partial}{\partial x^\nu} [\sqrt{(-g)} \hat{R}^{\underset{\vee}{\mu\nu}}], \tag{III.32}$$

represents the vacuum electric current density. The vacuum consists of particles and antiparticles. However, the total magnetic current density as defined by

$$S^\mu = \frac{1}{4\pi} \frac{\partial}{\partial x^\nu} [\sqrt{(-g)} F_*{}^{\mu\nu}] = 0, \tag{III.33}$$

vanishes, where we used the field equations (III.28), and (III.22) along with the equations

$$F_{\mu\nu,\rho} + F_{\nu\rho,\mu} + F_{\rho\mu,\nu} = 0, \tag{III.34}$$

or their dual form

$$\frac{\partial}{\partial x^\nu} [\sqrt{(-g)} F_*{}^{\mu\nu}] = 0. \tag{III.35}$$

Thus, there exists no free monopoles. The vacuum magnetic current density

$$s_*^\mu = \frac{1}{4\pi} \frac{\partial}{\partial x^\nu} [r_o^2 \sqrt{(-g)} \hat{R}_*^{\underset{\vee}{\mu\nu}}], \tag{III.36}$$

constitutes a magnetic charge neutral structure where

$$s^{\mu} + s_{*}^{\mu} = 0, \tag{III.37}$$

are the dual form of the field equation (III.22).

IV. SPHERICALLY SYMMETRIC FIELD

In the non-hermitian case for the spherically symmetric time-independent field variables we have the field equations,

$$\frac{1}{2} r_o^2 \frac{d}{d\beta} [S \exp(\rho) \frac{d\Phi}{d\beta}] = R_o^2 \cos \Phi + \ell_o^2 \sin \Phi , \tag{IV.1}$$

$$\frac{1}{2} r_o^2 \frac{d}{d\beta} [S \exp(\rho) \frac{d\rho}{d\beta}] = - R_o^2 \sin \Phi + \ell_o^2 \cos\Phi - \exp(\rho) , \tag{IV.2}$$

$$\frac{1}{2} r_o^2 \frac{d}{d\beta} [\frac{dS}{d\beta} \exp(\rho)] = \exp(\rho) [1 - \frac{\sin\Phi \exp(\rho)}{R_o^2 + r_o^2}], \tag{IV.3}$$

$$[\frac{d^2}{d\beta^2} + \frac{1}{4} (\frac{d\Phi}{d\beta})^2] \exp (\frac{1}{2}\rho) = 0 , \tag{IV.4}$$

where

$$dr = f\, d\beta, \;\; f = v \cosh\Gamma, \;\; \cosh\Gamma = \exp(-\rho) \sqrt{[\exp(2\rho) + \lambda_o^4)]}$$
$$= \exp(-\rho) (R_o^2 + r_o^2) \tag{IV.5}$$

and where the constants of integrations λ_o and ℓ_o can be expressed as

$$\ell_o^2 = \ell_s r_o , \quad \lambda_o^2 = \ell_w r_o . \tag{IV.6}$$

In order to obtain Coulomb force in the correspondence limit we must identify ℓ_o and λ_o as

$$\ell_o^2 = N\, g\, q_o^{-1} , \quad \lambda_o^2 = N\, e\, q_o^{-1} \tag{IV.7}$$

where e and g represent electric and magnetic charges, respectively, where the parameter N can assume values from 0 to 10^{60}, or more precisely, $N = 10^{\pm n}$, and n = 0, 1, 2, ..., 60, ...

For the $\Phi = \pm n \pi$ (n = 0, 1, 2, ...) value of the *mixing angle* we obtain $\ell^2 = r_o^2$ where

$$\ell^2 = \ell_s^2 + \ell_w^2 = \frac{2G}{c^4} N^2(e^2+g^2) \tag{IV.8}$$

and

$$\ell_s^2 = \frac{2G}{c^4} N^2 g^2, \quad \ell_w^2 = \frac{2G}{c^4} N^2 e^2 . \tag{IV.9}$$

For $\Phi = \frac{\pi}{2}$ the magnetic charge distribution decouples from the electric charge and we obtain $g = 0$, viz.,

$$g\,(r_c) = 0 , \tag{IV.10}$$

where r_c is the distance from the origin beyond which the coupling parameters

$$\alpha_s = \frac{\ell_o^2}{r_o^2} = \frac{g}{\sqrt{(e^2+g^2)}} \quad \alpha_w = \frac{\lambda_o^2}{r_o^2} = \frac{e}{\sqrt{(e^2+g^2)}}, \quad \alpha_s^2 + \alpha_w^2 = 1, \tag{IV.11}$$

assume the values $\alpha_s = 0$ and $\alpha_w = 1$. Hence, we see that the space dependence of the coupling parameters have their origins in the confined distribution of the magnetic charge leading to the concept of *running coupling constants.*

For the special case of a constant function Φ we obtain the results

$$r_o^2 = \frac{2G}{c^4} U^2, \quad \lambda_o^2 = \frac{2G}{c^4} NeU, \quad \ell_o^2 = \frac{2G}{c^4} NgU, \quad q_o^{-1} = \frac{2G}{c^4} U, \tag{IV.12}$$

where

$$U = N\sqrt{(\frac{g^2}{\cos^2\Phi} + e^2)} + (-1)^s\, Ng \tan\Phi , \qquad s=0,1 . \tag{IV.13}$$

The angular function Φ is the *mixing angle* for short-range coupling of the interactions. At $\Phi = \pi/2$ (i.e., $g=0$) the magnetic charge distribution decouples from the long-range interactions (electromagnetic, gravitational) and we obtain (see Reference 1)

$$\exp(u) = (1 + \frac{\lambda_o^4}{\beta^4}) \left\{1 - \frac{2MG}{c^2\beta} + \frac{\beta^2}{3\lambda_o^2}[1 - \sqrt{(1 + \frac{\lambda_o^4}{\beta^4})} + \frac{\lambda_o^3}{\beta^3} K]\right\} \tag{IV.14}$$

where $\lambda_o^2 = r_o^2 = \frac{2G}{c^4} N^2 e^2$ and where

$$K(\gamma, \frac{1}{\sqrt{2}}) = \int \frac{d\,\gamma}{\sqrt{(1-\frac{1}{2}\sin^2\gamma)}}, \qquad \gamma = \cos^{-1}\left[\frac{\beta^2-\lambda_o^2}{\beta^2+\lambda_o^2}\right], \tag{IV.15}$$

is an elliptic integral of the first kind. The function exp(u) is defined by

$$\exp(u) = S \cosh^2\Gamma$$

and represents generalized gravitational field where the short-range part just outside the magnetic charge core has an asymptotic form λ_o^4/β^4. The electric field of the *polarized vacuum* just outside magnetic core is obtained as (see equation 281 in Reference 1)

$$E_o = \frac{4r_o^2}{\beta^2} E_e \left[\frac{M\ G}{c^2\beta} + \frac{\beta^3+\lambda_o^4 K(\beta)-\beta\sqrt{(\beta^4+\lambda_o^4)}}{3\lambda_o^2\beta}\right] \tag{IV.16}$$

where the Coulomb field $E_e = \frac{\pm e(\pm 1)}{\beta^2}$ shows that the asymptotic behavior of the *polarized vacuum* electric field varies as β^{-5}. In (IV.16) $\beta = \pm$ r, and β and the mass M have the same sign. The result above for the electric field of the polarized vacuum is an exact calculation referring to the region outside the confined, neutral magnetic charge core of a quark, lepton or any charged particle. In reference 1 it has been shown that the vacuum behaves like a dielectric medium consisting of pairs of particles and antiparticles.

V. STRATEGIC TRIAD

An important triad, for most physicists, might entail the goals of :(i) finding the mass of the top quark, (ii) finding the mass of Higgs boson, (iii) looking for CP violation in K and B decays. However, a more general requirement is the unification of interactions mediated by the exchange of the *strategic triad* of spin 0, spin 1, and spin 2 bosons between families of quarks and leptons. In order to complete the unification of all interactions we shall need, besides the classical field equations (III.21)-(III.23) and their *supersymmetric* counterparts, *a generalized Dirac wave equation* describing quarks and leptons interactions with spin 0, spin 1, and spin 2 bosons. We consider the role of the *extremum value of the action function* S_o (see section 9.4 of reference 1) which is given by

$$S_o = -\frac{q^2}{4\pi c} \int [\sqrt{(-\hat{g})} - \sqrt{(-g)}]\, d^4x\,, \tag{V.1}$$

obtained by substituting the field equation (III.19) in the action function (III.13), where the square-root $\sqrt{(-\hat{g})} = \sqrt{(-g)}\,\sqrt{(1+q_o^{-2}\Omega - q_o^{-4}\Lambda^2)}$ can be shown to be of the form

$$\sqrt{(-\hat{g})} = \sqrt{(-g)} \sqrt{[(1+\frac{1}{2} q_o^{-2}\Omega)^2 - q_o^{-4} c^2 P_\mu P^\mu]}\,, \tag{V.2}$$

and where the momentum density P_μ is defined by

$$cP_\mu = T_{\mu\nu}\hat{v}^\nu, \quad c^2P_\mu P^\mu = \tfrac{1}{4}\Omega^2+\Lambda^2, \quad T_{\mu\rho}T^{\nu\rho} = \delta^\nu_\mu(\tfrac{1}{4}\Omega^2+\Lambda^2). \tag{V.3}$$

Hence $T_{\mu\nu}$ is, formally, of the same form as the stress-energy-momentum tensor of the electromagnetic field viz.

$$T_{\mu\nu} = \frac{1}{2}\Omega\, g_{\mu\nu} - \Phi_{\mu\rho}\,\Phi_\nu{}^\rho . \tag{V.4}$$

In this case the new $T_{\mu\nu}$ contains, in addition to the stress-energy-momentum of the electromagnetic field, as will be seen, the energy and momentum densities for massive and massless spin 0, 1, 2 fields. The unit vector $\hat{v}^\mu$ in (V.3) is defined by

$$\hat{v}^\mu = \frac{dx^\mu}{ds}, \quad \hat{v}^\mu\hat{v}_\mu = 1, \; ds^2 = g_{\mu\nu}\, dx^\mu dx^\nu . \tag{V.5}$$

By using the splitting of the field $\Phi_{\mu\nu}$ according to

$$\Phi_{\mu\nu} = \Phi_{o\mu\nu} + \Phi_{1\mu\nu}\,, \tag{V.6}$$

where $\Phi_{o\mu\nu}$ represents the short-range field inside the core with a neutral distribution of the magnetic charge, and $\Phi_{1\mu\nu}$, small compared to $\Phi_{o\mu\nu}$, represents a pure radiation field obeying Maxwell's equations

$$\Phi_{1\mu\nu,\rho} + \Phi_{1\nu\rho,\mu} + \Phi_{1\rho\mu,\nu} = 0, \quad [\surd(-g)\,\Phi_1^{\mu\nu}]_{,\nu} = 0. \tag{V.7}$$

Using (V.6) in the definition of $T_{\mu\nu}$ by (V.4) we obtain the decomposition

$$T_{\mu\nu} = T_{o\mu\nu} + T_{I\mu\nu} + T_{1\mu\nu}\,, \tag{V.8}$$

$$\Omega = \Omega_o + \Omega_I + \Omega_1\,, \tag{V.9}$$

where

$$\Omega_o = \frac{1}{2}\Phi_o^{\mu\nu}\Phi_{o\mu\nu}\,, \quad \Omega_I = \Phi_o^{\mu\nu}\Phi_{1\mu\nu}\,, \quad \Omega_1 = \frac{1}{2}\Phi_1^{\mu\nu}\Phi_{1\mu\nu}, \tag{V.10}$$

$$T_{o\mu\nu} = \tfrac{1}{2}\Omega_o g_{\mu\nu} - \Phi_{o\mu\rho}\Phi_{o\nu}{}^\rho, \quad T_{I\mu\nu} = \tfrac{1}{2}\Omega_I g_{\mu\nu} - (\Phi_{o\mu\rho}\Phi_{1\nu}{}^\rho + \Phi_{o\nu\rho}\Phi_{1\mu}{}^\rho), \tag{V.11}$$

$$T_{1\mu\nu} = \frac{1}{2}\Omega_1 g_{\mu\nu} - \Phi_{1\mu\rho}\,\Phi_{1\nu}{}^\rho, \quad [\surd(-g)T_{1\mu}{}^\nu]_{|\nu} = 0\,, \tag{V.12}$$

$$\frac{1}{4\pi}[\sqrt{(-g)}T_{o\mu}^{\ \nu}]_{|\nu} = \Phi_{o\mu\rho}J_e^{\ \rho} + f_{o\mu\rho}s^{\rho},\quad J_e^{\mu} = \frac{1}{4\pi}\frac{\partial}{\partial x^{\nu}}[\sqrt{(-g)}\Phi_o^{\ \mu\nu}],$$

$$\frac{1}{4\pi}[\sqrt{(-g)}T_{I\mu}^{\ \nu}]_{|\nu} = \Phi_{1\mu\rho}J_e^{\ \rho} + f_{1\mu\rho}s^{\rho}, \tag{V13}$$

and where the sign (|) indicates covariant derivative with respect to the metric $g_{\mu\nu}$.

The particle mass is generated by the scalar Ω_o according to

$$\frac{1}{4\pi c}\int_{\sigma}\frac{1}{2}\Omega_o\sqrt{(-g)}\,d\sigma = mc\,. \tag{V.14}$$

Expansion of the action function S_o in (V.1) to q_o^{-2} order yields

$$S_o \cong -\frac{1}{4\pi c}\int\frac{1}{2}(\Omega_o+\Omega_I+\Omega_1)\sqrt{(-g)}\,d^4x\,. \tag{V.15}$$

Using the definitions

$$\sqrt{(-g)}\,d^4x = \sqrt{(-g)}\,d\sigma\,ds,\ d\sigma = \hat{v}^{\mu}d\sigma_{\mu}\,,$$
$$\Phi_{1\mu\nu} = \partial_{\mu}A_{\nu} - \partial_{\nu}A_{\mu}\,,$$

we obtain

$$S_o \cong -mc\int ds - \frac{1}{c}\int A_{\mu}J_e^{\mu}\,d\sigma\,ds - \frac{1}{16\pi c}\int \Phi_1^{\mu\nu}\Phi_{1\mu\nu}\sqrt{(-g)}\,d^4x\,, \tag{V.16}$$

which is the same as the classical action function for a charged point particle moving according to Lorentz's equations of motion in an external electromagnetic field. Because of the Maxwell's equations (V.7) for electric charge free fields, the last term in the action function (V.16) vanishes. For a point like electric charge the electric current density in the second term of (V.16) can be replaced by $J_e^{\mu} = \frac{dx^{\mu}}{ds}e\delta(x-x_o)$, where $\delta(x-x_o)$ is a 4-dimensional delta function. The above is a crude approximation to a *non-linear classical action* function in order to show that it contains the usual equations of motion[9]. This is a new way of deriving the classical electrodynamics from an action function yielding, due to the magnetic charge origin of mass, the first term in (V.16) along with the coupling, as in the second term of (V.16), of the field to the particle. We shall in what follows show that the corresponding quantum action function is, in contrast to classical theory, linear, and contains new physics and a variety of new results.

By using the usual 4x4 Dirac matrices γ_{μ}, γ_5 defined by the anticommutation relations

$$\gamma_{\mu}\gamma_{\nu} + \gamma_{\nu}\gamma_{\mu} = -2\,g_{\mu\nu}I_o,\quad \gamma_5\gamma_{\mu}+\gamma_{\mu}\gamma_5 = 0\,, \tag{V.17}$$

where I_o is a 4x4 unit matrix, we can obtain from (V.1) the two action operators

$$S^F_{o1} = -\frac{1}{4\pi c}\int_F \sqrt{(-g)}\ [c\gamma^\mu P_\mu + \tfrac{1}{2}\Omega - 2\ u\_(\tfrac{1}{2}\Omega + q^2)]\ d^4x\ , \tag{V.18}$$

$$S^F_{o2} = -\frac{1}{4\pi c}\int_F \sqrt{(-g)}\ [c\gamma_5\gamma^\mu P_\mu + \tfrac{1}{2}\Omega - 2\ u\_(\tfrac{1}{2}\Omega + q^2)]\ d^4x, \tag{V.19}$$

where the subscript F implies the substitution of Fermi-like solutions (i.e., the solutions of the field equations for non-hermitian $\hat{g}_{\mu\nu}$) in the integrand as opposed to the substitutions of Bose-like solutions (i.e., the solutions of the field equations for the hermitian $\hat{g}_{\mu\nu}$) in the two action operators

$$S^B_{o1} = -\frac{1}{4\pi c}\int_B \sqrt{(-g)}[c\gamma^\mu P_\mu + \tfrac{1}{2}\Omega - 2\ u\_(\tfrac{1}{2}\Omega - q_o^2)]\ d^4x, \tag{V.20}$$

$$S^B_{o2} = -\frac{1}{4\pi c}\int_B \sqrt{(-g)}[c\gamma_5\gamma^\mu P_\mu + \tfrac{1}{2}\Omega - 2\ u\_(\tfrac{1}{2}\Omega - q_o^2)]\ d^4x, \tag{V.21}$$

where u_ is the projection operator

$$u\_ = \frac{1}{2}(1 - i\ \gamma_5)\ . \tag{V.22}$$

With the introduction of a wave function Ψ, the operator equations (V.18)-(V.21) can be replaced by the generalized Dirac wave equations

$$cM^F_1\Psi^F_1 = -\frac{1}{4\pi c}\int_F \sqrt{(-g)}[c\gamma^\mu P_\mu + \tfrac{1}{2}\Omega - 2u\_(\tfrac{1}{2}\Omega + q_o^2)]\ d\sigma\,\Psi^F_1, \tag{V.23}$$

$$cM^F_2\Psi^F_2 = -\frac{1}{4\pi c}\int_F \sqrt{(-g)}[c\gamma_5\gamma^\mu P_\mu + \tfrac{1}{2}\Omega - 2u\_(\tfrac{1}{2}\Omega + q_o^2)]\ d\sigma\,\Psi^F_2,$$

(V.24

$$cM^B_1\Psi^B_1 = -\frac{1}{4\pi c}\int_B \sqrt{(-g)}[c\gamma^\mu P_\mu + \tfrac{1}{2}\Omega - 2u\_(\tfrac{1}{2}\Omega - q_o^2)]\ d\sigma\,\Psi^B_1, \tag{V.25}$$

$$cM^B_2\Psi^B_2 = -\frac{1}{4\pi c}\int_B \sqrt{(-g)}[c\gamma_5\gamma^\mu P_\mu + \tfrac{1}{2}\Omega - 2u\_(\tfrac{1}{2}\Omega - q_o^2)]\ d\sigma\,\Psi^B_2, \tag{V.26}$$

where we used the substitutions

$$\frac{dS^\ell_{oj}}{ds}\Psi_j = cM_j^\ell\Psi_j\ , \tag{V.27}$$

and where j = 1,2 and ℓ = F,B. The role of the pseudo-scalar function Λ as defined by (III.6) is reflected by the appearance of the pseudo-scalar matrix operator γ_5 in the wave equations (V.23)-(V.26) above originating from Λ's presence in the square-root (V.2). The quantities M_j^ℓ are free particle masses relating to quark, lepton, and Planck masses. In the mass relation (V.14) the mass m is generated by F- and B-type fields. Hence, the integrals in (V.23)-(V.26) each yields a momentum $p_\mu = mc\hat{v}_\mu$, where m refers to the mass m_F or m_B. The *first quantization* of the equations (V.23)-(V.26) follows by making use of the operator representation of the momentum p_μ by $p_\mu = i\hbar \frac{\partial}{dx^\mu}$, thereby obtaining *generalized Dirac wave equations* describing the coupling of *photons*, and *massive gauge bosons* with spin 1/2 matter fields.

The volume integrals over the q_o^2 term in the wave equations (V.23)-(V.26) can be performed by using the fundamental relation (II.4) as

$$\frac{1}{4\pi}\int q^2 d\sigma = \int_o^{r_o} q^2 r^2 dr = \frac{c^4}{2G} r_o. \tag{V.28}$$

Thus, we obtain the mass term

$$M(r_o) = \frac{c^2}{2G} r_o = N\sqrt{(\frac{e^2+g^2}{2G})}, \tag{V.29}$$

where we see that for N=1 the mass $M(r_o)$ is of the order of *Planck Mass* $\sqrt{(\frac{\hbar c}{G})}$. However, if in (V.28) we set r_o equal to Planck length $\sqrt{(\frac{\hbar G}{c^3})}$, then we obtain the Planck mass $\frac{1}{2}\sqrt{(\frac{\hbar c}{G})}$.

The appearance of $T_{\mu\nu}$ in the wave equations (V.23)-(V.26) through the definition of $P_\mu = \frac{1}{c} T_{\mu\nu} \hat{v}^\nu$ (momentum density) provides a *hydrodynamical picture* of the interplay between the generations of quarks and leptons and unify the mediation of forces through the spin 0,1,2 bosons. The energy-momentum-stress densities represented by an electromagnetic type tensor $T_{\mu\nu}$ for massless and massive bosons is not accidental but the proper way to unify all the interactions. The tensor $T_{\mu\nu}$ with nine independent components (i.e., $T^\rho_\rho = 0$) can be decomposed into 0,1,2 spin parts as

$$T_{\mu\nu} = T^{(o)}_{\mu\nu} + T^{(1)}_{\mu\nu} + T^{(2)}_{\mu\nu}\,. \tag{V.30}$$

By using (V.8)-(V.13) it can easily be seen that for a point charge with $J^\mu_e = e\,\hat{v}^\mu\,\delta(x-\xi)$, any one of the wave equations (V.23)-(V.26) yield terms like

$$(p_\mu - \frac{e}{c} A_\mu)\, \gamma^\mu ,$$

and form the basis for the derivation of the usual gauge invariant Dirac wave equation describing the interaction of a photon with a point electric charge.

For distances large compared to Planck length (or r_o) and to order q_o^{-2}, the field equations (III.21) and their supersymmetric counterparts, with the assumption $g_{\mu\nu} = \eta_{\mu\nu}$, reduce to

$$(\nabla^2 - \frac{\partial^2}{c^2 \partial t^2} + \kappa^2)\, T^{(2)}_{\mu\nu}(F) = 0\, , \tag{V.31}$$

$$(\nabla^2 - \frac{\partial^2}{c^2 \partial t^2} - \kappa^2)\, T^{(2)}_{\mu\nu}(B) = 0\, , \tag{V.32}$$

where $\kappa^2 = 2\, r_o^{-2}$ and where we have used the relations (V.13) with $J^\mu_e=0$, $s^\mu=0$ to obtain four restrictions $[\sqrt{(-g)}\, T^\nu_\mu]_{|\nu} = 0$ to construct the spin 2 tensors $T^{(2)}_{\mu\nu}(F)$, $T^{(2)}_{\mu\nu}(B)$ so that each has five independent components only[1]. The magnetic charge current densities $s^\mu(F)$ and $s^\mu(B)$, with the same approximation, obey the equations (V.31) and (V.32), respectively. Despite *tachyon* type solutions of the linearized equation (V.31), its exact spherically symmetric form has no such solutions and yields magnetic charge distribution in stratified layers with alternating signs and decreasing amounts of magnetic charges g_n (n=0,1,2...), where $\sum_0^\infty g_n = 0$. The magnetic charge distribution for $s^\mu(B)$ has the form of a dipole with no restriction of the amounts of magnetic charges for each opposite sign poles, i.e., $g_+ + g_- = 0$.

VI. COSMOLOGICAL APPLICATIONS

The four most important advances, amongst others, in cosmology in the past five decades after Hubble's observation of the expanding universe, and after George Gamow's theory of the Big-Bang creation of the universe, include, in chronological order: (1) Ralph A. Alpher and Robert Herman[2]'s theoretical prediction in the 1940s of the cosmic microwave background radiation (CMBR) left over from the Big-Bang, (2) Arno A. Penzias and Robert W. Wilson's observation in 1964 of the residual heat detected as the CMBR, (3) Alan H. Guth[3]'s hypothesis in 1979 of an inflationary universe described in page 1 of the book[4] authored by Michael Riordan and David N. Schramm as "a discovery that was to change the course of modern physics," (4) the observation in 1992 of microwave anisotropies in the CMBR as seen through COBE by

[1]The search for short-range forces in this theory was inspired by a letter I received from P.A.M. Dirac over twenty years ago (see the appendix of reference 1).

physicists and cosmologists led by George Smoot[5]. There is, amongst physicists and cosmologists, a consensus that the description of the universe, and the elementary particles in the early phase of the universe and after requires a unified theory of the large and the small, i.e., of cosmology and elementary particle physics. Aside from the recent developments of quantum gravity, the problem of overproduction of monopoles in grand unified theories, accounting for the observed flatness of the universe, of the problem of "horizon" were amongst the issues that inspired Alan Guth in 1979 to hypothesize the inflationary behavior of the universe as an exceedingly brief glitch to precede all the events in the Big-Bang cosmology.

In this paper we also show that the theory resulting from the non-symmetrization of general relativity does in fact predict a flat universe, and the neutral magnetic charge confinement of monopoles as the constituents of integral spin particles (bosons of spins 0,1,2) and half-integral spin particles i.e., quarks and leptons. It was a serendipitous observation of the physical significance of a relation resulting from the non-symmetrization of general relativity, which is now more than forty years old[1] relation (II.4), that inspired the writing of this section of this paper, where c and G represent speed of light and gravitational constant, respectively, and where r_o is a "fundamental length" of the order of Planck length and q_o^2 is an energy density, with q_o having the dimensions of an electric field. The relation (II.4) was the result of the correspondence requirement, where in the $r_o = 0$ limit the generalized theory of gravitation reduces to general relativity in the presence of the electromagnetic field. The equation (II.4) can also be interpreted as an *equation of state*. The energy density q_o^2 in the region of size r_o (~ Planck length) is of the order of 10^{114}ergs. cm^{-3} and it represents the energy density at the beginning of time when the size of the universe was of the order of r_o. Thus, what was obtained more than 40 years ago as a fundamental length r_o only now turned out to also be the size, as shown below, of the universe as it evolved from the time of the Big-Bang to the present time. The fundamental relation (II.4) does in fact govern the general behavior of the universe including the process of nucleosynthesis and the elementary particles as described by the field equations (III.21)-(III.23) of the theory for the 16 non-symmetric hermitian field variables $\hat{g}_{\mu\nu} = g_{\mu\nu} + iq_o^{-1}\Phi_{\mu\nu}$, where $g_{\mu\nu}$ and $\Phi_{\mu\nu}$ represent generalized gravitational and generalized electromagnetic fields, respectively. The antisymmetric part $\Phi_{\mu\nu}$ results from the creation of electric and magnetic charges from the vacuum where $r_o \neq 0$, and $q_o^2 \neq \infty$. The symmetric part $g_{\mu\nu}$ alone, as in general relativity, is due to the absence of electric and magnetic charges where $r_o = 0$, and $q_o^2 = \infty$. The theory includes also the *supersymmetric* transition of the hermitian field variables into the 16 non-symmetric nonhermitian field variables $\hat{g}_{\mu\nu} = g_{\mu\nu} + q_o^{-1}\Phi_{\mu\nu}$, where both sets of corresponding field equations reducing in the limit $r_o = 0$, to general relativity.

The relation (II.4) can be written as an analogue of $E = mc^2$ in its density form by

$$q_o^2 = \rho_o c^2 , \qquad \text{(VI.1)}$$

where $\rho_o = (c^2/2G)r_o^{-2}$, not to be confused with the function ρ appearing in the field equations below, implies that mass density is proportional to the curvature r_o^{-2} of the space time. At the instant of creation of the universe from a region of size r_o with a curvature r_o^{-2} ($\sim 10^{66}cm^{-2}$), mass density ρ_o ($\sim 10^{95}$ gr.cm^{-3}), and energy density (10^{114}erg.cm^{-3}), the corresponding Big-Bang values, as follows by setting r_o=0, are infinite. For the present universe, by replacing r_o by r_t and q_o by q_t, the same relation (II.4) applies,

$$q_t^2 r_t^2 = c^4/2G, \quad q_t^2 = \rho_t c^2, \tag{VI.2}$$

which is satisfied for the present values of $q_t^2 \sim 10^{-8}$erg.cm^{-3}, $r_t \sim 2.58\times10^{10}$ light years. The total mass of the universe, which follows from integrating (VI.2),

$$M_u c^2 = \frac{1}{4\pi}\int q_t^2 \, dV = c^2 \int \rho_t dV = (c^4/2G)\int_0^{r_t}(r^2/r^2)dr = (c^4/2G)r_t \tag{VI.3}$$

with $r_t \sim 2.58\times10^{10}$ light years, yields the value $M_u \sim 8\times10^{22}$ solar mass. The relation (II.4) states that as the universe expands, its average energy density decreases. The result (VI.3) above when written in the form

$$1 - \frac{2MG}{c^2 r} = 0,$$

is reminiscent of the Schwartzschild singularity in general relativity where, in this case, M/r remains, as the universe expands, a constant equal to $c^2/2G$. The distance $d = \frac{2MG}{c^2}$ can be related to Hubble's law V=Hd by using d=r=Vt, where t^{-1}=H=Hubble's parameter, and t is assumed to be the age of the universe.

The field equations for the 16 hermitian field variables for the spherically symmetric case are given[1] by

$$\frac{1}{2} r_o^2 \frac{d}{d\beta}[S_B \Phi \exp(\rho)] = R_o^2 \sinh\Phi - (-1)^s \ell_o^2 \cosh\Phi, \tag{VI.4}$$

$$\frac{1}{2} r_o^2 \frac{d}{d\beta}[S_B \dot{\rho} \exp(\rho)] = R_o^2 \cosh\Phi - (-1)^s \ell_o^2 \sinh\Phi - \exp(\rho) \tag{VI.5}$$

$$\frac{1}{2} r_o^2 \frac{d}{d\beta}[\dot{S}_B \exp(\rho)] = \exp(\rho)\left[\frac{\cosh\Phi}{\cos\Gamma} - 1\right], \tag{VI.6}$$

$$\left[\frac{d^2}{d\beta^2} - \frac{1}{4}\dot{\Phi}^2\right] \exp(\tfrac{1}{2}\rho) = 0 \qquad \text{(VI.7)}$$

where $\dot{\Phi} = \frac{d\Phi}{d\beta},\ \dot{\rho} = \frac{d\rho}{d\beta},\ \dot{S}_B = \frac{dS_B}{d\beta}$, and where

$$s = 0{,}1,\quad R_o^2 - r_o^2 = [\exp(2\rho) - \lambda_o^4]^{\frac{1}{2}},\ S_B = \exp(u)/\cos^2\Gamma,\ 0 \le \Gamma \le \frac{\pi}{2} \qquad \text{(VI.8)}$$

$$\cos\Gamma = \exp(-\rho)\,(R_o^2 - r_o^2) = \lambda_o^2 \exp(-\rho)\,[\lambda_o^{-4}\exp(2\rho) - 1]^{\frac{1}{2}},\ -\infty<\Phi<\infty, \qquad \text{(VI.9)}$$

$$\exp(\rho)\sin\Gamma = \pm\lambda_o^2,\ \ell_o^2 = N\,g\,q^{-1},\ \lambda_o^2 = N\,e\,q^{-1} \qquad \text{(VI.10)}$$

where the constants of integration, λ_o^2 and ℓ_o^2, are identified as functions of electric charge and magnetic charge to obtain Coulomb's law of force. Hence N is a parameter assuming values from zero to 10^{60}. It is interesting to note that the three ratios R_u/r_p, M_u/m_p, and τ_u/τ_p are of the same order of magnitudes as the parameter N, where R_u, M_u, τ_u represent the size, mass, and age of the universe, respectively. The symbols r_p, m_p, and τ_p represent the Planck length, Planck mass, and Planck time, respectively. The approximate solutions of the field equations, as obtained in reference 1, yield, for the various lengths the results

$$r_o^2 = \frac{2G}{c^4} N^2(e^2+g^2),\qquad \lambda_o^2 = \frac{2G}{c^4} N^2\, e\,\sqrt{(e^2+g^2)}\,,\qquad \ell_o^2 = \frac{2G}{c^4} N^2\, g\,\sqrt{(e^2+g^2)} \qquad \text{(VI.11)}$$

The singularity at $\Gamma = \frac{\pi}{2}$ or $\exp(\rho) = \lambda_o^2$ for the function S_B is invariant and cannot be removed by a coordinate transformation, and the definition of λ_o^2 in (VI.11) shows that $\lambda_o^2 < r_o^2$ and, therefore, the surface of the singularity lies inside the region whose horizon has the dimension of r_o. The magnetic charge g differs from the electric charge e in acting as a source of a short-range interaction only. Furthermore, being space dependent, it is a *running coupling constant* with a *magnetic horizon* defined by $g(r_c) = 0$, where r_c is the short distance beyond which there is no field of which the magnetic charge is a source, no magnetic charge distribution[1]. The field equations (VI.4) - (VI.7) at the origin are solved by $\exp(\rho) = 0$, $\Phi(0) = 0$, $\lambda_o^2 = 0$. $\ell_o^2 = 0$, $r_o^2 = 0$. with the function S_B assuming arbitrary values. At the current epoch of the universe the parameter r_o can be assigned the value of the size of the universe, and therefore, dividing the field equations by r_o and noting its evolved large value of 10^{10} light years, we find that the field equations are, approximately, satisfied by the flat space-time values

$$\nu = \pm 1, \quad \Phi(r_t) = 0, \quad \exp(\rho) = \beta^2, \quad \Gamma = 0, \quad S_B = 1, \tag{VI.12}$$

where the remaining term $-(-1)^s \ell_o^2 r_o^{-2} = -(-1)^s \mathrm{g}\, q \frac{2G}{c^4}$ on the right hand side of the equation (VI.4), because of the small factor $\frac{2G}{c^4}$ ($\sim 10^{-49}$) and the small energy density q_t^2 ($\sim 10^{-8}$erg.cm^{-3}), is not affected by the size of the magnetic charge and, therefore, is negligibly small. Thus, *the universe is flat*. The solution (VI.12) if used in the line element

$$ds_B^2 = S_B \cos^2\Gamma\, dx_o^2 - S_B^{-1} d\beta^2 - \exp(\rho) \cosh\Phi\, d\Omega_o^2, \tag{VI.13}$$

where $d\Omega_o^2 = d\theta^2 + \sin^2\theta\, d\varphi^2$, and the variable β is related to the usual radial coordinate r by the definition $dr = f\, d\beta$, $f = \nu \cos\Gamma$, yields the flat space-time metric $ds^2 = dx_o^2 - dr^2 - r^2 d\Omega_o^2$. It is important to observe that flat space-time here results from the large size of the universe relative to the terms in the field equations (VI.4)-(VI.7), and it must not be confused with the flat space-time solutions of general relativity in the absence of matter and for zero cosmological constant. The flatness in the latter case results from the initial conditions and has no relation to flat universe. The function S_B in equation (VI.13) is defined by

$$S_B = \frac{\lambda_o^{-4} \exp(u + 2\rho)}{\lambda_o^{-4} \exp(2\rho) - 1}. \tag{VI.14}$$

The field equations (VI.4)-(VI.7) being invariant under the transformations,

$$\Phi \rightarrow -\Phi, \qquad \mathrm{g} \rightarrow -\mathrm{g}, \tag{VI.15}$$

have only magnetic dipole solutions with equal and opposite signs of magnetic charges, i.e., solutions with neutral magnetic charge distributions, where there is no limitation on the amount of magnetic charge g. There are in fact no free monopole solutions. A more general proof for the absence of free monopoles is given in reference 1, equation (184). However, prior to the time 10^{-43} sec. the primordial gas of positive and negative monopoles of all fractional spins, and sizes was in an equilibrium state. The primordial gas could have deviated from the equilibrium state by a *transient gravitational repulsion* arising from a singularity, as seen from the definition (VI.14) of S_B, inside the region of the dimension of r_o, which could have energized the explosive creation of the universe. Hence $-\lambda_o^2 \leq \exp(\rho) \leq \lambda_o^2$. The interior singularity is at $\exp(\rho)$

$= \lambda_o^2$, and S_B is negative in the interval of the singularity.

In the hot region of dimension r_o the primordial gas of monopoles and electric charges is not the only equilibrium state and a *supersymmetric transition* yielding a different neutral distribution of monopoles can be obtained uniquely by the transformations,

$$\Phi \to i\Phi + \frac{\pi}{2}, \quad \Gamma \to i\Gamma, \quad r_o \to ir_o \tag{VI.16}$$

$$q_o^{-1} \to iq_o^{-1}, \quad \ell_o^2 \to i\ell_o^2, \quad \lambda_o^2 \to i\lambda_o^2, \tag{VI.17}$$

of the field equations (VI.4)-(VI.7), leading to the field equations

$$\frac{1}{2} r_o^2 \frac{d}{d\beta} [S_F \exp(\rho)\, \dot{\Phi}] = R_o^2 \cos\Phi + \ell_o^2 \sin\Phi , \tag{VI.18}$$

$$\frac{1}{2} r_o^2 \frac{d}{d\beta} [S_F \exp(\rho)\, \dot{\rho}] = - R_o^2 \sin\Phi + \ell_o^2 \cos\Phi + \exp(\rho) \tag{VI.19}$$

$$\frac{1}{2} r_o^2 \frac{d}{d\beta} [\dot{S}_F \exp(\rho)] = \exp(\rho) \left[1- \frac{\sin\Phi}{\cosh\Gamma}\right] , \tag{VI.20}$$

$$\left[\frac{d^2}{d\beta^2} + \frac{1}{4} \dot{\Phi}^2\right] \exp(\tfrac{1}{2}\rho) = 0 , \tag{VI.21}$$

which are the same as (IV.1)-(IV.4), where $f = \nu \cosh\Gamma$, $\cosh\Gamma = \exp(-\rho)\, (R_o^2 + r_o^2)$, $R_o^2+r_o^2 = [\exp(2\rho) + \lambda_o^4]^{\frac{1}{2}}$ and where Φ is now an angle restricted by $0 \le \Phi \le \frac{\pi}{2}$. The new gravitational potential

$$S_F = \frac{\lambda_o^{-4}\exp(u+2\rho)}{\lambda_o^{-4}\exp(2\rho)+1} , \tag{VI.22}$$

is a positive function and differs from S_B, defined in (VI.14), by the +1 sign in the denominator versus -1 in that of S_B. In this case the neutral magnetic charge distribution is synthesized[1] in an infinitely layered form of an infinite number of monopoles of decreasing magnitudes with distance and of alternating signs, where $\sum_0^\infty g_n = 0$ and g_n representing the positive or negative magnetic charge in the n-th layer. Thus, a quark, or a lepton has a structure consisting of a neutral magnetic charge distribution like a *jigsaw puzzle* but requiring, in this case, an infinite number of pieces to put together to recreate the picture.

The cooling of the primordial gas and synthesizing of the monopoles by the short-range strong forces creates either quarks with spin 1/2, based on an Ansatz or a conjecture $\sum_0^\infty g_n^2 = \frac{1}{2}\hbar c$, or leptons, as solutions of the *generalized Dirac Wave equations* (V.23))--(V.26) obtained from this theory. The field equations (VI.18)-(VI.21), like the field equations (VI.4)-(VI.7), also in the limit $r_o = 0$ reduce to the *spherically symmetric* field equations of general relativity. In the limit of r_o evolving into the value at the present cosmological epoch we find that the field equations (VI.18)-(VI.21) have the flat space-time solutions,

$$\Phi = \frac{\pi}{2},\ \exp(\rho) = r^2,\ \Gamma = 0,\ S_F = 1, \tag{VI.23}$$

The fact that the universe began with a structure as described by its primordial size r_o and finite energy density q_o^2 can be related to temperature fluctuations in CMBR and to structure formation in the large. Finally, in view of the present theory's prediction of the magnetic charge structure of all matter, luminous or dark, the assumption of the ratio of dark matter to luminous matter being proportional to $\ell_o^2/\lambda_o^2 = g/e = \frac{eg}{e^2} = \frac{1}{2}n\frac{\hbar c}{e^2}$ may be reasonable with respect to dark matter cosmology, where we have used Dirac's relation $eg=\frac{1}{2}n\hbar c$. Alternatively we can obtain the same result from summing over of all fractional spins in

$$\sum_0^\infty (g_n^2/e^2) = \frac{1}{2}\frac{\hbar c}{e^2}. \tag{VI.24}$$

VII. CONCLUSIONS

A glance at the wave equations (V.23)-(V.26) together with the classical field equations (III.21)-(III.23) reveals a close resemblance to quantum electrodynamics and a new way of unifying general relativity and quantum theory. We have, actually, shown that the relativistic quantum theory can be derived from the generalized theory of gravitation. Some of the epistemological issues pertaining to conflict or clash between quantum theory and general relativity arising, for example, from the study of black holes as sinks of information regardless of their sustenance or of their fading away into oblivion, is not relevant in these discussions.

The eight spin 1 bosons, resulting from the decay of the four highly massive spin 2 bosons in the wave equations (V.23)-(V.26) despite a striking resemblance to the Standard Theory's eight gluons, may not, because of the absence of gravity, be related to the latter. The same wave equations contain the interactions of the four massive spin 1 bosons with quarks and leptons. The masses of these four weak bosons are generated by the four scalar bosons contained in the four wave equations (V.23)-(V.26). Thus, if there exist four weak bosons then we need to theorize the existence of a *fourth generation*. The current understanding of experiments does not include, for example, the existence of an additional neutral weak boson. The masses of the quarks and leptons, because of the q^2 term in the wave equations (V.23)-(V.26) depend on the electric and magnetic charges and the mixing angle Φ.

Just as in electrodynamics, there is a probability (though very small) proportional to strong and weak coupling for the quarks and leptons to radiate any member of the *strategic triad* of spins 0, 1, and 2. The *supersymmetry* presented in this theory, is the only possible way to unify the descriptions of particles with different spins. It is a symmetry whose basic operation is to transform particles or fields with a known spin into other particles or fields whose spins differ by the minimal unit $\frac{1}{2}\hbar$. The process of *supersymmetrizing* transforms bosons into fermions and vice versa. The presence of discrete symmetries induced by the terms containing γ_5 and the projection operator $\frac{1}{2}(1 - i\ \gamma_5)$ in the wave equations (V.23)-(V.26) should be relevant in seeking to explain *CP* violating weak decays of K and B bosons. The reference 1 contains in section 7.2 a discussion on the violation of C and *P* symmetries in the absence of electric charge. This property of symmetry violation in the absence of the electric charge together with the wave equations (V.23)-(V.26) may provide a fundamental mechanism for *CP* violation.

During the instant of the big bang creation of the universe all of the symmetries were unbroken. If at an appropriate temperature a phase transition occurred from the exact symmetry to the broken unified gauge symmetries SU(3,1)×U(1) and SO(3,1)×O(2), then *CP* violation must have begun in the prenucleosynthesis era to cause the creation of surplus matter over antimatter. However, this idea of *CP* violation to explain the origin of the excess matter content of the universe may not, like other proposals, be the final word on this important cosmological problem.

Finally, the physicists ought not to assume *ab initio* that an approach unrelated to the currently favored models cannot possibly be an acceptable effort. There are some new and useful ideas in the proposed theory. It unifies two long-range forces to create two short-range forces representing weak and strong interactions. Most important is the fact that the theory meets crucial dynamical test by yielding, without much effort, the Lorentz's equations of motion for a charged particle in an external electromagnetic field. If one pursues an idea for a long time, as is the case for the effort spent in this work, one develops a mind-set, just as others acquire their own mind-sets by investing all their time on what is fashionable for the present and ignoring other perfectly viable view points. Mind-sets are not necessarily obstacles to progress, but they can be, inherently, basic ingredients for creativity. In a conversation I had with Albert Einstein on November 19, 1953 (see the last Appendix in Reference 1), his parting remark was, "Your theory, because of r_0, is more general than mine, but time will show which one of us is right".

REFERENCES

[1] Behram N. Kursunoglu, *Journal of Physics Essays*, Vol. 4, No. 4, pp. 439-518, 1991. University of Toronto Press.

[2] Ralph A. Alpher and Robert Herman "Remembrance of Things Past" in proceedings of the Coral Gables Conference *"Unified Symmetry: In the small and in the large,"* 1993. Nova Science Publishers, New York eds. Behram N. Kursunoglu and Arnold Perlmutter.

[3] Alan A. Guth, Phys. Rev. D.23, 347 (1981).

[4] Michael Riordan and David N. Schramm, *The Shadows of Creation.* W.H. Freeman and Company, 1991 New York.

[5] George F. Smoot *et al.*, in *"After the First Three Minutes"* eds. S.S. Holt, C.L. Bennet, and Virginia Trimble (New York: AIP Conference Proc. 222), 95 (1991 C). See also J.C. Mather et al. in the same proceedings.

[6] J.C. Mather in the proceedings of the Coral Gables Conference "*Unified Symmetry: In the small and in the large*," 1993, Nova Science Publishers, New York, eds. Behram N. Kursunoglu and Arnold Perlmutter.

[7] See the generalized Dirac wave equation in B.N. Kursunoglu's paper, "The Ascent of Gravity" in proceedings of the Coral Gables Conference "*Unified Symmetry: In the small and in the large*," 1993. Nova Science Publishers, New York eds. Behram N. Kursunoglu and Arnold Perlmutter.

[8] For more references, advances, and recent developments, see Behram N. Kursunoglu, Journal of Physics Essays Vol. 4, No. 4, pp 439-518, 1991. Behram N. Kursunoglu, Phys Rev. 88, 1369 (1952); proceedings of Les Theories Relativistes de la Gravitation, edited by M.S. Lichnerowicz (Royaumont, 1959), p. 359; Rev. Mod. Phys. 29, 412 (1957); Nuovo Cimento 15, 729 (1960); Phys. Rev. D9, 2723 (1974); ibid. 12, 1850(E) (1975); Phys. Rev. D13 1538 (1976); ibid D14, 1518 (1974).

[9] L.D. Landau and E.M. Lifshitz, *The Classical Theory of Fields*, Addison Wesley Publishing Company, Inc., 1962, Mass., page 66.

RESULTS FROM QUANTUM COSMOLOGICAL GRAVITY

R. P. Woodard

Department of Physics
University of Florida
Gainesville, FL 32611

INTRODUCTION

This is a report on my work with Professor Nicholas Tsamis of the University of Crete on quantum gravity with a non-zero cosmological constant.[1–5] Because this theory differs so radically from conventional quantum gravity we have taken to calling it, "quantum cosmological gravity," or QCG for short. The Lagrangian of QCG is:

$$\mathcal{L} = \frac{1}{16\pi G}\Big(R - 2\Lambda\Big)\sqrt{-g} + \text{counterterms} \tag{1}$$

where G is Newton's constant, Λ is the cosmological constant, and we employ an infinite series of local counterterms to absorb ultraviolet divergences. The astute reader will note that our metric has spacelike signature and our Riemann tensor is $R^{\rho}{}_{\sigma\mu\nu} = \Gamma^{\rho}{}_{\nu\sigma,\mu} + \Gamma^{\rho}{}_{\mu\lambda}\Gamma^{\lambda}{}_{\nu\sigma} - (\mu \leftrightarrow \nu)$.

Nick and I study QCG using conventional perturbation theory. That is, we separate the metric into a classical solution and a quantum graviton field:

$$g_{\mu\nu} = g^{\text{class}}_{\mu\nu} + \kappa h_{\mu\nu} \tag{2}$$

where $\kappa^2 \equiv 16\pi G$ is the usual loop counting parameter of perturbative quantum gravity. Our analysis is done using real time evolution on a 3+1 dimensional manifold. We do not Euclideanize, nor do we permit topology change. There are no extra dimensions, nor any missing ones, although our results apply to any dimension in which gravity is dynamical.[1,2] We require no special matter content, although our results pertain to gravity plus any matter theory that does not require fine tuning to avoid conflicts with low energy phenomenology.[3,5]

We *do* make several assumptions. The first of these is that the cosmological constant is positive. Although negative Λ has great interest for us, we do not currently understand QCG well enough to say much about this case. Our second assumption is a locally de Sitter classical background on a manifold which admits flat 3-sections:

$$ds^2_{\text{class}} \equiv g^{\text{class}}_{\mu\nu} dx^\mu dx^\nu = -dt^2 + \exp(2Ht)\, d\vec{x} \cdot d\vec{x} \tag{3}$$

The parameter H is known as the Hubble constant, and in 3+1 dimensions its relation to the cosmological constant is: $H^2 = \frac{1}{3}\Lambda$. The exponential expansion of spacetime evident in these coordinates is known as inflation and our chief concern is what QCG

Unified Symmetry: In the Small and in the Large,
Edited by B.N. Kursunoglu *et al.*, Plenum Press, New York, 1995

has to say about inflation. Our third assumption is that the natural mass associated with inflation is at or below the GUT scale:

$$M \equiv \left(\frac{H}{\kappa}\right)^{\frac{1}{2}} < 10^{16}\ \mathrm{GeV} \tag{4}$$

One consequence of this is that the natural dimensionless parameter of QCG is very small:

$$\kappa^2 H^2 = \left(\frac{M}{M_{\mathrm{planck}}}\right)^4 < 10^{-12} \tag{5}$$

Not that this restriction is enormously above the bound of $\kappa^2 H^2 < 10^{-120}$ which is currently observed,[6] so we have not solved the problem of the cosmological constant by fiat. Restriction (5) can be realized naturally in some models of supergravity;[7] and if we are in error imposing it, we are not the only ones because (5) is the usual assumption of inflationary cosmology.[8] Our final assumption is that the various constants which occur in the finite parts of the local counterterms in (1) are not unreasonably large numbers when expressed in Planck units. This is also a widely held belief, and it would be very difficult to otherwise reconcile the observed success of classical general relativity.

Nick and I have reached the following conclusions concerning QCG:

(1) It can be used reliably, as a quantum theory, in the far infrared;[3,5]

(2) Its corrections become strong at late times in an inflating universe;[3–5]

(3) It dominates the physics of late time inflation with respect to any matter theory whose phenomenological viability does not require fine tuning;[3,5]

(4) If one assumes that the natural QCG vacuum suffers only perturbatively small corrections then asymptotic graviton scattering amplitudes are infrared divergent even at tree order;[4] and

(5) Modulo some tensor algebra, QCG corrections act to slow the inflationary expansion rate by an amount which becomes non-perturbatively large at late times.[5]

The first three results are simple to understand and I will explain them all in a single section. Although the fourth result is also simple to understand, its interpretation is sufficiently subtle as to merit a separate section. Our fifth result is the most important and also the most involved. I will allot one section to explaining how we derive it and another to elucidating the underlying physics. Three short final sections discuss, respectively, the few things we can say at this time about the case of negative Λ, why no previous studies have revealed our relaxation mechanism, and what our scheme may mean for cosmology.

THE FIRST THREE RESULTS

Quantum general relativity is not perturbatively renormalizable,[9] but it can still be purged of ultraviolet divergences, order by order in perturbation theory, if one allows arbitrary local counterterms. This seemingly *ad hoc* procedure does not give us a completely consistent quantum theory of gravitation but it does give us one which can be used reliably in the far infrared. To understand why, let us consider the standard exercise of using Fermi theory to compute quantum corrections to low energy weak interactions between fermions.

Suppose we integrate out the Higgs boson and the massive gauge particles of the electroweak model. What results is a non-local effective action which depends upon the electromagnetic vector potential and the various Fermi fields. In low energy processes it makes sense to expand this functional in powers of the derivative operator, whereupon we recover the action of QED and Fermi theory, plus an infinite series of local counterterms. Note that as long as one considers the scattering of only photons and fermions below the heavy particle thresholds, this model is indistinguishable from the electroweak theory. One consequence is that this action can be used to compute loop corrections of arbitrarily high orders. Some of the ultraviolet divergences that appear in loops can be absorbed into redefinitions of the parameters of QED-Fermi theory; the other divergences — the non-renormalizable ones — are subtracted off by divergences which occur in the coefficients of the local counterterms. This apparent miracle derives from the renormalizability of the electroweak action.

Note that we do not even require the electroweak action in order to know the most divergent parts of the counterterms; they are fixed by the ultraviolet divergence structure of QED-Fermi theory. What we cannot get from QED-Fermi theory is the finite parts of counterterms. Although these finite parts are suppressed by powers of heavy masses, they do make contributions. Because most of the quanta in QED-Fermi theory have non-zero masses, loop corrections involving these quanta can be expanded for low in-coming and out-going momenta into a series of ever higher derivative contact interactions. Such terms are not distinguishable from the finite parts of the electroweak induced counterterms. If we did not possess the electroweak action we could not predict the strength of these effects. However, a loop which involves massless quanta — photons or neutrinos — can give rise to corrections containing logarithms of the external momenta. *These* effects can never be simulated by the finite parts of the electroweak induced counterterms because the latter are analytic at zero momenta whereas the former are not. Even if we did not possess the electroweak action we could still trust the predictions of QED-Fermi theory about such logarithm terms. An illustration of this fact is the correct calculation of the long range force due to the exchange of two neutrinos, done in 1968 by Feinberg and Sucher.[10]

The connection to QCG should be obvious. Neither Fermi theory nor quantum general relativity is perturbatively renormalizable. However, just as the renormalizable electroweak theory underlies Fermi theory, so there must be *some* consistent quantum theory of gravitation in back of general relativity. We do not know this theory; it probably involves other quanta, and its fundamental dynamical variable may not even include the metric. Nonetheless, for purely geometrical issues below the heavy quantum threshold — which is presumably the Planck mass — the unknown true quantum theory of gravitation must be indistinguishable from the action of general relativity plus an infinite series of local counterterms. Some of the counterterms renormalize Λ and G; these we treat by imposing physical renormalization conditions the same as in any quantum field theory. The other counterterms subtract off the various non-renormalizable ultraviolet divergences of quantum general relativity; they also supply finite, higher dimension contact interactions which are suppressed by powers of the Planck mass. We do not know these finite terms because we do not know the fundamental quantum theory of gravitation, but they are guaranteed to give large corrections only on very small distance scales. Our procedure is therefore to make the necessary ultraviolet subtractions by hand and then to forget about the finite parts of counterterms. Because the graviton is massless the strongest infrared effects at any loop order will still be reliably predicted by QCG alone, without knowledge of the fundamental theory of quantum gravity.

So much for the first result. Of course the reliability of infrared effects in QCG would be uninteresting if these effects were not significant. That they are derives from two properties of the theory, the first of which is that it allows massless gravitons to self-interact through a coupling of dimension three:

$$\kappa\Lambda\left(\tfrac{1}{3}h^{\alpha}{}_{\beta}h^{\beta}{}_{\gamma}h^{\gamma}{}_{\alpha} - \tfrac{1}{4}hh^{\alpha}{}_{\beta}h^{\beta}{}_{\alpha} + \tfrac{1}{24}h^{3}\right)\sqrt{-g^{\mathrm{class}}} \tag{6}$$

Most particle theorists are familiar with the notion that the infrared is dominated by those massless particles which have the lowest dimension self-interaction. The rationale behind this piece of folk-wisdom is that infrared effects influence a local observation through the coherent superposition of distant interactions within the observer's past light cone. These interactions must be transmitted by *massless* quanta because massive propagators oscillate inside the light cone and so give destructive interference. Interactions of *low dimension* give the strongest infrared effects because increasing the dimension of an interaction means either adding derivatives or else adding more fields. Derivatives are bad because differentiated propagators show weaker long range correlation than undifferentiated ones. More fields are bad because this means that the interaction is transmitted by more propagators, each of which tends to suppress distant correlations; one can also have fewer interactions at a given loop order.

A few examples are illuminating. The long range force discovered by Feinberg and Sucher[10] comes from massless neutrinos interacting through the dimension six coupling of Fermi theory. This force falls off like the sixth power of the distance between sources. The lowest dimension coupling in $\Lambda = 0$ quantum gravity consists of three gravitons with two derivatives distributed among them. One would expect a stronger effect from this dimension five coupling and this is precisely what was found by Weinberg[11] in his study of the problem. (Note again the complete irrelevance of renormalizability for infrared effects involving massless particles.) The canonical example of a dimension four self-interaction is provided by QCD which has interaction vertices between three gluons with one derivative or four gluons with no derivatives. Because of the lower dimension coupling we expect infrared effects in QCD to be stronger than those of either $\Lambda = 0$ quantum gravity or Fermi theory. In fact they are so strong that perturbation theory breaks down, but the long range interaction is conjectured to obey a constant force law. The infrared effects of QCG ought to be even *stronger* because the theory allows massless gravitons to self-interact via the coupling (6) of dimension three.*

The second crucial property of QCG is that for positive Λ the curved background geometry *enhances* the tendency of low dimension self-interactions to produce strong infrared effects. This is because the inflationary redshift increases the population of soft gravitons. Of course the causal structure of an inflating universe differs from that of flat space and one must take account of the decoupling of modes whose physical wavelengths have redshifted beyond the Hubble radius. However, there is still an enhancement of the infrared because the number of modes whose physical wavelengths are *just reaching* the Hubble radius increases without bound as time progresses.** In fact the inflationary enhancement is so important that infrared effects in QCG become arbitrarily strong even when the spatial manifold is compact.

So much for the second result. Even strong infrared effects in QCG would not be interesting if they could be masked by yet stronger infrared effects due to matter. This does not happen for any currently observed matter theory because the quanta involved are either massive — and hence incapable of sustaining the long range correlations necessary to support an infrared effect — or else the interactions are conformally invariant in $3+1$ dimensions. A interesting property of the de Sitter geometry is that it allows coordinates to be chosen in which the invariant element is proportional to that of flat space:

$$g^{\text{class}}_{\mu\nu} dx^\mu dx^\nu = \Omega^2 \left(-du^2 + d\vec{x} \cdot d\vec{x} \right) \tag{7a}$$

$$\Omega \equiv \frac{1}{Hu} \equiv \exp(Ht) \tag{7b}$$

When a conformally invariant Lagrangian is expressed in this coordinate system using appropriately rescaled fields, all the factors of Ω cancel and one obtains the same

---

* One can demonstrate the graviton's masslessness in QCG either by noting the propagator's absence of oscillations inside the light cone [1] or by constructing the free Hamiltonian and showing there is no mass gap.[12] QCG shows a time dependent screening of the effective cosmological constant rather than gravitational confinement because infrared divergences break time translation invariance and change the relevant sense of "infrared" from large spatial separations at fixed time to large temporal separations at fixed position.[5]

** The attenuation due to causality is responsible for the curious fact that quantum corrections to expectation values of the metric can exceed the classical result only by powers of the conformal observation time,[5,13] whereas in-out matrix elements of the same operators — which receive contributions as well from the *future* light cone — experience infrared divergences proportional to powers of the conformal cutoff time.[3,4]

Lagrangian as in flat space. What this means physically is that conformally invariant quanta do not experience the inflationary redshift which enhances infrared effects in QCG. They also lack the dimension three coupling; the self-interactions of conformally invariant quanta have dimension four.

Gravitation is unique among the observed forces for being mediated by massless particles which are not conformally invariant. This means that even though the natural dimensionless coupling constant (5) of QCG is much smaller than α_{QCD}, quantum gravity must eventually dominate the infrared of an inflating universe with respect to quantum chromodynamics. A simple way to understand this is that in conformal coordinates and using conformally rescaled fields, the vertices of QCG contain positive powers of Ω which are not cancelled by propagators. On the other hand, the vertices of QCD are constant.* In the far future — in other words, as the conformal time u goes to zero — the excess conformal factors make up for the smallness of $\kappa^2 H^2$ and the interactions of QCG become much stronger than those of QCD.

I should mention two caveats before closing this section. The first is that the eventual dominance of QCG over observed matter refers only the physics of the far infrared. If one probes at scales intermediate between the inflationary mass (4) and the Planck mass then matter couplings can be much more important than gravitational ones for all time. For example, suppose the Bohr radius of a Hydrogen atom is much smaller than the Hubble radius. In this case the electron and the proton remain bound to one another, and electromagnetism is always a more significant determinant of the bound state structure than QCG.

The final caveat is that one can imagine non-conformally invariant quanta whose masses are much less than the inflationary mass scale M. Obvious candidates are axions and the various superpartners of the graviton in extended supergravity. This sort of matter might compete with pure QCG for a time if inflation occurs at high scales.

INFRARED DIVERGENCES IN TREE ORDER SCATTERING

One constructs invariant scattering amplitudes by integrating external wave functions against vertices which are connected by propagators. Although the final result is independent of the choice of field variable, the three components from which it is constructed are not. For QCG it turns out that these components are most efficiently represented if we conformally rescale the full metric:

$$g_{\mu\nu} \equiv \Omega^2 \widetilde{g}_{\mu\nu} \equiv \Omega^2 \Big(\eta_{\mu\nu} + \kappa\psi_{\mu\nu}\Big) \tag{8}$$

The full inverse of $\widetilde{g}_{\mu\nu}$ is denoted by $\widetilde{g}^{\mu\nu}$, i.e., $\widetilde{g}_{\mu\nu}\widetilde{g}^{\nu\rho} = \delta_\mu^{\ \rho}$. We call $\psi_{\mu\nu}$ the "pseudo-graviton" field, and its indices are raised and lowered with the Lorentz metric. After some judicious partial integrations the invariant Lagrangian of QCG can be cast in a form not too different from that of $H = 0$ quantum gravity:[1]

$$\begin{aligned} \mathcal{L}_{\rm inv} - S^{\nu}{}_{,\nu} = & -\tfrac{1}{2}\sqrt{-\widetilde{g}}\,\widetilde{g}^{\rho\sigma}\,\widetilde{g}^{\mu\nu}\,\psi_{\rho\sigma,\mu}\psi_{\nu}{}^{\alpha}\,(\Omega^2)_{,\alpha} \\ & + \sqrt{-\widetilde{g}}\,\widetilde{g}^{\alpha\beta}\,\widetilde{g}^{\rho\sigma}\,\widetilde{g}^{\mu\nu}\Big[\tfrac{1}{2}\psi_{\alpha\rho,\mu}\,\psi_{\nu\sigma,\beta} - \tfrac{1}{2}\psi_{\alpha\beta,\rho}\,\psi_{\sigma\mu,\nu} + \tfrac{1}{4}\psi_{\alpha\beta,\rho}\,\psi_{\mu\nu,\sigma} - \tfrac{1}{4}\psi_{\alpha\rho,\mu}\,\psi_{\beta\sigma,\nu}\Big]\Omega^2 \end{aligned} \tag{9}$$

(Commas denote ordinary differentiation in this and all subsequent formulae.) Breaking up covariant derivatives and making the various partial integrations combines the dimension five coupling of conventional quantum gravity with the dimension three one (6) unique to QCG. This makes for some confusion between infrared and ultraviolet but it does give simple interaction vertices. Except for the first term, and the factor

* There is a very weak dependence upon the conformal factor through ultraviolet regularization.

of Ω^2 on the second, this form for the Lagrangian is the same as that for perturbation theory around flat space in $H = 0$ quantum gravity!

The simplest gauge fixing term is $-\frac{1}{2} F_\mu F_\nu \eta^{\mu\nu}$, where:

$$F_\mu = \Omega\Big(\psi^\nu{}_{\mu,\nu} - \tfrac{1}{2}\psi_{,\mu} + 2\,\psi^\nu{}_\mu\,\Omega_{,\nu}\,\Omega^{-1}\Big) \tag{10}$$

At tree order it is most convenient to give the spatial fourier transform of the pseudo-graviton propagator:*

$$\int d^3x\, \exp(i\vec{k}\cdot\vec{x})\, i\Big[{}_{\mu\nu}\Delta_{\rho\sigma}\Big](u,\vec{x};u',0) = \frac{H^2 uu'}{2k}\exp\Big[-ik|\Delta u|\Big]\Big[2\,\eta_{\mu(\rho}\eta_{\sigma)\nu} - \eta_{\mu\nu}\eta_{\rho\sigma}\Big]$$
$$+\frac{H^2}{2k^3}\Big(1 + ik|\Delta u|\Big)\exp\Big[-ik|\Delta u|\Big]\Big[2\,\overline{\eta}_{\mu(\rho}\overline{\eta}_{\sigma)\nu} - 2\,\overline{\eta}_{\mu\nu}\overline{\eta}_{\rho\sigma}\Big] \tag{11}$$

where $\Delta u \equiv u' - u$, parenthesized indices are symmetrized, and a bar tensor indicates the suppression of temporal components:

$$\overline{\eta}_{ij} = \delta_{ij} \qquad , \qquad \overline{\eta}_{0\mu} = 0 \tag{12}$$

Except for the initial factor of H^2uu' the first term of the propagator (11) is precisely the same as that of the de Donder gauge propagator of conventional quantum gravity on a flat space background!

The external state wave functions for physical gravitons are:

$$\Psi_{\mu\nu}\Big(u,\vec{x};\vec{k},\lambda\Big) = \frac{Hu}{\sqrt{2k}}\Big(1 + \tfrac{i}{ku}\Big)\exp\Big[ik(u - 1/H) + i\vec{k}\cdot\vec{x}\Big]\,\epsilon_{\mu\nu}(\vec{k},\lambda) \tag{13}$$

where the polarization tensors are purely spatial, transverse, traceless, and canonically normalized:

$$\epsilon_{\mu 0}(\vec{k},\lambda) = k_i\,\epsilon_{ij}(\vec{k},\lambda) = \epsilon_{ii}(\vec{k},\lambda) = 0 \tag{14a}$$

$$\epsilon^*_{\mu\nu}(\vec{k},\lambda) = \epsilon_{\mu\nu}(-\vec{k},\lambda) \tag{14b}$$

$$\epsilon^*_{ij}(\vec{k},\lambda)\,\epsilon_{ij}(\vec{k},\lambda') = 2\,\delta_{\lambda\lambda'} \tag{14c}$$

These wave functions evolve according to the free field equations of QCG. They are canonically normalized:

$$-i\int d^3x\,\Psi^*_{\alpha\beta}\Big(u,\vec{x};\vec{k},\lambda\Big)\Big[{}^{\alpha\beta}\overleftrightarrow{W}{}^{\rho\sigma}_0\Big]\,\Psi_{\rho\sigma}\Big(u,\vec{x};\vec{k}',\lambda'\Big) = (2\pi)^3\delta^3(\vec{k} - \vec{k}')\,\delta_{\lambda\lambda'} \tag{15}$$

using the Wronskian of free QCG:

$$\Big[{}^{\alpha\beta}\overleftrightarrow{W}{}^{\rho\sigma}_\mu\Big] = \Omega\,(\overrightarrow{\partial}_\mu - \overleftarrow{\partial}_\mu)\,\Omega\,\Big[\tfrac{1}{2}\eta^{\alpha(\rho}\eta^{\sigma)\beta} - \tfrac{1}{4}\eta^{\alpha\beta}\eta^{\rho\sigma}\Big] \tag{16}$$

Note especially that the inner product (15) is conserved. Integrating the wave function (13) into an amputated Green's function inserts an in-coming graviton with spatial momentum $\vec{k}$ and polarization λ; the conjugate wave function removes an out-going graviton with the same quantum numbers.

---

* To overcome the inconvenient temporal inversion evident in (7b) we define the propagator as the expectation value in free de Sitter vacuum of the *anti*-time-ordered product of the two free field operators. The alternative usually followed in the literature is to define the conformal time as negative, but then one encounters the inconvenience of needing absolute values. Note that in our conformal coordinates there is also a minus sign in the canonical commutation relations.

It is now simple to understand why QCG scattering amplitudes harbor infrared divergences at tree order. The mathematics is that the various integrals over conformal time diverge at $u = 0$. This is because vertices supply powers of $\Omega = 1/Hu$ whereas propagators and external wave functions fail to vanish at $u = 0$. The physics is equally transparent: all gravitons are redshifted down to zero physical momentum in the asymptotic future, so the various plane waves are brought into ever stronger interaction with one another.

The appearance of an infrared divergence signals that there is something unphysical about the question being posed. The traditional problem is that one cannot discriminate between nearly degenerate states.[14] Note, however, that superposing degenerate ensembles of states can only absorb infrared divergences when the tree amplitudes are finite. What Nick and I found is that infrared divergences afflict even the tree amplitudes for vacuum into three gravitons, and for a single graviton decaying into two.[4] These processes contribute to the amplitude at order κ. There is simply nothing at lower order to which we can add soft gravitons in an attempt to cancel these divergences.

Nick and I agree that some unphysical feature of our scattering amplitudes must be responsible for their infrared divergences. When the same problems were long ago noted for light, minimally coupled scalars, Tagirov[15] concluded that such particles simply could not exist. The point might be debated, but no one can question that gravity exists; the problem must therefore be something we are assuming about it. Nick and I believe that the unphysical assumption is the indefinite persistence of the de Sitter background. It was also this assumption which led to the infrared divergences we found in computing two loop corrections to the in-out matrix element of the metric[3] and one loop corrections to the graviton self-energy[5]. One can see this from the fact that the most divergent contributions come in each case from virtual gravitons that stray far into the future of the last observed field.

There is a regrettable tendency upon reaching this conclusion to utterly discount the results of in-out matrix elements and scattering amplitudes. It is certainly true that precise statements about the evolution of the QCG background must be derived using a different formalism which we will describe in the next section. However, it is not correct that in-out results are devoid of content. For example, the fact that we find a non-zero amplitude for the vacuum to go into three gravitons implies the vacuum's decay. The fact that this amplitude contains an infrared divergence indicates that the vacuum must suffer non-perturbatively large corrections at late times. One cannot use in-out amplitudes to infer the *rate* at which changes in the background become non-perturbatively large, but one can rule out the notion that they remain small. For if causal evolution results in only perturbatively small corrections to the QCG vacuum then the assumption of de Sitter in the asymptotic future would not lead to infrared divergences.

In fact, as soon as one is confronted by a guess as to the evolution of the background this guess can be either verified or debunked by the simple device of using it to fix the asymptotic vacua and then computing the associated in-out matrix elements.* If the guess is correct at late time — at least to within perturbatively small corrections — then the associated in-out matrix elements will be free of infrared divergences. In this way one can, for example, rule out the oft-made suggestion that the background eventually settles down into de Sitter but with a somewhat different cosmological constant.

Before closing this section I should mention that not all our colleagues share the interpretation Nick and I have given to our fourth result. Some people believe that the problem can be corrected with a different choice of vacuum state but without changing the de Sitter background. Others reject the S-matrix and in-out matrix elements as valid observables in QCG. I will summarize why Nick and I do not find these objections to be convincing.

* The designation, "in-out" becomes a little fuzzy at this point because I am considering changes in the asymptotic vacua. I will try to be clear about this by using the qualifier, "associated" to describe matrix elements between states which are not asymptotically de Sitter.

First consider the issue of the vacuum. Our choice is the obvious one dictated by local correspondence with the known result of flat space and by de Sitter invariance.* We are perfectly willing to consider other alternatives, but it is difficult to see how this can do any good. Changing the vacuum alters the real part of the propagator. It also reshuffles the external state wave functions; in the new vacuum it will generally be a linear combination of Ψ and Ψ^* which inserts an in-coming graviton. Unfortunately the problem cannot be cured by changing the propagator because infrared divergences show up even in the 3-graviton tree, which involves no propagators. Neither is anything accomplished by forming different linear combinations of the wave functions. The old basis was complete; as long as the new basis is also complete it can be used to access the same divergent amplitudes. And using a less than complete basis is even worse because it puts perturbative unitarity at risk. In this regard it is important to note that the full Fock space must be present if QCG is to agree with conventional quantum gravity in the limit that the Hubble constant H goes to zero.

Considerably more can be said for the rejection of in-out matrix elements and the S-matrix as QCG observables. An inflating universe possesses causal horizons which would seem to preclude the observability of all but a tiny portion of the final state. On the other hand, the initial vacuum certainly evolves to *something*. If we had this final state then the associated in-out matrix elements would be the same as expectation values. Even if our final state were only perturbatively close to the true one, there would be no infrared divergences. Further, the same objections about observability can be made regarding matter theories on a de Sitter background. Yet the tree order S-matrix of QCD is not afflicted by infrared divergences.[4] We therefore sympathize with doubts about the observability of the S-matrix and in-out matrix elements on a de Sitter background, but we do not agree that this explains why these objects harbor infrared divergences in QCG. The correct explanation is, as I have stated, that the background suffers non-perturbatively large corrections at late times.

RELAXATION IN PERTURBATIVE TIME EVOLUTION

The appearance of infrared divergences for even tree order asymptotic scattering amplitudes means that the locally de Sitter geometry (3) is not even close to the true background of the far future. Instead of using the in-out formalism to check another guess we shall *derive* the answer by following the evolution of the metric's expectation value in the presence of a prepared initial state. The sort of initial state we seek corresponds not to the mathematical ideal of a spatial section of the full de Sitter manifold but rather to a simple model of conditions that might realistically be expected to prevail when inflation begins after causal evolution from a Big Bang. We will make a brief survey of this prehistory in order to motivate our choice of model.

The temperature shortly after the Big Bang must be so much greater than the inflationary mass scale (4) that thermal contributions to the stress-energy dwarf those due to the cosmological constant. This means there is no inflation. There is also no possibility for strong infrared effects because thermal fluctuations disrupt the long range correlations needed to sustain them. During this period it is just as if there were no cosmological constant: the universe is radiation dominated and its background geometry expands and cools classically.

Causality precludes the maintenance of thermal equilibrium throughout the entire universe during this period. One should think rather of many patches of local thermal equilibrium, none of them extending further than the causal horizon. The usual view is that the currently observed universe descends from a portion of one such patch. When the temperature of this region eventually falls below M the cosmological constant begins to dominate the stress-energy tensor and inflation commences. Thereafter points in the region's interior rapidly lose causal contact with the outside

* There is a trivial breaking of de Sitter invariance due to the zero mode.[1,16] However, this can be ignored in tree order scattering by merely choosing the momenta to be non-zero and off resonance.[4]

universe, and we can ignore the other patches. We can also ignore thermal effects because even a few e-foldings of inflation redshift the residual temperature to practically zero. The appropriate focus of our study is therefore the evolution of zero temperature QCG on a single patch, starting from a homogeneous and isotropic initial state.

Since interior points rapidly lose causal contact with the boundary we can simplify the problem further by imposing periodic boundary conditions. To be explicit, we assume each of the three spatial coordinates lies within the range:

$$-\frac{1}{2H} < x_i \leq \frac{1}{2H} \tag{17}$$

with the endpoints identified. This means we are specializing to the manifold $T^3 \times R$, which does admit the local de Sitter background (3) even though topological obstructions prevent it from possessing the full de Sitter group of isometries. Very little changes from the implicitly $R^3 \times R$ analysis of the previous section. The spatial momenta become discrete of course, so momentum integrals go over to sums, and the continuum state normalization (15) becomes discrete. There are also zero modes whose evolution in free QCG is analogous to that of free particles in quantum mechanics rather than harmonic oscillators.

There are many homogeneous and locally isotropic states in QCG. The simplest choice for perturbative computations — and the one we used — is free de Sitter vacuum for the non-zero modes and minimum uncertainty Gaussians for the zero modes. We do *not* believe that the true initial state was actually free any more than we believe the topology of our universe is really $T^3 \times R$. We can afford to make these simplifying assumptions because the inflationary expansion of spacetime quickly imposes its own order on local physics. The finite number of initially low momentum modes rapidly decouple as their wave lengths redshift beyond the Hubble radius. (This is why the zero modes are irrelevant.) Any infrared effect which persists to late times must therefore derive from redshifting the infinite reservoir of initially high momentum modes. These reach equilibrium among themselves in about a Planck time. Since we must in any case assume that all observations come much later than the ultraviolet length cutoff, which is also the Planck time, the high modes have plenty of time to equilibrate.

The exact position space propagator is an uninteresting mode sum which has been given elsewhere.[4] Note, however, that we require this propagator only for the narrow range of spatial coordinates given by (17). A similar restriction on the range of conformal times derives from the fact that the onset of inflation at $t = 0$ corresponds to $u = 1/H$, while the infinite future at $t = \infty$ corresponds to $u = 0$. Within this very small range of conformal coordinates the mode sums comprising the propagator can be excellently approximated as follows:[4]

$$i\Big[{}_{\mu\nu}\Delta_{\rho\sigma}\Big](u,\vec{x};u',\vec{x}\,') \approx \frac{1}{4\pi^2}\frac{H^2 uu'}{\Delta x^2 - \Delta u^2 + i\epsilon}\Big[2\,\eta_{\mu(\rho}\eta_{\sigma)\nu} - \eta_{\mu\nu}\eta_{\rho\sigma}\Big]$$
$$- \frac{H^2}{8\pi^2}\ln\Big[H^2\Big(\Delta x^2 - \Delta u^2 + i\epsilon\Big)\Big]\Big[2\,\overline{\eta}_{\mu(\rho}\overline{\eta}_{\sigma)\nu} - 2\,\overline{\eta}_{\mu\nu}\overline{\eta}_{\rho\sigma}\Big] \tag{18}$$

where $\Delta u \equiv u' - u$ and $\Delta x \equiv \|\vec{x}\,' - \vec{x}\|$, and I remind the reader that a barred tensor has its natural zero components suppressed as in (12).

Schwinger[17] has developed a diagrammatic formalism for computing expectation values almost as simply as the usual Feynman rules give in-out matrix elements. The idea is that one first evolves forward from the initial state to some time in the future of the last observation. This is accomplished by a functional integral with the conventional action — over the desired time interval — evaluated at fields $\psi^{+}_{\mu\nu}$. One then evolves back to the initial state by a functional integral over the complex conjugate action evaluated at fields $\psi^{-}_{\mu\nu}$. Vertices can have either all + lines or all − lines, and propagators can link any two kinds of line. The ++ propagator is given in (18), while the −− is its complex conjugate. The +− propagator is obtained from

(18) by replacing the $i\epsilon$ terms with $i\text{sgn}(u-u')\epsilon$, where $\text{sgn}(x)$ is the signum function. The $-+$ propagator is obtained by the replacement: $i\epsilon \longmapsto -i\text{sgn}(u-u')\epsilon$.

To compute the expectation value of an operator one merely sums all the relevant diagrams. Schwinger's formalism has the important property that the various $+$ and $-$ contributions add so as to give a real result for the expectation value of a Hermitian operator such as the metric. The two kinds of vertices also interfere destructively whenever an interaction lies outside the past light cone of the operator being observed. This means that the evolution we are studying is causal. It also means that we can ignore the topological peculiarities of $T^3 \times R$. For example, the fact that topological obstructions prevent invariance under continuous spatial rotations is not locally observable. The local field equations are isotropic, so if the initial background is locally isotropic — which we impose by fiat — then the background has this property for all time.

The object we actually compute is the amputated expectation value of the pseudo-graviton field. By virtue of the initial state's homogeneity and local isotropy we can express the expectation value using just two functions of the conformal time:

$$D_{\mu\nu}{}^{\rho\sigma} \left\langle 0 \right| \kappa\psi_{\rho\sigma}(u,\vec{x}) \left| 0 \right\rangle = a(u)\, \overline{\eta}_{\mu\nu} + c(u)\, \delta_\mu{}^0 \delta_\nu{}^0 \tag{19}$$

The symbol $D_{\mu\nu}{}^{\rho\sigma}$ stands for the pseudo-graviton kinetic operator:[1]

$$D_{\mu\nu}{}^{\rho\sigma} \equiv \left[\tfrac{1}{2}\overline{\delta}_\mu{}^{(\rho}\overline{\delta}_\nu{}^{\sigma)} - \tfrac{1}{4}\eta_{\mu\nu}\eta^{\rho\sigma} - \tfrac{1}{2}\delta_\mu{}^0\delta_\nu{}^0\delta_0{}^\rho\delta_0{}^\sigma\right]\mathrm{D}_A + \delta_{(\mu}{}^0\overline{\delta}_{\nu)}{}^{(\rho}\delta_0{}^{\sigma)}\mathrm{D}_B + \delta_\mu{}^0\delta_\nu{}^0\delta_0{}^\rho\delta_0{}^\sigma\mathrm{D}_C \tag{20}$$

where $\mathrm{D}_A \equiv \Omega(\partial^2 + 2/u^2)\Omega$ is the kinetic operator for a massless, minimally coupled scalar and $\mathrm{D}_B = \mathrm{D}_C \equiv \Omega\,\partial^2\Omega$ is the kinetic operator for a conformally coupled scalar. When single external lines emanate from a vertex in Schwinger's formalism they are always retarded propagators, so the result for the expectation value is:[5]

$$\left\langle 0 \right| \kappa\psi_{\mu\nu}(u,\vec{x}) \left| 0 \right\rangle = A(u)\, \overline{\eta}_{\mu\nu} + C(u)\, \delta_\mu{}^0\delta_\nu{}^0 \tag{21a}$$

$$A(u) = -4\, G_A^{\text{ret}}\left[a\right](u) + G_C^{\text{ret}}\left[3a + c\right](u) \tag{21b}$$

$$C(u) = G_C^{\text{ret}}\left[3a + c\right](u) \tag{21c}$$

The retarded Green's functions of the massless minimally coupled and conformally coupled scalars are respectively:

$$G_A^{\text{ret}}\left[u^{-4}\,(Hu)^\varepsilon\right](u) = \frac{H^2}{3\varepsilon(1-\frac{1}{3}\varepsilon)}\left\{(Hu)^\varepsilon - (1-\tfrac{1}{3}\varepsilon) - \tfrac{1}{3}\varepsilon(Hu)^3\right\} \tag{22a}$$

$$G_C^{\text{ret}}\left[u^{-4}\,(Hu)^\varepsilon\right](u) = \frac{H^2}{2(1-\varepsilon)(1-\frac{1}{2}\varepsilon)}\left\{-(Hu)^\varepsilon + 2(1-\tfrac{1}{2}\varepsilon)Hu - (1-\varepsilon)(Hu)^2\right\} \tag{22b}$$

Note that four powers of $1/u$ are lost in attaching the external line. Since the pseudo-graviton is part of a rescaled metric $\widetilde{g}_{\mu\nu} = \Omega^{-2}\, g_{\mu\nu}$ whose classical value is constant, the functions $a(u)$ and $c(u)$ must grow faster than u^{-4} if quantum corrections are to exceed the classical result.

Because of the special symmetries of our initial state it is simple to find an invariant measure of the rate at which spacetime is expanding. First write the invariant element in co-moving coordinates:

$$\left\langle 0 \right| g_{\mu\nu}(u,\vec{x})\, dx^\mu dx^\nu \left| 0 \right\rangle = -dt^2 + R^2(t)\, d\vec{x}\cdot d\vec{x} \tag{23a}$$

then define the effective Hubble constant as the logarithmic derivative of the scale factor:

$$H_{\text{eff}}(t) \equiv \frac{d}{dt} \ln\Big[R(t)\Big] \tag{23b}$$

To express this in terms of the functions $A(u)$ and $C(u)$ we merely compare (23a) with the quantum-corrected invariant element in conformal coordinates:

$$\Big\langle 0 \Big| g_{\mu\nu}(u, \vec{x})\, dx^\mu dx^\nu \Big| 0 \Big\rangle = \Omega^2 \Big\{ -\Big[1 - C(u)\Big] du^2 + \Big[1 + A(u)\Big] d\vec{x} \cdot d\vec{x} \Big\} \tag{24}$$

This allows us to identify the scale factor and the quantum-corrected coordinate transformation which relates u to t:

$$R(t) = \Omega \sqrt{1 + A(u)} \tag{25a}$$

$$dt = -\Omega \sqrt{1 - C(u)}\, du \tag{25b}$$

We then apply (23b) to obtain:

$$H_{\text{eff}}(t) = \frac{H}{\sqrt{1 - C(u)}} \left\{ 1 - \frac{\frac{1}{2} u \frac{d}{du} A(u)}{1 + A(u)} \right\} \tag{26}$$

Note that this is in principle a non-perturbative result if one could somehow evaluate $A(u)$ and $C(u)$ non-perturbatively.

Renormalization is accomplished by subtracting primary divergences using local counterterms. We choose the cosmological counterterm to remove primary divergences in the amputated 1-point function and to enforce the condition that the effective Hubble constant be initially time independent:

$$\frac{d}{dt} H_{\text{eff}}(t) \Big|_{t=0^+} = 0 \tag{27a}$$

The condition that $H_{\text{eff}}(0) = H$ is enforced by the choice of initial state, and is indeed obeyed by $|0\rangle$. Since both the coefficients $A(u)$ and $C(u)$, and their first derivatives, vanish at onset we can express the condition for $\delta\Lambda$ in terms of the initial values of the coefficients $a(u)$ and $c(u)$:

$$\left(\frac{d}{du}\right)^2 A(u)\Big|_{u=H^{-1}} = a(H^{-1}) - c(H^{-1}) = 0 \tag{27b}$$

If desired we can enforce the vanishing of the other linearly independent combination of $a(H^{-1})$ and $c(H^{-1})$ by using time dependent changes in the coordinate system. We choose δG to remove primary divergences of dimension two in the 2-point function. Since the imaginary part of our propagator gives the correct classical response to a point mass[1] we know that Newton's Law is true at onset. We could use the finite part of δG to enforce the initial *constancy* of Newton's law, but it turns out that the finite part of δG is not relevant to the leading infrared contributions to the amputated 1-point function. The finite parts of the higher dimension counterterms are even less relevant.

To analyze the actual u dependence of the functions $a(u)$ and $c(u)$ we begin by extracting factors of κ and H. The n-point vertex contributes a factor of $\kappa^{n-2} H^{-2}$. From (18) we see that each propagator is H^2 times functions which depend only very weakly on H. If an ℓ-loop contribution consists entirely of 3-point interactions there will be $V = 2\ell - 1$ vertices and $P = 3\ell - 2$ propagators. Since we include an extra factor of κ in the definition (19), the ℓ-loop contributions to $a(u)$ and $c(u)$ can be written as:

$$\kappa^{V+1}\, H^{2P-2V} = \Big(\kappa H\Big)^{2\ell}\, H^{-2} \tag{28}$$

times the integral of a function which depends only weakly on H — and not at all on κ — over the $2\ell - 2$ free vertices. To see that this result is correct even if higher point vertices are involved, note that diagrams involving the higher point vertices at the same loop order can be obtained by contracting propagators.* Suppose we make such a contraction between an m-point and an n-point vertex. Before the contraction the structure contributed a factor of:

$$\kappa^{m-2} H^{-2} \times H^2 \times \kappa^{n-2} H^{-2} = \kappa^{m+n-4} H^{-2} \tag{29}$$

Afterwards we are left with a single vertex having $m + n - 2$ legs, so the factor is unchanged. What *does* change is the number of vertex integrations.

Recall that κ has the dimensions of length, while H and the pseudo-graviton field have the dimensions of inverse length. It follows that the coefficient functions $a(u)$ and $c(u)$ have the same dimensions as the pseudo-graviton kinetic operator: namely, inverse length squared. At ℓ loops these functions can be written as $(\kappa H)^{2\ell} H^{-2}$ times the integral of a function which depends only weakly on H. This integral has dimension of length^{-4} and it can depend only upon the observation time u, the ultraviolet regularization parameter ϵ — which has dimensions of length squared — and the Hubble constant H. The only strong source of dependence upon H is through the upper limit — H^{-1} — of the conformal time integrations. Now it happens that our conformal coordinate patch actually conceals most of its infinite invariant volume near $u = 0$. We must therefore expect that infrared effects derive mostly from the *lower* limits, and since they must be ultraviolet finite, it follows that the leading infrared effects at any order go like u^{-4} times dimensionless functions of u. These dimensionless functions can only be logarithms, and there can be at most one for every vertex integration. At ℓ loops there are at most $2\ell - 2$ free vertices, so we expect the ℓ-loop coefficient functions to behave as:

$$a_\ell(u) \;,\; c_\ell(u) \quad \sim \quad \Big(\kappa H\Big)^{2\ell} H^{-2} u^{-4} \Big\{\# \ln^{2\ell-2}(Hu) + (\text{subdominant})\Big\} \tag{30}$$

Since the inclusion of higher point vertices at the same loop order results in fewer free vertices, the leading term at any order is composed entirely of 3-point vertices.

One loop effects cannot make significant corrections to $H_{\text{eff}}(t)$ because the strongest growth they can give to the amputated 1-point function is proportional to u^{-4}. From (21-22) — and expanding (26) to lowest order — we see that the resulting corrections to $H_{\text{eff}}(t)$ approach a constant in the far future. (This result was previously obtained by Ford[18] using a different technique.) Since this constant must be a pure number times $(\kappa H)^2 \sim 10^{-12}$, any one-loop effect has to be irrelevant.

One way to understand the absence of a one loop effect is that free interaction vertices can contribute from anywhere within the observer's past light cone. Infrared effects derive from the fact that the invariant volume of this region grows — like $\ln(u)$ in fact — as the observer evolves into the future. The longer he evolves, the greater the invariant volume over which free interaction vertices can contribute, and the larger the effect. But the vertex at the amputation point is not free, it is fixed at the observer's position. There is only this single vertex at one loop, so there can be no growth in time. In fact, with our renormalization condition (27) we ought properly to subtract off even the possible time independent effect.

The first time dependence comes at two loops. One can make a strong case that both of the two possible logarithms occur at this order:[5]

$$a(u) = \frac{\kappa^4 H^2}{u^4} \Big\{a_0 \ln^2(Hu) + (\text{subdominant})\Big\} + O(\kappa^6 H^4) \tag{31a}$$

* One can view the higher interaction vertices as filling out certain portions of the Schwinger parameter space needed to decouple unphysical polarizations from the primitive diagram.

$$c(u) = \frac{\kappa^4 H^2}{u^4} \left\{ c_0 \ \ln^2(Hu) + (\text{subdominant}) \right\} + O(\kappa^6 H^4) \tag{31b}$$

where the "subdominant" terms diverge less strongly as u approaches zero. Substituting into (21-22) and from there into (26) gives:

$$H_{\text{eff}}(t) = H \left\{ 1 - \kappa^4 H^4 \left[(\tfrac{1}{12} a_0 + \tfrac{1}{4} c_0)\,(Ht)^2 + (\text{subdominant}) \right] + O(\kappa^6 H^6) \right\} \tag{32}$$

We do not yet know the numerical values of a_0 and c_0 but it is probably possible to obtain them by using a symbolic manipulation program to perform the tedious tensor algebra which is the bane of quantum gravity. The loop integrals by themselves are not significantly more difficult than ones we have already computed[5] for the analogous decay in massless ϕ^3 theory on flat $R^3 \times R$. If the combination $a_0 + 3c_0$ comes out positive then it will prove that QCG corrections act to break inflation by an amount which becomes non-perturbatively large at late times.

A PHYSICAL MODEL OF RELAXATION

Nick and I feel the simplest way to understand the beginning of relaxation is that dimension three self-interactions always generate negative vacuum energy, which is in turn a source for the expansion of spacetime. The time dependence — and the dominance of cubic self-interactions — comes about because the amount of negative self-energy generated in this way depends upon the size of the loops involved. Loops where the various interaction vertices explore a larger invariant volume of spacetime contribute more than smaller loops. A single loop has no free interaction vertices so it simply contributes a divergent constant that must be renormalized away — or absorbed into a coordinate redefinition — if the initial observer is to see inflation. The size of higher loops is constrained by the requirement that contributing interaction vertices must occur after the onset of inflation and in the past light cone of the observer. As time goes on these loops explore larger and larger invariant volume, so their impact on the expansion of spacetime becomes stronger and stronger.

Many people find this explanation unsatisfying because they feel that any quantum instability must have a classical antecedent. However, no one has ever found a reason why the local de Sitter background (3) should be unstable in classical general relativity.[12,19] In fact a classical mechanism *does* act to destabilize de Sitter space, and it is easy to understand both why this effect is not strong enough to do the job and why its quantum descendant is.

The classical mechanism I referred to is simply the presence of some gravitational radiation in the initial configuration. We can use QCG to study the evolution of the metric by taking the initial state to be not free vacuum but rather a suitably chosen coherent state. In perturbation theory one simply writes the coherent state as an exponential of creation operators acting on the free vacuum, and then expands to obtain an infinite series of Fock space states. The matrix element of the pseudo-graviton field between any two of these n graviton states can be computed using the reduction formalism Nick and I developed at the end of our last paper.[5] Some of the Schwinger diagrams which contribute to these expectation values contain no loops; summing only these terms gives the purely classical effect.

Now consider the pseudo-graviton matrix element between n and m gravitons of fixed momenta. What one finds is that the effect of any mode depends upon its physical wave length at the observation time. The dominant effect comes from modes whose physical wave lengths are just reaching the Hubble radius when the observation is made. When the physical wave length is much below the Hubble radius the interaction overlap with the constant background is weak; and causality cuts off the interaction from modes whose physical wave length has redshifted beyond the Hubble radius. Since all gravitons eventually redshift beyond the causal horizon, there is no strong classical effect at late times.

The connection between these classical processes and the quantum loop effect we see in pure vacuum comes via Feynman's tree theorem.[20] This is a technique in which one decomposes loops of Feynman propagators, connected by local interactions, into sums of tree diagrams. When the method is applied to the Schwinger diagrams that give our two loop effect, the result is just a sum of expectation values of the pseudo-graviton field in the presence of various two graviton states. The crucial difference from the classical process we just considered is that the decomposition of the tree theorem involves sums over gravitons whose initial spatial momenta are arbitrarily large. The quantum effect never cuts off because we never run out of wave lengths which are just redshifting to the Hubble radius. In fact, the number of these *increases* with time as Ω^2 by a simple phase space effect.

That relaxation is a quantum effect deriving from the uncertainty principle can be seen by considering the Fourier transforms of the initial configuration and its first time derivative:

$$q_{\mu\nu}(\vec{k}) \equiv \int d^3x \, \exp\left(i\vec{k}\cdot\vec{x}\right) \psi_{\mu\nu}\left(\tfrac{1}{H},\vec{x}\right) \tag{33a}$$

$$p_{\mu\nu}(\vec{k}) \equiv \int d^3x \, \exp\left(i\vec{k}\cdot\vec{x}\right) \dot{\psi}_{\mu\nu}\left(\tfrac{1}{H},\vec{x}\right) \tag{33b}$$

For these to serve as initial value data for a well defined classical solution they must fall off at high momentum. This means that for high k both $q_{\mu\nu}$ and $p_{\mu\nu}$ are being confined to near zero. We can do this in classical general relativity, in fact we *must* do it if the initial configuration is to have finite curvature. However, this is precisely the sort of thing that quantum field theory does not allow. In quantum field theory one cannot localize a $q_{\mu\nu}$ mode and its conjugate momentum to within a phase space volume smaller than $2\pi\hbar \times H^3$. This is what gives quantum cosmological gravity the unending supply of ultraviolet modes which the locally de Sitter background redshifts to the infrared.

We do not have a good understanding of the non-perturbative regime which prevails at very late times. As the rate of expansion slows, so too does the redshift of high momentum modes. This slows down the growth in the number of modes at any given physical wave length, which would seem to weaken relaxation. However, a competing effect is that the effective Hubble radius increases, restoring causal contact with some of the modes whose wave lengths had redshifted beyond the larger Hubble radius of earlier times. Which of the two effects wins is not presently clear to us.

Before closing this section I must debunk two misconceptions about relaxation. First, Nick and I are often asked the question, "if the local de Sitter vacuum of QCG is unstable, why can't we simply start the theory out in the true vacuum, whatever that is, and dispense with all this talk of time evolution from plausible initial conditions?" This appeals to the widespread prejudice that there should be a time independent vacuum state. The analysis presented above shows that there is no such state in QCG; relaxation goes on as long as perturbation theory remains valid. Although we do not understand the behavior at late times there is nothing to suggest that relaxation ever stops.

The final misconception concerns the renormalization group. Nick and I are continually asked, "why not integrate out the short distance physics and then deal with the resulting effective field theory of long distances?" I will not attempt to prove the negative that there is no useful version of this scheme, but I certainly don't see one. Let me first explain why the procedure seems ambiguous in QCG. I will then deal with the most common friendly suggestion for implementing it and the most common unfriendly one.

The heart of the problem is that QCG lacks a time independent notion of ultraviolet and infrared. Although free QCG evolves plane waves (13) onto plane waves of the same coordinate momenta, the associated *physical* momenta redshift with the inflating universe. This cannot be ignored because it has dynamical consequences; for example, one reason that relaxation occurs is the growth in the number of modes whose momenta are redshifting below H at any given time. We should therefore think not of integrating out high frequency space*time* modes but rather of obtaining an effective theory for the evolution of those modes whose initial spatial momenta are below a certain scale.

The sort of effective field equations just described can be derived but it is not clear that this is simple to do, or that there is any point to it. Any mode whose initial wave length is finite must eventually redshift beyond the causal horizon and decouple. At late times the initially infrared modes are not in interaction with each other but rather with the effective forces generated by the originally ultraviolet modes that were integrated out. The evolution equations of the initially infrared modes all become linear and the forces are $\mathbb{C}$-number functions of time. If we follow the evolution of the expectation value of a given mode we can find the forcing functions directly, as we did in (31) for two of the zero modes. This seems much simpler than trying to erect an effective action for the infrared.

People who like the idea of relaxation often suggest a variant of the renormalization group in an attempt to avoid the breakdown of perturbation theory which occurs when relaxation becomes significant. The idea is to evolve for a only brief time dt, during which perturbation theory remains reliable. One then attempts to mock up the subsequent evolution for another brief period by restarting the process in a modified version of QCG with the bare cosmological constant reset to its effective value at the end of the previous evolution:

$$\Lambda \longmapsto 3H^2_{\rm eff}(dt) \tag{34}$$

Something like this might work in the late stages of relaxation — at least we cannot rule it out — but it does not capture the beginning of relaxation if $a_0 + 3c_0$ is non-zero. To see this divide the time interval t up into N steps, $dt = t/N$, each of which is still of long enough duration that we may neglect Hdt relative to $(Hdt)^2$. Then the n-th iteration gives:

$$H^{(n)}_{\rm eff}(dt) = H^{(n-1)}_{\rm eff}(dt)\left\{1-\kappa^4H^4\left[\left(\tfrac{1}{12}a_0+\tfrac{1}{4}c_0\right)(Hdt)^2+(\text{subdominant})\right]+O(\kappa^6H^6)\right\} \tag{35}$$

The result of iterating N times should agree with (32), but in fact we get:

$$H^{(N)}_{\rm eff}(dt) = H\left\{1-\kappa^4H^4\left[\left(\tfrac{1}{12}a_0+\tfrac{1}{4}c_0\right)\frac{(Ht)^2}{N}+(\text{subdominant})\right]+O(\kappa^6H^6)\right\} \tag{36}$$

It is simply not true that a universe whose past includes long lived virtual gravitons is the same as one without the virtual gravitons but with a somewhat smaller cosmological constant.

Of course not everyone likes relaxation, and some of those who do not, attempt to debunk it by invoking the renormalization group in a different form. Their arguments start with the contention that integrating out the modes above a certain momentum scale — whatever this is taken to mean — must produce the local Einstein equations, with suitably renormalized values for Λ and G, plus a series of irrelevant higher curvature terms. The next step is to note that one can arrange the various higher curvature terms so that they vanish when the metric is locally de Sitter:

$$R_{\mu\nu\rho\sigma} = H^2\left(g_{\mu\rho}g_{\nu\sigma} - g_{\mu\sigma}g_{\nu\rho}\right) \tag{37}$$

Given the first assumption, this step is certainly correct; in fact, if we did *not* arrange things thusly the result would be just a trivial redefinition of the bare cosmological constant. The third step is to note that since the locally de Sitter background (3) solves the renormalized field equations for all time, there is no relaxation. The problem with this argument is that the first assumption is not correct. Theories such as QCG, which contain infrared divergences in gauge invariant, off shell Green's functions,[5] cannot possess effective actions of the stated form.

We do not have to fight this point out in the ethereal context of quantum gravity because it can be demonstrated past the point of reasonable dispute in the much simpler model of a massless scalar with cubic self-interaction in flat space:

$$\mathcal{L} = -\phi_{,\mu}\phi_{,\nu}\eta^{\mu\nu} - \frac{1}{6}\lambda\phi^3 + \text{counterterms} \tag{38}$$

Veneziano proved long ago that infrared divergences preclude the existence of a perturbative S-matrix for this theory when the massless limit is taken in the $\phi = 0$ vacuum,[21] One can also see that off shell Green's functions are infrared divergent at two loops and higher,[5] so the theory cannot possess an effective action of the conjectured form. However, let us ignore this fact and attempt anyway to follow the analog of QCG relaxation. That is, we will study the perturbative evolution of the expectation value of the field in the presence of the state that is free vacuum at $t = 0$. If our critics are right this must obey a field equation of the form:

$$\Box\phi = \frac{1}{2}\lambda'\phi^2 + g\phi^3 + \text{local higher dimension terms} \tag{39}$$

It turns out that the first secular contribution comes at two loops and from diagrams with precisely the same topology as the dominant two loop contributions of QCG:[5]

$$\widehat{\phi}(t) = \frac{-\lambda^3}{2^{10}\pi^4}t^2\left\{\ln^2(2\mu t) - \ln(2\mu t) + \frac{1}{2}\right\} + O(\lambda^5) \tag{40}$$

The conjectured field equation must consist of more than just a renormalization of the scalar potential — the typically assumed form — because although the solution and its first time derivative vanish at $t = 0$, it does not remain zero. In fact even allowing higher derivatives does not help as long as corrections are constrained to be local, Poincaré invariant and perturbative. The assumption that integrating out the high momentum modes produces such an effective field equation is simply wrong. It is equally wrong in QCG, and for the same reason.

THE CASE OF A NEGATIVE COSMOLOGICAL CONSTANT

The maximally symmetric background with a negative cosmological constant is anti-de Sitter space. As the name suggests, it is in many ways the opposite of de Sitter space. Whereas even a locally de Sitter patch rapidly loses causal contact with the outside universe, information in anti-de Sitter space requires only a finite time to flow out to — or in from — spatial infinity. Even for a mathematically perfect initial condition this raises the problem of choosing a sensible spatial boundary condition. The usual choice is a condition in which information from inside is reflected back.[22] This has the unphysical consequence that one cannot make a small disturbance without making an opposing anti-disturbance at the antipodal point — which is inside the manifold for Lorentz signature anti-de Sitter space. We prefer the transmissive condition studied recently by Kleppe.[23] However, the point is really moot when set against the larger issues arising from how the universe might actually enter an anti-de Sitter phase.

Suppose first that the bare cosmological constant is negative and that deflation begins, as we conjecture for positive Λ, when the temperature ceases to dominate the stress-energy tensor. It makes no sense to assume that this occurs simultaneously throughout the universe; what should really happen is the onset of deflation in a region of local thermal equilibrium. However, whereas the portion beyond an inflating patch rapidly becomes irrelevant, the opposite is true for deflation. In this case the outside universe *dominates*. The trouble is that we don't understand very well what to assume about this outside. If our patch is the first of its neighbors to have begun deflation then the others have unknown temperatures — all presumably different and somewhat larger than the deflationary mass scale. If other nearby patches cooled first then we know even less.

The most physically interesting periods of deflation are triggered by matter phase transitions which occur after the observed universe has relaxed from a GUT scale inflation. This carries the additional complication of heavy dependence on the previous history of relaxation. The reason for this is that relaxation means long wave length phenomena in QCG are suppressing a very large and positive bare cosmological constant. As explained previously, this is not a once-and-for-all process; it requires the constant participation of long wave length modes. If the local geometry begins deflating then the redshift of new modes down from the ultraviolet ceases, but old modes from the very far infrared resume interaction. If the first effect dominates then deflation cannot last long because only a small percentage decrease in the suppression of a GUT scale positive cosmological constant is required to overcome the negative effective cosmological constant that might be induced by the electroweak or QCD phase transitions.

WHY THIS MECHANISM HAS NOT BEEN SEEN BEFORE

So many people have searched for an instability of de Sitter space that one might wonder why the mechanism Nick and I have proposed was not studied sooner. As near as I can understand there are three reasons. The first is that our mechanism involves quantum gravity, not quantum matter theory. There is a very widespread prejudice against invoking quantum gravity to do anything because we do not possess a fundamentally consistent theory of it. Few people thought to look for relaxation in QCG.

The second reason is that exotic formulations of QCG have — it seems to us — misled the few people who attempted to study it. For example, there is the Euclidean version, with its infamous conformal rotation and the option of topology change. Then there is the seductive influence of de Sitter invariance. We exploit conformal flatness on a tiny portion of the full de Sitter manifold, thereby abandoning even the pretense of de Sitter invariance.[4,16] Our effect derives from causal correlations which contribute over longer and longer times. However, if one attempts to work on the full de Sitter then there is a very subtle error through which it can seem that a de Sitter invariant vacuum exists and that its propagator is plagued by acausal correlations which grow with increasing spatial separation.[1,24]

The final reason is that our effect derives from interactions, not from the free theory. At the level of diagrams it arises because the un-fixed interaction vertices in the amputated 1-point function can explore an increasing invariant volume of spacetime as the observation time increases. There are no un-fixed interaction vertices at one loop, which is why no effect was seen there.[18] Very few people thought to look in higher loops because these are much harder to evaluate and there is another widespread prejudice that if something doesn't occur by one loop then it cannot occur at all.

WHAT OUR SCHEME MEANS FOR COSMOLOGY

If Nick and I are right then the bare cosmological constant is not unreasonably small. Further, there is no need for fine tuning of scalar potentials. Our view is that inflation happened in the early universe for no other reason than that the bare cosmological constant is large and positive. The current observed universe derived from a patch of about one Hubble volume which began inflation when the local temperature dropped below the scale M.

The reason our universe is not observed to be inflating today — at least not rapidly — is that infrared processes in QCG tend to screen the cosmological constant. There are two reasons why the effect is strong:

(1) QCG allows massless gravitons to interact through a coupling of dimension three; and

(2) The inflationary redshift continually increases the number of gravitons whose physical wavelength is about one Hubble length.

The reason QCG acts to null inflation is that dimension three self-interactions always induce a negative vacuum energy, which is a source, along with the bare cosmological constant, for the expansion of spacetime. The reason we get any inflation at all is that infrared effects require time to build up. One can see from (32) and condition (5) that a significant diminution in the effective Hubble constant requires a very large number of e-foldings:

$$N = Ht \approx \frac{1}{G\Lambda} > 10^{12} \tag{41}$$

In this respect it is the smallness of the natural dimensionless parameter of QCG which provides the lag needed for a long period of inflation.

All experimentally confirmed matter quanta are either massless or else conformally invariant at the classical level. This means that they give only negligibly small corrections to the classical geometry of an inflating universe. Although certain conjectured light matter quanta may be competitive with QCG for a time, in the end only the graviton is left. This means that QCG makes *unique* predictions for relaxation.

Unfortunately, the most interesting predictions of QCG are not easy to obtain because perturbation theory breaks down. One can see from (32) that when the two loop correction is of order one then the three loop effect — which ought to go like $\kappa^6 H^6 (Ht)^4$ — is at least twelve orders of magnitude larger. We do not worry that these higher order effects might reverse relaxation because the tendency of dimension three couplings to generate negative self-energy is true at every order. However, it does mean that non-perturbative methods must be used to gain control over the late stages of inflation. This is very frustrating because the end of rapid inflation should tell us what reheating temperature was reached; and the last sixty e-foldings governs the magnitude and spectrum of observable density perturbations.

Finally, we would like to suggest two more points of potential experimental contact. The first arises from the fact that the universe should still be experiencing a residual expansion due to its not completely screened bare cosmological constant. It is conceivable that a modified form of perturbation theory could be used to predict this residual expansion rate. There is some evidence that a suitably time varying effective cosmological constant can explain the apparent discrepancy between the age of the oldest galaxies, as inferred by models of stellar evolution, and the age of the universe as inferred by the currently measured Hubble constant.[25]

The second point of potential experimental contact derives from electroweak and QCD chiral symmetry breaking phase transitions. If these occurred after inflation and relaxation at some high scale then they should have induced an effectively negative cosmological constant. We do not understand the process by which deflation is relaxed — or even if it is relaxed — but one would think that the time scale of this gravitational process should be longer than that of either phase transition. If so then there would have been substantial periods of deflation, which would almost inevitably have affected low scale baryogenesis and nucleosynthesis.

Acknowledgements

This work was partially supported by DOE contract DE-FG05-86-ER40272 and by NATO Collaborative Research Grant CRG-910627.

REFERENCES

1. N. C. Tsamis and R. P. Woodard, *Commun. Math. Phys.* **162** (1994) 217.
2. N. C. Tsamis and R. P. Woodard, *Phys. Lett.* **B292** (1992) 269.
3. N. C. Tsamis and R. P. Woodard, *Phys. Lett.* **B301** (1993) 351.

4. N. C. Tsamis and R. P. Woodard, "The Physical Basis For Infrared Divergences In Inflationary Quantum Gravity," University of Florida preprint UFIFT-HEP-93-17, July 1993.
5. N. C. Tsamis and R. P. Woodard, "Strong Infrared Effects In Quantum Gravity," University of Florida preprint UFIFT-HEP-92-24, February 1994.
6. A Sandage, *Observatory* **88** (1968) 91.
7. C. Kounnas, M. Quiros and F. Zwirner, *Nucl. Phys.* **B302** (9188) 403.
8. E. W. Kolb and M. S. Turner, *The Early Universe* (Addison-Wesley, Redwood City, CA, 1990).
9. G. 't Hooft and M. Veltman, *Ann. Inst. Henri Poincare* **20A** (1974) 69;
S. Deser and P. van Nieuwenhuizen, *Phys. Rev.* **D10** (1974) 401; *Phys. Rev.* **D10** (1974) 411;
S. Deser, H. S. Tsao and P. van Nieuwenhuizen, *Phys. Rev.* **D10** (1974) 3337;
M. Goroff and A. Sagnotti, *Phys. Lett.* **B160** (1985) 81; *Nucl. Phys.* **B266** (986) 709.
10. G. Feinberg and J. Sucher, *Phys. Rev.* **166** (1968) 1638.
11. S. Weinberg, *Phys. Rev.* **140** (1965) B516.
12. L. F. Abbott and S. Deser, *Nucl. Phys.* **B195** (1982) 76.
13. A. D. Dolgov, M. B. Einhorn and V. I. Zakharov, "On Infrared Effects In de Sitter Background," University of Michigan preprint UM-TH-94-11, March 1994.
14. T. D. Lee and M. Nauenberg, *Phys. Rev.* **133** (1964) B1549.
15. E. A. Tagirov, *Ann. Phys.* **76** (1973) 561.
16. G. Kleppe, *Phys. Lett.* **B317** (1993) 305.
17. J. Schwinger, *J. Math. Phys.* **2** (1961) 407; *Particles, Sources and Fields* (Addison-Wesley, Reading, MA, 1970);
R. Jordan, *Phys. Rev.* **D33** (1986) 444.
18. L. H. Ford, *Phys. Rev.* **D31** (1985) 710.
19. P. Ginsparg and M. J. Perry, *Nucl. Phys.* **222** (1983) 245.
20. R. P. Feynman, *Acta Phys. Pol.* **24** (1963) 697; in *Magic Without Magic*, edited by J. Klauder (Freeman, New York, 1972), p. 355.
21. G. Veneziano, *Nucl. Phys.* **B44** (1972) 142.
22. S. J. Avis, C. J. Isham and D. Storey, *Phys. Rev.* **D18** (1978) 3565.
23. G. Kleppe, "Greens Functions for Anti de Sitter Space Gravity," University of Alabama preprint, UAHEP-94-03, April 1994.
24. A. D. Dolgov, M. B. Einhorn and V. I. Zakharov, "The Vacuum of de Sitter Space," University of Michigan preprint UM-TH-94-14, May 1994.
25. T. S. Olson and T. F. Jordan, *Phys. Rev.* **D35** (1987) 3258;
P. J. E. Peebles and B. Ratra, *Astrophys. J. Lett.* **325** (1988) L17.

SECTION II

THE VERY EARLY UNIVERSE

BARYOGENESIS FROM ELECTROWEAK STRINGS

Manuel Barriola

Harvard-Smithsonian CFA

60 Garden st.

Cambridge, MA 02138

USA

INTRODUCTION

One implication of the standard hot big bang cosmology is that from the relics of the early universe we should in principle expect the universe to contain the same abundance of baryons and antibaryons. However there exist compelling empirical evidence that the universe is made out of matter with relative few antimatter. From the present asymmetry we can extrapolate that at the energy scale of the electroweak phase transition there was one part in 10^8 more matter than antimatter in the universe. The goal of baryogenesis is to explain why there is this asymmetry between the amount of matter and antimatter i.e. there is a nonbanishing ratio between the net baryon number density and entropy density (n_B/s) in the universe.

Sakharov[1] realized that there are three necessary conditions for any baryogenesis mechanism. First we need the existence of interactions that violate baryon number. The second condition is that the theory in consideration biases the processes that violate baryon number towards the production of a net baryon number. This requires interactions that break the symmetries of charge conjugation (C) and the product of charge conjugation and parity (CP). Finally the interactions have to be in the presence of processes that are out of thermal equilibrium.

Baryogeneses at GUT scales is however affected by two major problems that are in conflict with the standard cosmological picture. The first is that in the standard electroweak model baryon and lepton currents are anomalous and only the difference between them (B-L) is conserved. For temperatures above the electroweak scale baryon and lepton violation processes proceed in thermal equilibrium so that they will equilibrate all the asymmetry produced at the GUT scales and only B-L will be conserved. The second problem comes from the prediction in GUT theories of the existence of relic particles whose mass density is incompatible with the matter density of the universe.

Unified Symmetry: In the Small and in the Large,
Edited by B.N. Kursunoglu *et al.*, Plenum Press, New York, 1995

In particular in any GUT theory we will get that the density of magnetic monopoles will close the universe[2]. The most elegant solution to this problem (together with the horizon and flatness problem) is to invoke that the universe went through a period of inflation[3]that diluted the extra abundances of the relics. However by the same token inflation will also erase any baryon asymmetries produced. The reheating mechanism will not help because the temperature of reheating has to be well bellow the GUT scale in order not to reproduce the unwanted relics, avoiding also the production of baryon asymmetry.

In order to circumvent these problems it has been suggested that baryogenesis takes place at the electroweak phase transition (for a review see Refs.4,5and 6). The three main ingredients of this picture are:

First in the electroweak model we know that, in the field configuration space, the sphaleron[7]is the saddle point configuration connecting two different minima. These minima correspond to vacuum configurations with different baryon number. Transitions among the minima will change the baryon number.

Second, in order to produce a net baryon number asymmetry we need that the electroweak interactions violate C and CP. Extensions of the Minimal Standard Model (MSM) are needed because the MSM violates C and CP in amounts too small to explain the baryon to antibaryon asymmetry.

Finally if the phase transition is first order, bubbles produced at the phase transition will nucleate and expand, near the expanding bubbles we shall have processes that depart from thermal equilibrium.

Different calculations of the characteristics of the phase transition[9]seem to show that the transition is first order for small Higgs masses. The strength of the transition decreases as the Higgs mass increases to the point in which the transition is second order. In particular for the minimal standard electroweak model the present lower bound of the Higgs mass (60 Gev) suggests that the phase transition is weakly first order. It also seems likely that in this range for the Higgs mass the amount of anomalous baryon number violation after the phase transition will wash out any baryon number that was created during the phase transition. Finally we point out that if the transition is second order baryogenesis will not take place since the transition will be in thermal equilibrium.

Recently Brandenberger and Davis[10]have proposed a new mechanism that uses strings produced at the electroweak phase transition[12]. At the core of the strings we have trapped false vacuum in which baryon number changing transitions can occur through configurations that connect two vacua with different baryon number. The collapsing strings will be a source for the interactions that violate CP and of processes that are out of thermal equilibrium. In order for their mechanism to work they need that the thickness of the string is bigger than the magnetic screening length of the field theory at high temperatures, so that the transitions that change baryon number can take place. This leads to the condition that the core of the string has to be thick so that the Higgs mass is small. As we saw before this implies that the phase transition will be more likely first order. Ref.11also studies strings but this ones are produced at a symmetry breaking whose energy is slightly higher than the electroweak scale. The electroweak symmetry is unbroken at the core of the cosmic strings and transitions that change baryon number proceed as before.

In this article we would like to briefly review electroweak strings (in particular Z-strings[12]) and different mechanisms that imply magnetic fields at the time of the electroweak phase transition. We shall show how anomalous currents occur in the electroweak model. We shall examine the interaction of the magnetic Z-flux along the core of the string with the background magnetic field. This interaction will introduce baryon and lepton anomalous currents. Finally we shall study their implications as a possible mechanism for baryogenesis.

ELECTROWEAK STRINGS AND MAGNETIC FIELDS

It has been shown[12]that in the standard electroweak model there are solutions that look like Nielsen-Olesen strings with magnetic Z-flux along their cores. Nambu[12]has also shown that these strings can connect monopole-antimonopole pairs. The strings are metastable for some range in parameter space even though the standard electroweak theory is topologically trivial. However they are not stable for the physical value of the Weinberg angle θ_w. We should expect that Z-string solutions should survive for extensions of the standard electroweak model. For example Refs.13and 14have shown that this is the case in the two-Higgs model extension. Ref.13has also studied the stability of this solutions to small perturbations. They find that for realistic values of the Higgs mass and the Weinberg angle the string is unstable.

Dvali and Senjanovic[16]have also shown that if we consider the electroweak model with two Higgs doublets and we add an extra global $U(1)_{gl}$ symmetry to the theory, then there are topologically stable string solutions that also carry magnetic Z-flux along their core. This result also suggest that this Z-flux tubes may be metastable when the extra $U(1)_{gl}$ is an approximate symmetry. In Ref.17these authors have also shown that the minimal supergravity extension of the standard model has these topological solution if the hidden sector has an exact R-symmetry. These strings have different characteristics than 'ordinary' embedded Z-strings produced at the time of the phase transition. We shall not consider them in this paper.

In what follows we will consider the embedded Z-strings solutions of Ref.12. We shall assume extensions of the minimal standard model in which these strings are stable or meta-stable. We should also point out that these strings will be stabilized by the nature of this baryogenesis mechanism. As we shall see, at the core of the strings massless baryons will be produced while outside they would be massive. Ref.15has shown that bound states stabilize non-topological solitons.

Since the electroweak model has monopole solutions we expect that when our string network is produced we will have $(M - \bar{M})$ pairs connected by strings and small string loops. The $M - \bar{M}$ and strings will interact with the background plasma and the monopoles will contract with some speed v_c. If l is the typical length of the strings in the network then after the time interval $\delta t_d = l/v_c$, the strings will decay. The strings will also have some transverse velocity with respect to the background that we will denote by v_t.

There are several mechanism that will generate primodial magnetic fields in the early universe. Some of them relay on physical effects during an inflationary phase transition[18]. It has also been suggested[19]that in the electroweak phase transition the fields have to be uncorrelated at distances larger than the initial correlation length(ξ). Therefore the gradients of the Higgs fields can not be compensated by the gauge fields for scales bigger than ξ. These gradients imply that we will have electromagnetic fields.

Recently Ref.20have used the Savvidy vacuum[21]to generate magnetic fields at the scale of grand unification. They propose that magnetic field fluctuations at the GUT scales produce a phase transition to a new ground state with a non zero magnetic field. All these mechanisms imply strong magnetic fields at the time of the electroweak phase transition.

ANOMALOUS CURRENTS AND BARYOGENESIS

Now that we have reviewed electroweak strings and the existence of magnetic fields at the time of the phase transition, we will show how the anomalous currents form in this model. In the electroweak model all the currents that couple with the gauge fields are free of anomalies so that the theory is renormalizable. However the baryon and lepton currents have anomalies of the form

$$\partial_\mu J^\mu_B = \partial_\mu J^\mu_L = N_f \left(\frac{g^2}{32\pi^2} W^{\mu\nu}_a \tilde{W}^a_{\mu\nu} - \frac{g'^2}{32\pi^2} Y_{\mu\nu} \tilde{Y}^{\mu\nu} \right). \tag{1}$$

Where N_f is the number of families, $W^{\mu\nu}_a$ $a = 1, 2, 3$ and $Y_{\mu\nu}$ are the $SU(2)$ and $U_Y(1)$ field strengths respectively, with the tilde referring to the usual definition of the dual of the field strength. g and g' are the associated gauge couplings. Note that there is not an anomaly for the difference of the baryon to the lepton currents.

If we integrate both sides of Eq.(1)over a volume V and assume that the currents vanish at the surface of V we obtain that the baryon number $B = \int d^3x J^0_B$ changes in some time interval by,

$$\Delta B = \frac{N_f}{32\pi^2} \int dt \int d^3x \left(g^2 W^{\mu\nu}_a \tilde{W}^a_{\mu\nu} - g'^2 Y_{\mu\nu} \tilde{Y}^{\mu\nu} \right). \tag{2}$$

We see that if the r.h.s. of Eq.(2)is non zero we shall have a change in the baryon number. Another way to interpret this result is realizing that the integrand can be expressed as a total divergence so if we neglect surface effects and define the Chern-Simons numbers N_{cs} and n_{cs}

$$\begin{aligned} N_{cs} &= \frac{g^2}{32\pi^2} \int d^3x \epsilon^{ijk} \left(W^a_{ij} W^a_k - \frac{1}{3} g \epsilon_{abc} W^a_i W^b_j W^c_k \right) \\ n_{cs} &= \frac{g'^2}{32\pi^2} \int d^3x \epsilon^{ijk} Y_{ij} Y_k \end{aligned} \tag{3}$$

we can perform the time integral to obtain

$$\Delta B = N_f (\Delta N_{cs} - \Delta n_{cs}). \tag{4}$$

Therefore a change in the Chern-Simons number introduces a change in the baryon number. However the Chern-Simons number is not a meaningful physical quantity since it is gauge dependent and only changes in the Chern-Simons number will be gauge invariant.

After the phase transition the W^μ_3 and Y^μ mix to form the massive Z^μ and the massless electromagnetic A^μ vector bosons.

In terms of these new vector fields Eq.(2)takes the form,

$$\Delta B = \frac{N_f}{32\pi^2}\int d^4x[\quad g^2\vec{E}^{\bar{a}}_W\cdot\vec{B}^{\bar{a}}_W + \alpha^2\cos 2\theta_w\vec{E}_Z\cdot\vec{B}_Z + \frac{\alpha^2}{2}\sin 2\theta_w(\vec{E}_A\cdot\vec{B}_Z + \vec{E}_Z\cdot\vec{B}_A) + i.t.] \tag{5}$$

where the electric and magnetic fields are defined respectively,

$$\begin{aligned} E^{\bar{a}i}_W &= W^{0i}_{\bar{a}}, \quad B^{\bar{a}i}_W = \frac{1}{2}\epsilon^i_{jk}W^{jk}_{\bar{a}} \\ E^i_A &= A^{0i}, \quad B^i_A = \frac{1}{2}\epsilon^i_{jk}A^{jk} \\ E^i_Z &= Z^{0i}_{\bar{a}}, \quad B^i_Z = \frac{1}{2}\epsilon^i_{jk}Z^{jk} \end{aligned} \tag{6}$$

and $\bar{a} = 1,2$. The last term of Eq.(5)represent interaction terms of the electric and magnetic components of $\vec{Z}$ and $\vec{A}$ with $\vec{W}_{\bar{a}}$ and self-interactions of $\vec{W}_{\bar{a}}$. From Eq.(5)we see that there is no coupling between the electric and magnetic components of $\vec{A}$.

Once the W^1, W^2 and Z bosons become massive, unless there is some topological obstructions in their configurations, we should expect that they are exponentially suppressed. Equation(5)can be reduced to

$$\Delta B = \frac{N_f}{32\pi^2}\int d^4x\left[\alpha^2\cos 2\theta_w\vec{E}_Z\cdot\vec{B}_Z + \frac{\alpha^2}{2}\sin 2\theta_w(\vec{E}_A\cdot\vec{B}_Z + \vec{E}_Z\cdot\vec{B}_A)\right]. \tag{7}$$

$\vec{B}_Z$ and $\vec{E}_Z$ survive because we assume that after the phase transition we have Z-strings. At the core of the strings we have false vacuum and the Z bosons will be massless. From Eq.(7)we can infer a neat physical interpretation for the change in the baryon number. The first term in the r.h.s. of Eq.(7)is the change in the helicity of the string network[22]. The second term can be easily understood if we realize that the background magnetic fields transform to electric fields in the frame of the moving strings. Therefore this term reflects the interactions of the magnetic Z flux along the core of the Z-strings with these electric fields. The last term will have the inverse interpretation of the previous one. The transformation of $\vec{B}_Z$ from the string to the background frame will give us an $\vec{E}_Z$ field that will interact with the background magnetic field.

To quantify the effect of the first term of Eq.(7)would require first to know, using a computer simulation, the percentage of loops that are linked in the string network. We shall not do this calculation[23]. In what follows we shall assume that the contribution from this term is not greater than the contributions coming from the other two terms.

BIAS MECHANISM

Now that we have established a new mechanism to introduce anomalous currents that will violate baryon number, we need a process that biases these currents towards the production of a net baryon number. We are going to consider the two-Higgs model as an example. We shall show that the classical equations of motion for the strings are modified by the quantum corrections coming from the effective potential. For collapsing strings, this corrections will favor the production of baryons over antibaryons.

The relevant term in the effective action that biases the baryon number is given by the coupling of the relative phase θ of the two Higgs doublets with the W^a gauge bosons through the triangle diagram[24]. For slowly varying θ this term can be expanded in a power series in the momenta[25]. We also do the same diagram but replacing the W^a by the Y gauge fields. When we include both diagrams we find that the effective action contains the term,

$$\Delta S = \int d^4x\kappa\theta\frac{m_t^2}{T^2}\left(g^2W_a^{\mu\nu}\tilde{W}_{\mu\nu}^a - g'^2Y_{\mu\nu}Y^{\mu\nu}\right) \tag{8}$$

where $\kappa = \frac{14}{3\pi^2\zeta(3)}$ and the expansion is up to quadratic order in m_t/T. Using the standard definitions of Z^μ, A^μ vector bosons and assuming that the W^1 and W^2 are exponentially suppressed we obtain

$$\Delta S \sim \frac{N_f\alpha^2}{64\pi^2}\kappa\sin 2\theta_w\frac{m_t^2}{T^2}\int d^4x\theta\left(\vec{E}_A\cdot\vec{B}_Z + \vec{E}_Z\cdot\vec{B}_A\right). \tag{9}$$

This term in the action will modify the classical equations of the string network interacting with the background electromagnetic fields introducing a bias towards the production of a net baryon number. This term violates C and P but conserves CP. However the evolution of θ will not conserve CP because of explicit breaking of CP in the two-Higgs potential. Using the approach of Ref.24we can consider that θ is homogeneous in space and integrate by parts the first term in the r.h.s. of Eq.(8), we obtain

$$\Delta S \sim \int dt\frac{d\theta}{dt}(N_{CS} - n_{CS}). \tag{10}$$

This correspond to a potential for the Chern-Simons number giving a net change in the baryon number.

As the string network is formed Eq.(9)will bias the evolution of the interaction of the magnetic flux of the strings with the electromagnetic background. But in order for the mechanism to work we need that the angle θ changes in time in a definite direction so we favor the production of a net baryon number. This directionality is given by the fact that the string network is collapsing. In particular we have seen that we will have strings bounded by monopole-antimonopole pairs and small loops that will be collapsing. At the moving ends of the strings, θ and the Higgs fields that describe the strings will evolve in a definite way in time. The $M - \bar{M}$ pairs and string loops will collapse at relativistic speeds, this will give production of baryons that are out of thermal equilibrium.

ESTIMATE OF THE BARYON ASYMMETRY

The magnetic Z-flux along the core of the string is $\sim 2gT^2$. We also know that the strings will be moving in a background magnetic field. To be definite we will consider the magnetic fields generated from the random fluctuations of the Higgs field gradients at the time of the electroweak phase transition[19]. These gradients will generate electromagnetic fields with electric and magnetic components that will be smooth in scales of the order of the correlation length $\xi \sim 2gT$ and with a strength of the order of $E \sim B \sim \frac{1}{2}gT^2$. At this temperature the scale of ξ corresponds to the inter-particle separation of the background plasma. As the correlation length grows the electromagnetic fields will start to be affected by the fact that the plasma is a very good conductor. So we would expect that the electric fields will be erased and the magnetic fields will be frozen in the plasma.

The transformation of these magnetic fields under a Lorentz transformation from the frame of the background plasma to the system of the string moving with velocity v_t with respect of the background is of the order of $E_A \sim \gamma v_t B_A$. As a first approximation we have neglected the electric fields assuming that the plasma is a very good conductor. The rate of baryon number produced per unit time and volume is

$$\Gamma \sim E_A \cdot B_Z \sim \gamma v_t g^2 T^4 \tag{11}$$

This equation has to be modified to include the effect of the corrections that come from the effective potential. As we saw before this contribution will be significant only at the ends of the strings or in the collapsing loops. This will give that the volume in which the correction is effective is $\sim \delta^3$ where δ is the width of the string ($\delta \sim \lambda^{-1/2}\eta^{-1}$). The rate of baryon generation per string is

$$\frac{dN_B}{dt} \sim \gamma v_t g^2 T^4 \epsilon \delta^3 \tag{12}$$

where ϵ is a dimensionless constant that gives the numerical contribution of Eq.(9)to the rate of baryon generation. To find the rate of increase in the baryon number density we have to multiply the previous expression by the density of strings. We are going to assume one string per correlation volume at the time of the phase transition ($\xi^3 \sim \lambda^{-3}\eta^{-3}$). We obtain

$$\frac{dn_B}{dt} \sim \gamma v_t g^2 \lambda^{1/2} \epsilon T^4. \tag{13}$$

This has to be integrated by the time that lasts the string network ($t_d \sim d/v_c \sim \lambda^{-1}\eta^{-1}$) so that

$$n_B \sim \gamma v_t g^2 \lambda^{1/2} \epsilon T^3. \tag{14}$$

The entropy density at the time of the phase transition is

$$s = \frac{\pi^2}{45} \mu T^3, \tag{15}$$

where μ is the number of spin states. We obtain that the ratio between the baryon and entropy density is,

$$\frac{n_B}{s} = \frac{45}{\pi^2}\frac{1}{\mu} \gamma v_t g^2 \lambda^{1/2} \epsilon. \tag{16}$$

For $\lambda \sim 1$ we need $v_t \epsilon \sim 10^{-7}$ in order to obtain the observational value of $n_B/s \sim 10^{-8}$.

Acknowledgements

I would like to thank R. Brandenberger, G. Field, T. Vachaspati and A. Vilenkin for helpful discussions. I also thank the Ministry of Education of Spain for financial support.

REFERENCES

1. A. D. Sakharov JETP Lett. 5:24 (1967).
2. J. Preskill, Phys. Rev. Lett. P**43**, 1365 (1979).
3. A. H. Guth, Phys. Rev. D **23**, 347 (1981).
4. A.G. Cohen, D.B. Kaplan and A.E. Nelson, Annu. Rev. Nucl. Part. Sci **43** 27 (1993).
5. N. Turok, in 'Perspectives on Higgs Physics', ed. G. Kane (World Scientific, Singapore, 1992).
6. A.D. Dolgov UM-AC.93-91 preprint (1993).
7. N. S. Manton Phys. Rev. D **28**, 2019 (1983): F. R. Klinkhammer and N. S. Manton, Phys Rev. D **30**, 2212 (1984).
8. R. H. Brandenberger and A.C. Davis, Phys. Lett. **B308**, 79 (1993).
9. See 4and references therein.
10. R. Brandenberger and A. C. Davis, Phys. Lett. **B308**, 79 (1993).
11. R. Brandenberger, A. C. Davis and M. Trodden, BROWN-HET-935 preprint (1994).
12. T. Vachaspati, Phys. Rev. Lett. **68**, 1977 (1992). For earlier work see Y. Nambu, Nucl. Phys. **B130**, 505 (1977); M.B. Einhorn and A. Savit, Phys. Lett. **77B**, (1978) 295; N.S. Manton, Phys. Rev. **D28**, (1983) 2018.
13. M. A. Earnshaw and M. James Phys. Rev. **d48**, 5818 (1993).
14. L. Perivolaropoulos Phys. Lett. **B316**, 528 (1993).
15. T. Vachaspati and R. Watkins, Phys. Lett. **B318**, 163 (1993); T.D. Lee, Phys. Rep. **23C**, 254 (1976).
16. G. Dvali and G. Senjanovic, Phys. Rev. Lett. **71**, 2376 (1993).
17. G. Dvali and G. Senjanovic IC/94 preprint (1994).
18. C.J. Hogan, Phys. Rev. Lett. **51** (1983) 1488; M.S. Turner and L.W. Widrow, Phys. Rev. **D37** (1988); W. Garretson, G.B. Field and S.M. Carroll, Phys. Rev. **D46** (1992) 5346.
19. T. Vachaspati, Phys. Lett. **B265**, 258 (1991).
20. K. Enqvist and P. Olesen NORDITA-94/6 preprint (1994).
21. G.K. Savvidy, Phys. Lett. **B71** (1977) 133.
22. I want to thank George Field for pointing this out when I showed him the calculation.
23. T. Vachaspati and G. Field, TU/94 preprint (1994).
24. N. Turok and J. Zadrozny, Phys. Rev. Lett. **65**, 2331, (1990).
25. M. Dine and S. Thomas SCIPP 94/01 preprint (1994).

RECONSTRUCTING THE INFLATON POTENTIAL

Edward W. Kolb,[1,2] Mark Abney,[2]
Edmund J. Copeland,[3] Andrew R. Liddle,[3] and James E. Lidsey[1]

[1]NASA/Fermilab Astrophysics Center
Fermi National Accelerator Laboratory, Batavia, IL 60510 USA

[2]Department of Astronomy and Astrophysics, Enrico Fermi Institute
The University of Chicago, Chicago, IL 60637 USA

[3]School of Mathematical and Physical Sciences
University of Sussex
Brighton BN1 9QH, United Kingdom

INTRODUCTION

Inflation involves a period of rapid growth of the Universe. This is most easily illustrated by considering a homogeneous, isotropic Universe with a flat Friedmann–Robertson–Walker (FRW) metric described by a scale factor $a(t)$. Here, "rapid growth" means a positive value of $\ddot{a}/a = -(4\pi G_N/3)(\rho + 3p)$ where ρ is the energy density and p the pressure. It is useful to identify the energy density driving inflation with some sort of scalar "potential" energy density $V > 0$ that is positive, and results in an effective equation of state $\rho \simeq -p \simeq V$, which satisfies $\ddot{a} > 0$. If one identifies the potential energy as arising from the potential of some scalar field ϕ, then ϕ is known as the *inflaton* field.

The prime observational consequences of inflation derive from the stochastic spectra of density (scalar) perturbations and gravitational wave (tensor) modes generated during inflation. Each stretches from scales of order centimeters to scales well in excess of the size of the presently observable Universe. Once within the Hubble radius, gravitational waves redshift away and so their main influence is on the large-scale microwave background anisotropies, such as those probed by COBE [1]. Advanced gravitational wave detectors such as the proposed beam-in-space experiments may be able to detect the gravitational waves on a much shorter (about 10^{14}cm) wavelength range. The density perturbations are thought to lead to structure formation in the Universe. They produce microwave background anisotropies across a much wider range of angular scales than do the tensor modes, and constraints on the scalar spectrum are also available from the clustering of galaxies and galaxy clusters, peculiar velocity flows and a host of other measurable quantities [2].

Broadly speaking, inflation predicts a very nearly Gaussian spectrum of density perturbations that is *scale dependent*, i.e., the amplitude of the perturbation depends upon the length scale. Such a dependence typically arises because the Hubble expansion rate during the inflationary epoch changes, albeit slowly, as the field driving the expansion rolls towards the minimum of the scalar potential. This implies that the amplitude of the fluctuations as they cross the Hubble radius will be weakly time-dependent.

Within the context of slow-roll inflation, any scale dependence for density perturbations is possible if one considers an arbitrary functional form for the inflaton potential, $V(\phi)$. In this sense, inflation makes no unique prediction concerning the form of the density perturbation spectrum and one is left with two options. Either one can aim to find a deeper physical principle that uniquely determines the potential, or observations that depend on $V(\phi)$ can be employed to limit the number of possibilities. One such observation is the amplitude of the tensor perturbations produced by inflation.

Recently, we provided a formalism which allows one to reconstruct the inflaton potential $V(\phi)$ directly from a knowledge of these spectra [3]. This developed an original but incomplete analysis by Hodges and Blumenthal [4]. An important result that follows from our formalism is that knowledge of the scalar spectrum alone is insufficient for a unique reconstruction. Reconstruction from only the scalar spectrum leaves an arbitrary integration constant, and since the reconstruction is nonlinear, different choices of this constant lead to different functional forms for the potential. A minimal knowledge of the tensor spectrum, say its amplitude at a single wavelength, is sufficient to lift this degeneracy.

The most ambitious aim of reconstruction is to employ observational data to deduce the inflaton potential over the range corresponding to microwave fluctuations and large-scale structure, although at present the observational situation is some way from providing the quality of data that this would require [3].

In this talk I will discuss the promise of potential reconstruction assuming one knows 1) the amplitude of the tensor spectrum at one point from microwave background fluctuations, presumably on quadrupole scales corresponding to $3000h^{-1}$Mpc, and 2) the scalar spectrum from microwave background fluctuations and the large-scale structure investigations from quadrupole scales down to scales of several Mpc.

PERTURBATIONS FROM SLOW-ROLL INFLATION

We are interested in the perturbations resulting from inflation. The "density" perturbations are usually described in terms of fluctuations in the local value of the mass density. In a Universe with density field $\rho(\mathbf{x})$ and mean mass density ρ_0, the density contrast is defined as

$$\delta(\mathbf{x}) = \frac{\delta\rho(\mathbf{x})}{\rho_0} = \frac{\rho(\mathbf{x}) - \rho_0}{\rho_0}. \tag{1}$$

It is convenient to express this contrast in terms of a Fourier expansion:

$$\delta(\mathbf{x}) = A \int \delta_{\mathbf{k}} \exp(-i\mathbf{k}\cdot\mathbf{x}) d^3k, \tag{2}$$

where A is an overall normalization constant, interesting only for those who enjoy keeping track of factors of 2π. What is usually meant by the density perturbation on a scale λ, $(\delta\rho/\rho)_\lambda$, is related to the square of the Fourier coefficients $\delta_{\mathbf{k}}$:

$$\left(\frac{\delta\rho}{\rho}\right)^2_\lambda \equiv A' \frac{k^3|\delta_k|^2}{2\pi^2}\bigg|_{\lambda=k^{-1}}, \tag{3}$$

where again we have included an overall normalization constant A'. The perturbations are normally taken to be (statistically) isotropic, in the sense that the expectation of $|\delta_{\mathbf{k}}|^2$ averaged over a large number of independent regions can depend only on $k = |\mathbf{k}|$. The dependence of $\delta\rho/\rho$ as a function of λ is the spectrum of the density perturbations.

In a spatially flat isotropic Universe the Hubble expansion rate is $H(t) = \dot{a}/a$, and its inverse $H^{-1}(t)$ (the Hubble radius) is the scale beyond which causal processes no longer operate. Of crucial importance is the relative size of a scale λ to the Hubble radius. The *physical* length between two points of coordinate separation d is $\lambda(t) = a(t)d$, so that a length scale comoving with the expansion will grow in proportion to $a(t)$. The condition for inflation to occur is precisely the condition for physical scales to grow more rapidly than the Hubble length; that is, for the comoving Hubble radius H^{-1}/a to shrink. Thus, a given scale can start sub-Hubble radius, $\lambda < \lambda_H$, pass outside the Hubble radius during inflation, and finally re-enter the Hubble radius long after inflation. Thus, perturbations can be imparted on a given length scale in the inflationary era as that scale leaves the Hubble radius, and will be present as that scale re-enters the Hubble radius after inflation in the radiation-dominated or matter-dominated era.

Microphysics cannot affect the perturbation while it is outside the Hubble radius, and the evolution of its amplitude is *kinematical*, unaffected by dissipation, the equation of state, instabilities, and the like. However, for super-Hubble-radius sized perturbations one must take into account the freedom in the choice of the background reference space-time, i.e., the gauge ambiguities. As usual when confronted with such a problem, it is convenient to calculate a *gauge-invariant* quantity. For inflation it is convenient to study the Bardeen potential ζ [5]. In the uniform Hubble constant gauge ζ is particularly simple. It is related to the background energy density and pressure, ρ_0 and p_0, and the perturbed energy density ρ_1 by $\zeta \equiv \delta\rho/(\rho_0 + p_0)$, where $\delta\rho = \rho_1 - \rho_0$ is the density perturbation.

In the standard matter-dominated (MD) or radiation-dominated (RD) phase, ζ at Hubble radius crossing is equal (up to a factor of order unity) to $\delta\rho/\rho$. Thus, the amplitude of a density perturbation when it crosses back inside the Hubble radius after inflation, $(\delta\rho/\rho)_{\rm HOR}$,[1] is given by ζ at the time the fluctuation crossed outside the Hubble radius during inflation.

As inferred from the adoption of ζ, the convenient specification of the amplitude of density perturbations on a particular scale is when that particular scale just enters the Hubble radius, denoted as $(\delta\rho/\rho)_{\rm HOR}$. Specifying the amplitude of the perturbation at Hubble radius crossing evades the subtleties associated with the gauge freedom, and has the simple Newtonian interpretation as the amplitude of the perturbation in the gravitational potential. Of course, when one specifies the fluctuation spectrum at Hubble radius crossing, the amplitudes for different lengths are specified at *different* times.

Now let us turn to the scalar field dynamics during inflation. Consider a minimally coupled, spatially homogeneous scalar field ϕ, with Lagrangian density

$$\mathcal{L} = \partial^\mu\phi\partial_\mu\phi/2 - V(\phi) = \dot{\phi}^2/2 - V(\phi). \tag{4}$$

With the assumption that ϕ is spatially homogeneous, the stress-energy tensor takes the form of a perfect fluid, with energy density and pressure given by $\rho_\phi = \dot{\phi}^2/2 + V(\phi)$ and $p_\phi = \dot{\phi}^2/2 - V(\phi)$. The classical equation of motion for ϕ is

$$\ddot{\phi} + 3H\dot{\phi} + V'(\phi) = 0, \tag{5}$$

[1]The notation "HOR" follows because often in the literature the Hubble radius is referred to (incorrectly) as the horizon.

and the expansion rate in a flat FRW spacetime is given by ($\kappa^2 = 8\pi G_N$)

$$H^2 = \frac{\kappa^2}{3}\left(\frac{1}{2}\dot{\phi}^2 + V(\phi)\right). \tag{6}$$

Here dot and prime denote differentiation with respect to cosmic time and ϕ respectively. We assume that inflation has already provided us with a flat universe by the time the largest observable scales cross the Hubble radius.

By differentiating Eq. (6) with respect to t and substituting in Eq. (5), we arrive at the "momentum" equation

$$2\dot{H} = -\kappa^2\dot{\phi}^2. \tag{7}$$

All minimal slow-roll models are examples of sub-inflationary behavior, which is defined by the condition $\dot{H} < 0$. Super-inflation, where $\dot{H} > 0$, cannot occur here, though it is possible in more complex scenarios [6, 7]. We may divide both sides of this equation by $\dot{\phi}$ if this quantity does not pass through zero. This allows us to eliminate the time-dependence in the Friedmann equation [Eq. (6)] and derive the first-order, non-linear (Hamilton-Jacobi) differential equations

$$(H')^2 - \frac{3}{2}\kappa^2 H^2 = -\frac{1}{2}\kappa^4 V(\phi); \qquad \kappa^2\dot{\phi} = -2H'. \tag{8}$$

We now consider the production of density perturbations that arise as the result of quantum-mechanical fluctuations of fields in de Sitter space. First, let's consider scalar density fluctuations produced if we assume that the inflaton field ϕ is a massless, minimally coupled field. (Later we will include the corrections due to the fact that the inflaton field has a potential.)

Just as fluctuations in the density field may be expanded in a Fourier series as in Eq. (1), the fluctuations in the inflaton field may be expanded in terms of its Fourier coefficients $\delta\phi_{\mathbf{k}}$: $\delta\phi(\mathbf{x}) \propto \int \delta\phi_{\mathbf{k}} \exp(-i\mathbf{k}\cdot\mathbf{x})d^3k$. During inflation there is an event horizon as in de Sitter space, and quantum-mechanical fluctuations in the Fourier components of the inflaton field are given by [8]

$$k^3 \left|\delta\phi_{\mathbf{k}}\right|^2 / 2\pi^2 = (H/2\pi)^2, \tag{9}$$

where $H/2\pi$ plays a role similar to the Hawking temperature of black holes. Thus, when a given mode of the inflaton field leaves the Hubble radius during inflation, it has impressed upon it quantum mechanical fluctuations. In analogy to Eq. (3), what is called the fluctuation in the inflaton field on scale k is proportional to $k^{3/2}|\delta\phi_{\mathbf{k}}|$, which by Eq. (9) is proportional to $H/2\pi$. Fluctuations in ϕ lead to perturbations in the energy density:

$$\delta\rho_\phi = \delta\phi(\partial V/\partial\phi). \tag{10}$$

Now considering the fluctuations as a particular mode leaves the Hubble radius during inflation, we may construct the gauge invariant quantity ζ using the fact that during inflation $\rho_0 + p_0 = \dot{\phi}^2$:

$$\zeta = \delta\phi\left(\frac{\partial V}{\partial\phi}\right)\frac{1}{\dot{\phi}^2}. \tag{11}$$

Now using Eqs. (8), the amplitude of the density perturbation when it crosses the Hubble radius *after* inflation is

$$\left(\frac{\delta\rho}{\rho}\right)_\lambda^{\mathrm{HOR}} \equiv \frac{m}{\sqrt{2}}A_S(\phi) = \frac{m\kappa^2}{8\pi^{3/2}}\frac{H^2(\phi)}{|H'(\phi)|} \propto \frac{V^{3/2}(\phi)}{m_{Pl}^3 V'(\phi)}, \tag{12}$$

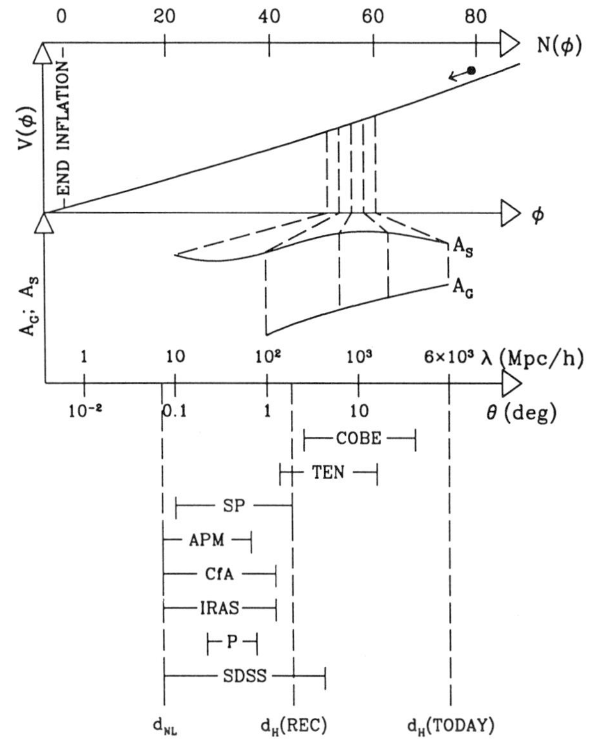

Fig. 1: The basic idea of inflation is that as the field evolves in the potential, quantum fluctuations in the inflaton field produce scalar density perturbations A_S, while fluctuations in the transverse, traceless metric components produce tensor gravitational wave perturbations, A_G. For reconstruction the two main steps involve converting the observations (lower half of figure) into the primordial scalar (A_S) and tensor (A_G) fluctuation spectra and then working in reverse to reconstruct the potential $V(\phi)$. The main observational information from the cosmic microwave background arises through the Cosmic Background Explorer (COBE) *satellite [9], and the Tenerife (TEN) [10] and South Pole (SP) [11] collaborations. Galaxy surveys (APM [12], CfA [13], IRAS [14,15]) may provide useful information up to* $100h^{-1}$ *Mpc, while the Sloan Digital Sky Survey (SDSS) [16] should extend to the lowest scales measured by* COBE. *Peculiar velocity measurements using the POTENT (P) [17]* methods *are important on intermediate scales. The angle θ measures angular scales on the CMBR in degrees, and length scales λ are in units of h^{-1} Mpc. d_H refers to the horizon size today and at recombination and $d_{\rm NL} \approx 8h^{-1}$ Mpc is the scale of non-linearity. Perfect observations will only reconstruct a small portion of the inflaton potential corresponding to between $53 \le \Delta N \le 60$ e-foldings before the end of inflation.*

where $H(\phi)$ and $H'(\phi)$ are to be evaluated when the scale λ crossed the Hubble radius *during* inflation. The constant m equals 2/5 or 4 if the perturbation re-enters during the matter or radiation dominated eras respectively.[2] Now we wish to know the λ-dependence of $(\delta\rho/\rho)_\lambda$, while the right-hand side of the equation is a function of ϕ when λ crossed the Hubble radius during inflation. We may find the value of the scalar field when the scale λ goes outside the Hubble radius in terms of the number of e-foldings of growth in the scale factor between Hubble radius crossing and the end of inflation.

It is quite a simple matter to calculate the number of e-foldings of growth in the scale factor that occur as the scalar field rolls from a particular ϕ to the end of inflation ϕ_e:

$$N(\phi) \equiv \int_{te}^{t} H(t')dt' = -\frac{\kappa^2}{2}\int_{\phi}^{\phi_e} \frac{H(\phi')}{H'(\phi')}d\phi'. \tag{13}$$

The total amount of inflation is given by $N_{\rm TOT} \equiv N(\phi_i)$, where ϕ_i is the initial value of ϕ at the start of inflation (when $\ddot{a}$ first becomes positive). In general, the number of e-folds between when a length scale λ crossed the Hubble radius during inflation and the end of inflation is given by [18]

$$N(\lambda) = 45 + \ln(\lambda/\mathrm{Mpc}) + \frac{2}{3}\ln(M/10^{14}\,\mathrm{GeV}) + \frac{1}{3}\ln(T_{\rm RH}/10^{10}\,\mathrm{GeV}), \tag{14}$$

where M is the mass scale associated with the potential and $T_{\rm RH}$ is the "re-heat" temperature. Relating $N(\lambda)$ and $N(\phi)$ from Eq. (13) results in an expression between ϕ and λ.

In addition to the scalar density perturbations caused by de Sitter fluctuations in the inflaton field, there are gravitational mode perturbations, $g_{\mu\nu} \to g_{\mu\nu}^{\rm FRW} + h_{\mu\nu}$, caused by de Sitter fluctuations in the metric tensor [19,20]. Here, $g_{\mu\nu}^{\rm FRW}$ is the Friedmann–Robertson–Walker metric and $h_{\mu\nu}$ are the metric perturbations. That de Sitter space fluctuations should lead to fluctuations in the metric tensor is not surprising, since after all, gravitons are the propagating modes associated with transverse, traceless metric perturbations, and they too behave as minimally coupled scalar fields. The dimensionless tensor metric perturbations can be expressed in terms of two graviton modes we will denote as h. Performing a Fourier decomposition of h, $h(\vec{\mathbf{x}}) \propto \int \delta h_k \exp(-i\vec{\mathbf{k}}\cdot\vec{\mathbf{x}})d^3k$, we can use the formalism for scalar field perturbations simply by the identification $\delta\phi_{\mathbf{k}} \to h_{\mathbf{k}}/\kappa\sqrt{2}$, with resulting quantum fluctuations [cf. Eq. (9)]

$$k^3|h_{\mathbf{k}}|^2/2\pi^2 = 2\kappa^2(H/2\pi)^2. \tag{15}$$

While outside the Hubble radius, the amplitude of a given mode remains constant, so the amplitude of the dimensionless strain on scale λ when it crosses the Hubble radius after inflation is given by

$$\left|k^{3/2}h_{\mathbf{k}}\right|_\lambda^{\rm HOR} \equiv A_G(\phi) = \frac{\kappa}{4\pi^{3/2}}H(\phi) \sim \frac{V^{1/2}(\phi)}{m_{Pl}^2}, \tag{16}$$

where once again $H(\phi)$ is to be evaluated when the scale λ crossed the Hubble radius *during* inflation.

---

[2] The 4 for radiation is appropriate to the uniform Hubble constant gauge. One occasionally sees a value 4/9 instead which is appropriate to the synchronous gauge. The matter domination factor is the same in either case. Note also that it is exact for matter domination, but for radiation domination it is only strictly true for modes much larger than the Hubble radius, and there will be corrections in the extrapolation down to the size of the Hubble radius.

RECONSTRUCTION EQUATIONS TO SECOND ORDER

To some extent all inflationary calculations rely on the use of the slow-roll approximation. In the form we present here, the slow-roll approximation is an expansion in terms of quantities defined from derivatives of the Hubble parameter H. In general there are an infinite hierarchy of these which can in principle all enter at the same order in an expansion.

The slow-roll approximation arises in two separate places. The first is in simplifying the classical inflationary dynamics of expansion, with the lowest-order approximation ignoring the contribution of the inflaton's kinetic energy to the expansion rate. The second is in the calculation of the perturbation spectra; the standard expressions are true only to lowest-order in slow-roll. In the expressions in the previous section, we utilized the Hamilton-Jacobi approach [21] to treat the dynamical evolution exactly.

A very elegant calculation of the perturbation spectra to next order in slow-roll has now been provided by Stewart and Lyth [22]. The slow-roll approximation can be specified by parameters defined from derivatives of $H(\phi)$. There are in general an infinite number of these as each derivative is independent, but usually only the first few enter into any expressions. We shall require the first two, which are all of the same order when defined by

$$\epsilon(\phi) = \frac{2}{\kappa^2}\left[\frac{H'(\phi)}{H(\phi)}\right]^2 ; \qquad \eta(\phi) = \frac{2}{\kappa^2}\frac{H''(\phi)}{H(\phi)} . \tag{17}$$

The slow-roll approximation applies when these slow-roll parameters are small in comparison to unity. The condition for inflation, $\ddot{a} > 0$, is precisely equivalent to $\epsilon < 1$.

The lowest-order expressions for the scalar (A_S) and tensor (A_G) amplitudes assume $\{\epsilon,\ \eta\}$ are negligible compared to unity. Improved expressions for the scalar and tensor amplitudes for finite but small $\{\epsilon,\ \eta\}$ were found by Stewart and Lyth [22]:

$$\begin{aligned} A_S &\simeq -\frac{\sqrt{2}\kappa^2}{8\pi^{3/2}}\frac{H^2}{H'}\left[1-(2C+1)\epsilon + C\eta\right] \\ A_G &\simeq \frac{\kappa}{4\pi^{3/2}} H \left[1-(C+1)\epsilon\right] , \end{aligned} \tag{18}$$

where $C = -2 + \ln 2 + \gamma \simeq -0.73$ is a numerical constant, $\gamma \approx 0.577$ being the Euler constant. The right hand sides of these expressions are evaluated when the scale in question crosses the Hubble radius during inflation, $2\pi/\lambda = aH$. The spectra can equally well be considered to be functions of wavelength or of the scalar field value.

The standard results to lowest-order are given by setting the square brackets to unity. Historically it has been common even for this result to be written as only an approximate equality (the ambiguity arising primarily because of a vagueness in defining the precise meaning of the density perturbation), though the precise normalization to lowest-order was established some time ago by Lyth [23] (see also the discussion in [2]).

The improved expressions for the spectra in Eqs. (18) are accurate in so far as ϵ and η are sufficiently slowly varying functions that they can be treated adiabatically as constants while a given scale crosses outside the Hubble radius. Corrections to this would enter at next order. This differs from the usual situation in which H is treated adiabatically. For the standard calculation to be strictly valid H must be constant, but provided it varies sufficiently slowly (characterized by small ϵ and $|\eta|$), it can be evaluated separately at each epoch. This injects a scale dependence into the spectra. There is a special case corresponding to power-law inflation for which ϵ and η are precisely constant and equal to each other. In this case there are exact expressions for the perturbation [22,24]. Furthermore, the corrections to each spectrum are the same and they cancel when the ratio is taken. In the general case ϵ and η may be treated as different constants if it is assumed that the timescale for their evolution is much longer than the timescale for perturbations to be imprinted on a given scale. This assumption worsens as η is removed from ϵ, which would be characterized by the next order terms becoming large.

It is useful to define the dimensionless quantities $\tilde{\phi}$ and v, and a dimensionless derivative denoted by a dot:

$$\tilde{\phi} \equiv \frac{\kappa}{\sqrt{2}}\phi; \qquad v(\tilde{\phi}) \equiv \frac{\kappa^4}{48\pi^3}V(\phi)\,; \qquad \dot{X} \equiv \frac{dX}{d\tilde{\phi}}\,. \tag{19}$$

In addition, we can use the identity $\eta = \epsilon + \dot{\epsilon}/2\sqrt{\epsilon}$ and adopt as the expansion variables $\sum_{n=0} \epsilon^{-n/2} d^n\epsilon/d\tilde{\phi}^n$. In terms of these variables, the expressions for $A_S(\lambda)$, $A_G(\lambda)$, v, and the $\tilde{\phi}$–λ relation become

$$\begin{aligned}
A_G &= \frac{\kappa}{4\pi^{3/2}} H\left[1-(C+1)\epsilon\right] \\
A_S &= -\frac{\kappa}{4\pi^{3/2}} H \frac{1}{\sqrt{\epsilon}}\left[1-(C+1)\epsilon + \frac{C}{2}\frac{\dot{\epsilon}}{\sqrt{\epsilon}}\right] \\
v &= A_G^2\left[1+\left(2C+\frac{5}{3}\right)\epsilon\right] = A_S^2\epsilon\left[1+\left(2C+\frac{5}{3}\right)\epsilon - C\frac{\dot{\epsilon}}{\sqrt{\epsilon}}\right] \\
\frac{d\tilde{\phi}}{d\lambda} &= \frac{\sqrt{\epsilon}}{\lambda}(1+\epsilon).
\end{aligned} \tag{20}$$

In the third expression, v depends upon $A_G(\lambda)$ and ϵ. Since we anticipate that we will only have information about $A_G(\lambda)$ at the largest scales, we have to use the "consistency" equation (also called the evolution equation) to relate $A_G(\lambda)$ to the more experimentally accessible $A_S(\lambda)$ at the expense of introducing the additional $\dot{\epsilon}/\sqrt{\epsilon}$ term.[3] This was done through the identity

$$\frac{A_G^2}{A_S^2} = \epsilon\left[1 - C\frac{\dot{\epsilon}}{\sqrt{\epsilon}}\right]. \tag{21}$$

which follows from the expressions for A_S and A_G in Eq. (20). Now we develop the evolution equation by taking the derivative of $A_S(\phi)$:

$$\begin{aligned}
\frac{\dot{A}_S}{A_S} &= \frac{1}{\epsilon^{1/2}}\left(\epsilon - \frac{1}{2}\frac{\dot{\epsilon}}{\sqrt{\epsilon}}\right) + \frac{1}{\epsilon^{1/2}}\left[\frac{C}{2}\frac{\ddot{\epsilon}}{\epsilon}\epsilon - (C+1)\frac{\dot{\epsilon}}{\sqrt{\epsilon}}\epsilon - \frac{C}{4}\left(\frac{\dot{\epsilon}}{\sqrt{\epsilon}}\right)^2\right] \\
&\quad \times \left[1+(C+1)\epsilon - \frac{C}{2}\frac{\dot{\epsilon}}{\sqrt{\epsilon}}\right].
\end{aligned} \tag{22}$$

Now in addition to the expansion in ϵ and its derivatives, a truncation is necessary. The truncation here is to assume $\ddot{\epsilon}/\epsilon \ll \dot{\epsilon}/\sqrt{\epsilon}$. With this truncation, to second order

$$\frac{\dot{A}_S}{A_S} = \frac{1}{\epsilon^{1/2}}\left[\epsilon - \frac{1}{2}\frac{\dot{\epsilon}}{\sqrt{\epsilon}}\right]. \tag{23}$$

Now we can express $\dot{A}_S/A_S$ in terms of the spectral index $1-n \equiv d\ln A_S^2/d\ln\lambda$, and the evolution equation becomes

$$\begin{aligned}
\frac{\dot{\epsilon}}{\sqrt{\epsilon}} &= 2\epsilon - (1-n) \\
\frac{d\epsilon}{d\lambda} &= \frac{\epsilon}{\lambda}(1+\epsilon)\left[2\epsilon-(1-n)\right],
\end{aligned} \tag{24}$$

[3] Of course if the consistency equation is only used to evolve ϵ, it can not be used as a check of inflation as discussed in [3].

where for the second equality we have used the $d\tilde{\phi}/d\lambda$ expression. This evolution equation serves two purposes. It removes $\dot{\epsilon}/\sqrt{\epsilon}$ from the equation for v, and it is a differential equation that can be evolved to give ϵ as a function of λ. To solve the equation it is necessary to know the spectral index as a function of λ, along with the initial condition $\epsilon(\lambda_0)$ as a function of $1 - n_0$, $A_G(\lambda_0)$ and $A_S(\lambda_0)$.

So the system to be solved can be expressed as

$$
\begin{aligned}
v[\tilde{\phi}(\lambda)] &= A_S^2(\lambda)\epsilon(\lambda)\left\{1 + \frac{5}{3}\epsilon(\lambda) + C\left[1 - n(\lambda)\right]\right\} \\
\frac{d\tilde{\phi}}{d\lambda} &= \frac{\epsilon^{1/2}(\lambda)}{\lambda}\left[1 + \epsilon(\lambda)\right] \\
\frac{d\epsilon}{d\lambda} &= \frac{\epsilon}{\lambda}(1+\epsilon)\left[2\epsilon - (1-n)\right].
\end{aligned}
\tag{25}
$$

In the next section we discuss the simplification of the above expressions obtained by dropping the second-order terms and working to first order. In the section after that we solve an example first-order problem.

FIRST-ORDER APPROXIMATION TO RECONSTRUCTION

To first order in the slow-roll expansion variables the expressions simplify considerably. For example, to first order, $\epsilon = A_G^2/A_S^2$, $v = A_G^2$, and $\lambda d\tilde{\phi}/d\lambda = A_G/A_S$. The evolution–consistency equation is also quite simple. It can be written as

$$
\frac{\lambda}{A_G(\lambda)}\frac{dA_G(\lambda)}{d\lambda} = \frac{A_G^2(\lambda)}{A_S^2(\lambda)}. \tag{26}
$$

Again, the procedure will be the same as in the second-order case. The potential depends upon $A_G(\lambda)$, about which we will have information only on the largest scales (possibly only on one scale), so we specify the initial value of $A_G(\lambda)$, and use the consistency–evolution equation to evolve $A_G(\lambda)$ in terms of $A_S(\lambda)$. We can thus express the system to be solved in terms of two equations and a single first-order differential equation which can easily be solved in terms of the initial value $A_G(\lambda_0)$, yielding:

$$
\begin{aligned}
v[\tilde{\phi}(\lambda)] &= \left[A_G^{-2}(\lambda_0) - 2\int_{\lambda_0}^{\lambda}\frac{d\lambda'}{\lambda'}\frac{1}{A_S(\lambda')}\right]^{-1} \\
\pm\tilde{\phi} &= \int_{\lambda_0}^{\lambda}\frac{d\lambda'}{\lambda'}\frac{v^{1/2}[\tilde{\phi}(\lambda')]}{A_S(\lambda')}
\end{aligned}
\tag{27}
$$

A WORKED EXAMPLE

Let's assume a simple power-law potential of the form $V(\phi) = \lambda_\phi\phi^4$ with $\lambda_\phi = 4\times10^{-14}$. This generates perturbation spectra of the form (evaluated at horizon crossing after inflation)

$$
\begin{aligned}
A_S(\lambda) &= 4\times10^{-8}\left[50 + \ln(\lambda/\lambda_0)\right]^{3/2} \\
A_G(\lambda) &= 4\times10^{-8}\left[50 + \ln(\lambda/\lambda_0)\right].
\end{aligned}
\tag{28}
$$

On any scale, the number of statistically independent sample measurements of the spectra that can be made is finite. Given that the underlying inflationary fluctuations are stochastic, one obtains only a limited set of realizations from the complete probability distribution function. Such a subset may insufficiently specify the underlying distribution, which is the quantity predicted by an inflationary model. The cosmic variance is an important matter of principle, being a source of uncertainty which remains even if perfectly accurate experiments could be carried out. At any stage in the history of the Universe, it is impossible to specify accurately the properties (most significantly the variance, which is what the spectrum specifies assuming gaussian statistics) of the probability distribution function pertaining to perturbations on scales close to that of the observable Universe.

Even assuming "perfect" observations, cosmic variance sets a lower limit on the uncertainty at any one scale. Assuming that the only errors come from cosmic variance, the determination of the spectra might look like in Fig. 2. In the realization generated

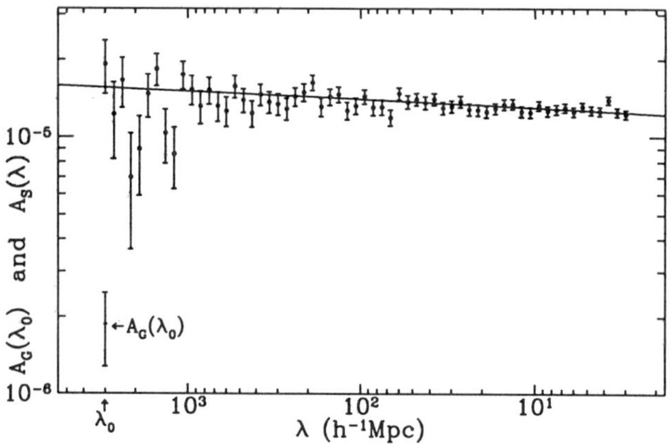

Fig. 2. An illustration of an anticipated data set limited by cosmic variance. The data was generated with a $\lambda_\phi\phi^4$ potential with $\lambda_\phi = 4 \times 10^{-14}$. The upper points are $A_S(\lambda)$, while the single lower point is $A_G(\lambda)$. The solid line is the mean $A_S(\lambda)$, while the mean $A_G(\lambda_0)$ is 2×10^{-6}.

by the random number generator, the value of $A_G(\lambda_0)$ is 1.87×10^{-6}, slightly below the ensemble mean of 2×10^{-6}.

As a first exercise, we simply perform a first-order reconstruction by doing a simple trapezoidal integration, and making the naïve assumption that the errors are uncorrelated. If we do that we obtain the reconstructed potential shown in Fig. 3. Also shown in Fig. 3 by the solid curve is the actual potential used to generate the synthetic data from which the potential was reconstructed.

There are several things we can notice in Fig. 3. First of all, reconstruction works: the true potential is within the error bars. The second obvious feature is that the slope of the reconstructed data is better than one might expect given the errors.

This feature can be explored by taking another approach to the uncertainty introduced in $A_G(\lambda_0)$ by cosmic variance. Let's ignore that error, and pick three realizations

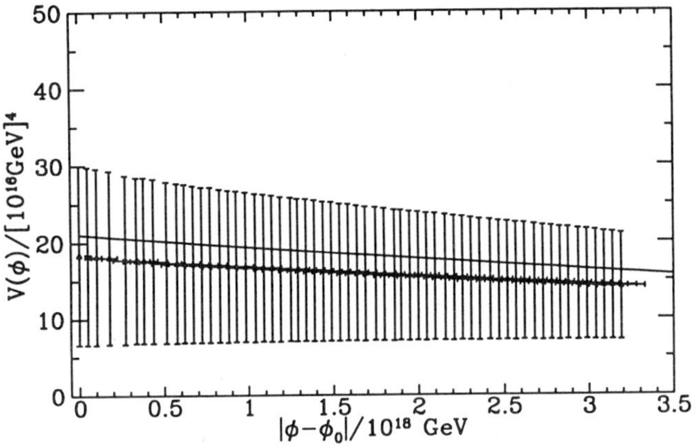

Fig. 3. First-order reconstruction of the example $\lambda_\phi \phi^4$ potential. The solid line is the actual potential, while the points and associated errors were generated from the data of Fig. 2.

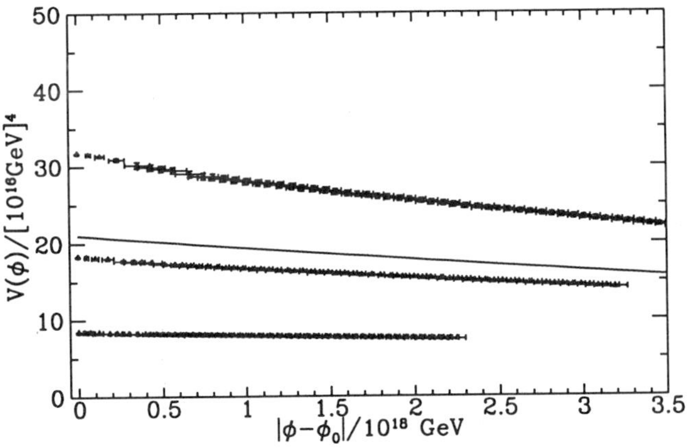

Fig. 4. The reconstructed potential ignoring uncertainty in $A_G(\lambda_0)$ for three choices of $A_G(\lambda_0)$ corresponding to the midpoint and $\pm 1\sigma$.

of $A_G(\lambda_0)$, one at the "measured" value, one 1σ above the measured value, and one 1σ below the measured value. (Here "σ" is the value determined by cosmic variance.) If we do that, we generate the three curves shown in Fig. 4. Although we can't tell which of the curves is the true potential, we know that the true potential is one of a family of curves bounded by the two extremes in the figure.

We can understand why this occurs, because if we look at the slope of v^{-1}, the initial value of $A_S(\lambda)$ drops out, and the contribution comes from adding together a large number of different $A_S(\lambda)$. Since we are combining a large number of data points, the central limit theorem tells us that the errors in the reconstructed potential will become small.

CONCLUSIONS

The quantum-mechanical fluctuations impressed upon the metric during inflation depend upon the inflaton potential. During inflation the Hubble expansion takes microscopic fluctuations of wavelength of order 10^{-28}cm and stretches them to super-Hubble-radius size where they are frozen. Today they appear on scales as large as the observable Universe, 10^{+28}cm. It is possible to read the fossil record of the fluctuations by observing cosmic microwave background fluctuations and the power spectrum of large-scale structure.

If the tensor perturbations are large enough to be identified, and if the scalar power spectrum is determined, the inflaton potential may be reconstructed.

Hence cosmology and astrophysics may provide the first concrete piece of the potential of energy scales of 10^{16}GeV or so.

ACKNOWLEDGMENTS

EJC and JEL are supported by the PPARC. EWK and JEL are supported at Fermilab by the DOE and NASA under Grant NAGW–2381. ARL is supported by the Royal Society. We would like to thank David Lyth and Michael Turner for helpful discussions.

REFERENCES

1. L. M. Krauss and M. White, Phys. Rev. Lett. **69**, 869 (1992); R. L. Davis, H. M. Hodges, G. F. Smoot, P. J. Steinhardt and M. S. Turner, Phys. Rev. Lett. **69**, 1856 (1992); A. R. Liddle and D. H. Lyth, Phys. Lett. **291B**, 391 (1992); J. E. Lidsey and P. Coles, Mon. Not. R. astr. Soc. **258**, 57P (1992); D. S. Salopek, Phys. Rev. Lett. **69**, 3602 (1992); F. Lucchin, S. Matarrese, and S. Mollerach, Astrophys. J. Lett. **401**, 49 (1992); T. Sourdeep and V. Sahni, Phys. Lett. A **7**, 3541 (1992).

2. A. R. Liddle and D. H. Lyth, Phys. Rep **231**, 1 (1993).

3. E. J. Copeland, E. W. Kolb, A. R. Liddle and J. E. Lidsey, Phys. Rev. Lett. **71**, 219 (1993); Phys. Rev. D**48**, 2529 (1993).

4. H. M. Hodges and G. R. Blumenthal, Phys. Rev. D**42**, 3329 (1990).

5. J. M. Bardeen, Phys. Rev. D**22**, 1882 (1980).

6. J. E. Lidsey, Phys. Lett. **273B**, 42 (1991).

7. F. Lucchin and S. Matarrese, Phys. Rev. D**32**, 1316 (1985); Phys. Lett. **164B**, 282 (1985).

8. T. Bunch and P. C. W. Davies, *Proc. Roy. Soc. London* **A360**, 117 (1978).

9. G. F. Smoot *et al.*, Astrophys. J. Lett. **396**, L1 (1992); E. L. Wright *et al.*, Astrophys. J. Lett. **396**, L13 (1992).

10. A. A. Watson *et al*, *Nature* **357**, 660 (1992).

11. T. Gaier, J. Schuster, J. Gunderson, T. Koch, M. Seiffert, P. Meinhold and P. Lubin, Astrophys. J. Lett. **398**, L1 (1992).

12. S. J. Maddox, G. Efstathiou, W. J. Sutherland and J. Loveday, *Mon. Not. R. astr. Soc.* **242**, 43p (1990).

13. M. J. Geller and J. P. Huchra, *Science* **246**, 879 (1989); M. Ramella, M. J. Geller and J. P. Huchra, Astrophys. J. **384**, 396 (1992).

14. W. Saunders *et al*, *Nature* **349**, 32 (1991); N. Kaiser, G. Efstathiou, R. Ellis, C. Frenk, A. Lawrence, M. Rowan-Robinson and W. Saunders, *Mon. Not. R. astr. Soc.* **252**, 1 (1991).

15. K. B. Fisher, M. Davis, M. A. Strauss, A. Yahil and J. P. Huchra, Astrophys. J. **389**, 188 (1992).

16. J. E. Gunn and G. R. Knapp, "The Sloan Digital Sky Survey," Princeton preprint POP-488 (1992); R. G. Kron in *ESO Conference on Progress in Telescope and Instrumentation Technologies*, ESO conference and workshop proceeding no. 42, p635, ed. M.-H. Ulrich (1992).

17. E. Bertschinger and A. Dekel, Astrophys. J. Lett. **336**, L5 (1989); A. Dekel, E. Bertschinger and S. M. Faber, Astrophys. J. **364**, 349 (1990); E. Bertschinger, A. Dekel, S. M. Faber, A. Dressler and D. Burstein, Astrophys. J. **364**, 370 (1990).

18. P. J. Steinhardt and M. S. Turner, Phys. Rev. D**29**, 2162 (1984); E. W. Kolb and M. S. Turner, *The Early Universe*, (Addison-Wesley, New York, 1990).

19. V. A. Rubakov, M. Sazhin, and A. Veryaskin, Phys. Lett. **115B**, 189 (1982); R. Fabbri and M. Pollock, Phys. Lett. **125B**, 445 (1983); B. Allen, Phys. Rev. D **37**, 2078 (1988); and L. Abbott and M. Wise, Nucl. Phys. **B244**, 541 (1984).

20. A. A. Starobinsky, *Sov. Astron. Lett.* **11**, 133 (1985).

21. D. S. Salopek and J. R. Bond, Phys. Rev. D**42**, 3936 (1990); J. E. Lidsey, Phys. Lett. **273B**, 42 (1991).

22. E. D. Stewart and D. H. Lyth, Phys. Lett. **302B**, 171 (1993).

23. D. H. Lyth, Phys. Rev. D**31**, 1792 (1985).

24. D. H. Lyth and E. D. Stewart, Phys. Lett. **274B**, 168 (1992).

BEYOND POTENTIAL DOMINATED INFLATION

Katherine Freese[1] and Janna Levin[2]

[1]*Physics Department, University of Michigan*
Ann Arbor, MI 48109
[2]*Physics Department, Massachusetts Institute of Technology*
Cambridge, MA 02139;
present address: CITA, 60 St. George St.,Toronto, Canada

ABSTRACT

A year and a half ago we[4] proposed a solution to the horizon and monopole problems that differed from many models of inflation in that it does not involve potential domination. Instead, we proposed that a cosmology with a dynamical Planck mass might resolve these problems.

Recently Brustein and Veneziano[13] studied superstring inspired Lagrangians and found that the type of behavior required may be found in these models. As there is renewed interest in this subject, we wish to review some of the things we found. For instance, we discuss the horizon, flatness, and monopole problems in this non-potential dominated approach below.

The basic idea of our work was to attempt to use the behavior of gravity alone, without a potential, to solve the horizon and monopole problems of the standard Hot Big Bang cosmology. Here we discuss some of the problems we initially encountered and proposals we made to move beyond them.

INTRODUCTION

In the standard hot Big Bang cosmology there are finite growing particle horizons. As the particle horizon grows, regions which previously could not have communicated become able to see each other. Differences in properties imprinted during an earlier era should become apparent across these regions. Today, our observable universe would be comprised of many regions which could not have been in causal contact when the universe was still hot and thermalized. We would therefore expect these regions to differ in the temperature of the cosmic background radiation for instance or to differ in the density and composition of the matter distribution. Thus the large-scale homogeneity and isotropy of our observable universe defies a causal explanation.

Any dynamical explanation of the large-scale homogeneity and isotropy must explain how a causally connected region early in the history of the universe could be large enough to encompass our observable universe today. In the inflationary scenario proposed by Guth[1], the universe passed through a stage where potential energy dominated the energy density of the universe. During this era, a causally connected region that was small at the beginning of inflation grew large enough to contain our observed universe. Subsequently, entropy was produced as the potential energy converted to radiation. The presence of potentials provides the key element to the inflationary cosmology.

We propose that a cosmology with a dynamical Planck mass might resolve the horizon problem without a period of potential domination[2]. In a model of modified gravity, for instance, the structure of gravity might slow the Hubble expansion, thus slowing the cosmic evolution. The universe would then linger at a given temperature longer than in the standard model of cosmology and thus grow old at a high temperature. An old universe allows information more time to travel from one point to another – enough time for the entire observable universe to be in causal contact. Subsequently, gravity must change so as to imitate Einstein gravity by today. We call this early epoch of varying gravity the MAD era.

It is the intent of this paper to present an overview of the advances and the limitations of the MAD proposal. We work in the context of massless scalar theories of gravity although the MAD prescription may more generally be a feature of any modified gravity theory. In Ref [2] we found the cosmological solutions to the equations of motion for massless scalar theories of gravity during the radiation dominated era. We introduced there the suggestion that modified gravity could be used to find a new resolution to the horizon problem. In this paper, we review the role of gravity in a solution to the horizon problem and move toward completing the model. We extend the analysis to more general cases and provide a hopefully useful explanation of the roles of changing gravity and entropy production. We also illustrate how the MAD prescription can solve the monopole problem, though not the flatness problem.

We will consider a variety of approaches. First, as an extreme case, we study an adiabatic cosmology i.e, a cosmology in which no entropy is produced. For an adiabatic cosmology we will break the analysis into two regimes: the limit of a slowly moving Planck mass and the limit of a rapidly moving Planck mass. We study more closely in this paper problems with completing the slow roll adiabatic MAD model and indicate directions to pursue to move beyond these obstacles. The possibility of allowing for entropy production is then considered. We note that the horizon problem can be treated with a combination of modified gravity and entropy production. Although we have not worked out the details for a specific model which produces entropy, we discuss nonadiabatic versions toward the close of this paper. Also, any model which solves the horizon problem will necessarily accelerate the scale factor at some time prior to today. We will end by looking for accelerations in the context of general scalar gravity.

CONTENTS

We begin by defining the causality condition, the condition which must be satisfied for a causal explanation of the smoothness of our obervable universe in §I. We also give an introduction to inflation in §IA and to the MAD prescription in §IB. The next section (§II) gives a quick introduction to the action and equations of motion for a general scalar theory of gravity. Those readers unfamiliar with this material, are referred to Ref [2] where the solutions to the equations of motion are worked out in detail. The slow roll limit is discussed in §IIIA and obstacles to completing the MAD solution in the slow roll limit are discussed in §IIIB. §IIIC moves beyond the slow roll limit and gives a quick treatment of the case of a rapidly moving Planck mass in the limit of a large deviation from Einstein gravity. A discussion of the monopole problem is given in §IV. A discussion of the flatness problem is given in §V. In §VI we discuss mechanisms to accelerate the scale factor in scalar gravity theories. Extensions of the MAD proposal to allow for entropy production are discussed in §VII. Lastly, §VIII summarizes the results of this paper.

I. THE CAUSALITY CONDITION

We begin by quantifying the causality condition that would be required by any dynamical mechanism that seeks to explain the smoothness of our observable universe. As well, we give a simple introduction to inflation and MAD gravity. By observable universe we mean that portion of the total universe that we can see, i.e. the universe out to the surface of last scattering. Thus an estimate of the size of the observable universe today is given by the distance light could travel between photon decoupling and now, $d_{obs} \simeq R(t_o)\int_{t_{\rm dec}}^{t_o} dt/R(t) = O(1) \times (t_o - t_{\rm dec})$ for $R \propto t^p$ and $p = O(1)$ between t_{dec} and t_o. Here R is the scale factor. Thus, the size of the observable universe is roughly t_o, the age of the universe today.

We can compare the comoving size of the observable universe to the comoving size of a causally connected region at some earlier time t_c: $d_{hor}(t_c)/R(t_c) = \int_0^{t_c} \frac{dt}{R(t)}$ [In principle it is possible that the integral diverges for $t \to 0$ and there is no horizon problem subsequent to the era of quantum gravity. We are assuming that this is not the case and so are looking for dynamical solutions.] The observable universe today fits inside a causally connected region at t_c if

$$\frac{d_{hor}(t_c)}{R(t_c)} \geq \frac{d_{obs}}{R(t_o)} \,. \tag{1}$$

Here, subscript o refers to today and subscript c to some early time. If condition (1) is met, then the horizon size at t_c (before nucleosynthesis) is large enough to allow for a causal explanation of the smoothness of the universe today.

For power-law expansion of the scale factor both before t_c and after t_{dec} we can take $t \sim H^{-1}$ during these periods. Here $H(t) = \dot{R}/R$ is the Hubble parameter. The causality condition (1) then becomes

$$\frac{1}{R_c H_c} \geq \frac{1}{H_o R_o} \,. \tag{2}$$

Condition (2) is equivalent to the statement that $\dot{R}_o > \dot{R}_c$ and therefore is equivalent to the statement that the universe accelerates, $\ddot{R} > 0$ at some point between t_c and today. Therefore any dynamical solution to the horizon problem involves an epoch of cosmic acceleration. As a result of this period of accelerated expansion, in any model that does satisfy the causality condition (2), there is the possibility of causally generating density fluctuations as scales cross outside the Hubble radius and reenter while the expansion decelerates. In addition, anisotropies and inhomogeneities of the metric can be diluted. In most of the paper, we will use the simpler eqn (2), since our results are often the same as if we use eqn (1). However, we will encounter one case in which $d_{\rm horiz} \neq H^{-1}$ and so these conditions are not identical.

Notice that condition (2) is never satisfied in the standard model since the product $HR = \dot{R}$ always decreases with age. To see this recall that the acceleration of the scale factor is given by

$$\frac{\ddot{R}}{R} = -\frac{4\pi}{3M_o^2}(\rho + 3p) \ , \tag{3}$$

where the Planck mass is $M_o = 10^{19}$ GeV. In the standard hot big bang model, both ρ and p are always greater than or equal to zero. Consequently $\ddot{R}$ is always less than or equal to zero and the scale factor always decelerates.

I A. An Inflationary Epoch

The inflationary model proposed by Guth provides a dynamical resolution of the horizon problem. As a general class of early universe models, inflation suggests that our universe passes through an era when the energy density of the universe is dominated by the energy density in a potential. The energy momentum tensor for the potential is given by $T_f^{\mu\nu} = -Vg^{\mu\nu}$ which leads to a period of de Sitter expansion. The energy density is $\rho_f = V$, while the pressure is $p_f = -V$. As a result of the unusual equation of state, $p_f = -\rho_f$, the de Sitter cosmology leads to a cosmic acceleration given by (see eqn (3))

$$\ddot{R} = \frac{8\pi}{3M_o^2} VR > 0 \tag{4}$$

Consequently, during the era of potential domination, the negative pressure drives the scale factor to accelerate. The accelerated growth of the scale factor inflates a region which was initially subhorizon-sized and therefore in causal contact. If the scale factor grows sufficiently, our observable universe fits inside one of these blown up causally connected volumes. During inflation, the universe supercools. The temperature falls classically as R^{-1} until it reaches the Gibbons-Hawking temperature $H/(2\pi)$, at which point it is stabilized by quantum effects.[3] Therefore, the next crucial ingredient for a successful inflationary model is a period of entropy production which reheats the universe to some high temperature. Entropy production is precisely what is expected as the phase transition which drives inflation completes and the energy previously trapped in the potential is released.

To see explicitly how a given dynamical model meets the causality condition, we begin by rewriting condition (2) in terms of the temperature instead of the scale factor. The universe evolves adiabatically before time t_c so that $R_c T_c \propto (S_c/g_*(t_c))^{1/3}$, where S is the entropy and g_* counts the number of degrees of freedom participating in equilibrium. Therefore we can write R_c in terms of T_c. Similarly, the universe has resumed adiabatic evolution today so that $R_o T_o \propto (S_o/g_*(t_o))^{1/3}$, and we can write R_o in terms of T_o. We allow for a nonadiabatic period between time t_c and today so that $S_o \neq S_c$. With these substitutions, eqn (2) becomes

$$\frac{H_o}{H_c} \gtrsim \frac{T_o}{T_c}\left(\frac{S_c}{S_o}\right)^{1/3} . \tag{5}$$

For the case of inflation, where the Planck mass remains constant in time, one can see that a growth in entropy is needed. To demonstrate, we can use the expression for the Hubble constant from standard cosmology, which is valid both before and after inflation. At time t_c the standard model Hubble constant is given by $H_c = \gamma_c^{1/2} T_c^2 / m_{pl}(t_c)$ where $\gamma = (8\pi^3/90) g_*(t)$. Although in the standard model the Planck mass $m_{pl}(t_c)$ is constant at the value 10^{19} GeV, we leave the Planck mass general for now since it will give us an indication of how a dynamical Planck mass model might accommodate the causality condition. The Hubble constant today can be written as $H_o = \alpha_o^{1/2} T_o^2 / M_o$ where $\alpha_o = \gamma(t_o)\eta_o = (8\pi/3)(\pi^2/30) g_*(t_o)\eta_o$, and $\eta_o \sim 10^4 - 10^5$ is the ratio today of the energy density in matter to that in radiation. The temperature of the cosmic background radiation today is $T_o = 10^{-13}$ GeV and the Planck mass today is $M_o = 10^{19}$ GeV. The causality condition (2) then becomes

$$\frac{m_{pl}(t_c)}{M_o} \left(\frac{S_o}{S_c}\right)^{1/3} \gtrsim \beta \frac{T_c}{T_o} \ , \tag{6}$$

where

$$\beta = (\gamma(t_c)/\alpha_o)^{1/2} (g_*(t_c)/g_*(t_o))^{-1/3} \sim 10^{-2} \ . \tag{7}$$

Since $m_{pl}(t_c) = M_o$ for Einstein gravity, (6) becomes

$$\left(\frac{S_o}{S_c}\right)^{1/3} \gtrsim \beta \frac{T_c}{T_o} \ . \tag{8}$$

For example if T_c is on the order of the scale of Grand Unified Theories (GUTs), $T_c \sim 3 \times 10^{16}$ GeV, then $S_o/S_c \gtrsim 10^{71}$. Indeed, inflation satisfies condition (2) by generating a large ratio of the entropy after inflation to the entropy before inflation. If the evolution were always adiabatic and the Planck mass constant, then (8) reduces to the condition $T_o \gtrsim T_c$. If the universe is forever expanding it can only become cooler in the absence of entropy production and therefore the requirement that the universe is hotter today than at some earlier time can never be met.

Suppose inflation ends at time denoted t_e. We can also rewrite condition (8) in terms of R and T as

$$\frac{R_e}{R_c} \gtrsim \beta \frac{T_c^2}{T_e T_o} \ , \tag{9}$$

where the subscript e denotes quantities evaluated at the end of inflation. We knew condition (2) meant that $\ddot{R} > 0$ for some time between t_c and t_o. This form of the equation tells us just how much R must grow during inflation to satisfy the causality condition. For the example above of $T_c \sim 3 \times 10^{16}$ GeV and $T_e \sim T_c$ then $R_e/R_c \gtrsim 10^{27}$, which is roughly a factor of e^{62}. If the horizon problem is to be resolved by such an inflationary epoch beginning at the GUT scale, then 62 e-foldings of inflation are needed and the entropy must grow by a factor of 10^{71}.

I B. A MAD Epoch

For the case of MAD gravity, the ratio H_o/H_c is very much larger than in the standard model, and can in principle satisfy eqn (5), $H_o/H_c \gtrsim (S_c/S_o) T_o/T_c$. The modified structure of gravity slows the Hubble expansion during some early epoch relative to the Hubble expansion today. As a result of the slowed expansion, the universe ages and causally connected regions have time to grow large. If condition (5) is satisfied, then a causally connected region is large enough for our entire observable universe to have been in causal contact at time t_c. Subsequent to t_c, the dynamic nature of gravity must speed up the Hubble expansion to H_o.

By way of introduction, for the rest of this section we consider the limit of a slowly rolling Planck mass which is a subset of the MAD proposal. Although eqn (6) is not of general validity, it does hold in the slow roll limi. We assume adiabaticity so that $S_o = S_c$. There could of course be a solution to eqn (3) with a combination of large H_o/H_c and some entropy generation; as an extreme case we take $S_o = S_c$ for the moment. Keeping in mind that in our MAD suggestion $m_{pl}(t_c) \neq M_o$, we note that a large ratio of Planck masses could replace the large ratio of entropies needed in inflation. In order to satisfy condition (6) for the extreme case of an adiabatic cosmology $S_c = S_o$ and in the slow roll limit the MAD model requires

$$\frac{m_{pl}(t_c)}{M_o} \geq \beta \frac{T_c}{T_o}. \tag{10}$$

Thus, it appears that a large ratio of Planck masses could replace the large ratio of entropies needed in inflation. If $T_c \sim 3 \times 10^{16}$ GeV then $m_{pl}(t_c) \gtrsim 10^{27} M_o$.

For the slow roll example shown here, subsequent to t_c the Planck mass must decrease so that the strength of gravity grows. While the Planck mass drops, the universe will cool. Notice from eqn (10) that the ratio $m_{pl}(t_c)/M_o$ must be greater than the ratio T_c/T_o, if adiabaticity is imposed. Thus, the causality condition in (2) requires that, subsequent to T_c, the Planck mass must drop faster than the temperature does. An additional mechanism will be needed to force the ratio m_{pl}/T to drop. Below we illustrate the difficulty in implementing such a mechanism. In principle, as the Planck mass outraces the cooling of the universe, the cosmological expansion necessarily accelerates.

To summarize, the causality condition can be satisfied in different ways. Taking the two extreme cases, the causality condition can be met with either a large ratio of Hubble constants or a large ratio of entropies. Take the extreme of no entropy production but allow for a variable Planck mass. Then, causally connected regions at time t_c during the early stages of the universe's history can be much larger than in the standard model simply because the universe can be much older than in the standard model. Thus there could have been enough time for the entire observable universe to be in causal contact. Of course, subsequent to time t_c, gravity must change so as to attain its present behavior. As gravity changes, the scale factor must accelerate.

In contrast, if the Planck mass is constant then a solution to the horizon condition relies totaly on generous entropy production. At time t_c, the scale factor begins to accelerate. The accelerated growth of the scale factor inflates a region which was initially subhorizon-sized and therefore in causal contact to a size large enough to encompass our observable universe. During this stage the universe supercools. Therefore, there must be a period of entropy violation which reheats the universe to some high temperature. This is the manner in which inflation meets the horizon condition. In an inflationary cosmology, there is an era of potential domination which powers the acceleration and subsequently generates entropy.

Inflation always requires entropy production in order to resolve the horizon problem. In MAD models the entropy production of inflation can be replaced in part or in full by a change in the strength or behavior of gravity. Notice before we proceed with a specific MAD model that eqn (6) indicates that both a large ratio of Planck masses and even a moderate ratio of entropies could be exploited in conjunction to meet the causality condition. To illustrate the case most different from standard inflation, in this paper we consider a universe where there is no entropy generation; thus $R \propto T^{-1}$ at all times as in the standard model. [In future work we will consider some change in the entropy as well].

In this paper we work in the context of scalar theories of gravity to search for a specific working model. Although the equation of motion for the scale factor $H^2 = 8\pi\rho/(3m_{pl}^2)$ is very simplistic, ultimately we find that for the slow roll limit it gives the correct form of the causality condition, namely expression (10). We elaborate in §IIIB on the difficulties in anchoring the Planck mass and indicate directions to pursue to move beyond the slow roll limit.

II. SCALAR GRAVITY

As a foundation we begin with scalar theories of gravity in which the Planck mass is some function of a scalar field. However, a MAD solution to the horizon problem could result more generally from other theories with a dynamical Planck mass. In a scalar theory of gravity, Newton's constant $G = M_o^{-2}$ in the Einstein action $A_{\rm einst} = \int d^4x\sqrt{-g}\left(-M_o^2/16\pi\right)\mathcal{R}$ (where $\mathcal{R}$ is the Ricci scalar) is replaced by a scalar field $\Phi(t) = m_{pl}^2(t)$. Thus the Planck mass, which dictates the strength of gravity, is determined dynamically by the behavior of Φ. The action describing the theory is given by

$$A = \int d^4x\sqrt{-g}\left[-\frac{\Phi}{16\pi}\mathcal{R} - \frac{\omega(\Phi)}{\Phi}\frac{\partial_\mu\Phi\partial^\mu\Phi}{16\pi} - V(\Phi) + \mathcal{L}_{\rm matter}\right] , \qquad (11)$$

where we have used the metric signature $(-,+,+,+)$, $\mathcal{R}$ is the scalar curvature, $\mathcal{L}_{\rm matter}$ is the Lagrangian density for all the matter fields excluding the field Φ, and $V(\Phi)$ is the potential for the field Φ. In the original proposal of Brans and Dicke, ω was constant.[4] More generally, $\omega(\Phi)$ may change as a function of time. A theory is specified by choosing the functional form of $\omega(\Phi)$. We point out that the action is built so that the energy-momentum tensor in matter is conserved independently of the energy-momentum tensor in the Φ sector. Consequently, the universe evolves adiabatically so that $R \propto T^{-1}$. [More generally there could be some entropy generation if one included a coupling of Φ directly to matter.]

From this action in a Friedman-Robertson-Walker (FRW) cosmology, we derive the equations of motion for the scale factor of the universe $R(t)$ and for $\Phi(t)$. These equations and their solutions can be found in [2]. In particular, the equation of motion for the scale factor is

$$H^2 + \kappa/R^2 = H_R^2 - \mu H + \omega\mu^2/6 \ , \qquad (12)$$

where

$$\mu = \frac{\dot{\Phi}}{\Phi} \qquad (13)$$

and

$$H_R^2 = \frac{8\pi}{3\Phi}(\rho + V(\Phi)) \ . \qquad (14)$$

Notice from the start that we have assumed that there is no direct coupling of Φ to $\mathcal{L}_{\text{matter}}$. Hereafter, we take the energy density of ordinary matter to be homogeneous and isotropic and composed solely of radiant energy and pressure. Since conservation of energy-momentum in ordinary matter does therefore not involve Φ, it follows that the entropy per comoving volume in ordinary matter is conserved. For convenience we define

$$\bar{S} = R^3 T^3, \tag{15}$$

where $S = \bar{S}(4/3)(\pi^2/30)g_S$. In practice we can take $g_S = g_*$. We have already done this implicitly in the introduction. Notice that the assumption of adiabaticity is easily removed if we allow for a direct coupling of the Planck mass to the radiation bath.

III. BRANS-DICKE GRAVITY WITH $V(\Phi) = 0$ AND $\Delta S = 0$

In Ref. [2] we presented solutions to these equations of motion during the radiation-dominated era with $V(\Phi) = 0$ and an adiabatic universe. We found three possibilities: the Planck mass is constant, initially zero and growing, or infinite and decreasing in time. In all three cases, the Planck mass asymptotically approaches a constant value which we called $\tilde{m}_{pl}$. During the initial phase, the Planck mass can change rapidly (fast roll limit) and the time evolution of the cosmology can be quite different from the standard model. However, once the asymptotic value $\tilde{m}_{pl}$ is approached (slow roll limit), the universe looks like an ordinary radiation-dominated cosmology except $M_o = 10^{19}$ GeV is replaced by $\tilde{m}_{pl}$. In particular, $R \propto \tilde{m}_{pl}^{-1/2} t^{1/2}$ and $H = 1/2t$. Thus, despite the underlying structure of the theory, in the asymptotic limit gravity appears to be described by general relativity with a static gravitational constant.

The investigation of the horizon problem can be broken into two regimes: the slow roll limit and the fast roll limit. To define these two limits we use the expression for H^2 in eqn (12). By slow roll limit we mean $\mu^2 \ll H_R^2$ where again $\mu^2 = (\dot{\Phi}/\Phi)^2$ and $H_R^2 = (8\pi/3\Phi)\rho$. In general scalar gravity, the Planck mass is slowly rolling when $\Phi \sim \tilde{\Phi}$, i.e., it nears its asymptotic value. The fast roll limit, by which we mean $\mu^2 \gg H_R^2$, can apply when the Planck mass is far from its asymptotic value, and hence Φ is far from $\tilde{\Phi}$.

III A. Causality Condition in the Slow Roll Limit with $V(\Phi) = 0$ AND $\Delta S = 0$:

To summarize the results of the previous papers, when the Planck mass approaches its asymptotic value, it changes very slowly; $\mu^2 \ll H_R^2$. In the slow roll limit the equation of motion for the scale factor is the simple $H \approx H_R$. The Hubble expansion is slowed if the Planck mass is large and the gravitational coupling strength, $G = 1/m_{pl}^2$, is therefore small. For Brans-Dicke gravity in the slow roll limit we find that our simple introductory demonstration gave a correct condition on the Planck mass

$$\frac{\tilde{m}_{pl}}{m_{pl}(T_o)} \gtrsim \beta \frac{T_c}{T_o} \quad , \tag{16}$$

where $m_{pl}(T_c) \sim \tilde{m}_{pl}$ and $m_{pl}(T_o) = M_o = 10^{19}$ GeV, $T_o = 2.74^o$ K, and $\beta \sim 10^{-2}$. For example, for $T(\tilde{m}_{pl}) = 3 \times 10^{16}$ GeV, the causality requirement becomes $\tilde{m}_{pl}/M_o \gtrsim 10^{27}$. In terms of t_{einst}, the age of the standard cosmology described by

Einstein gravity, the age of the universe $t(\tilde{\Phi}) = t_{\text{einst}}\,(\tilde{m}_{pl}/M_o)$. A universe which has an early MAD era with large Planck mass $\tilde{m}_{pl} >> M_o$ is therefore older at a given temperature than a universe which has today's value of the Planck mass M_o for all time. Thus, for the above example of $T_c = 3 \times 10^{16}$ GeV, the universe is 10^{-11} seconds old at T_c, a factor of 10^{27} older than in the standard model.

Thus it would seem that the horizon problem could be solved with a large early value of the Planck mass followed by a drop to today's value of the Planck mass, preferably by the time of nucleosynthesis. We see immediately that for Brans-Dicke gravity we grind to a halt. Since the Planck mass has reached its asymptotic value it will not move much away from this value during the era of radiation domination. During the era of matter domination the behavior of the solutions changes and the Planck mass will begin to move again. However, the effect of nonrelativistic matter on the equations of motion is not sufficient to drive the Planck mass down by the many orders of magnitude needed. An additional mechanism is needed to anchor the Planck mass and complete this MAD Brans-Dicke model.

III B. Problems with the Slow Roll Limit

We know that while the Planck mass drops and the strength of gravity grows, the universe will cool. Notice from (16) that the ratio $\tilde{m}_{pl}/M_o$ must be greater than the ratio $T(\tilde{\Phi})/T_o$. Thus, the causality condition in (16) requires that, subsequent to $T(\tilde{\Phi}) = T_c$, the Planck mass must drop faster than the temperature does. Note an amusing corollary. If, as we look backward in time, the Planck mass continues to grow faster than the temperature does, the two scales would never meet, and there would never be an epoch of quantum gravity.

Notice also that by construction the cosmological expansion is very slow. If the causality condition is satisfied and subsequently m_{pl} is to outrace the cooling, a period of accelerated growth of the scale factor will necessarily ensue. So, the key ingredient missing is a mechanism which drives down the ratio $\frac{m_{pl}}{T}$ until m_{pl} is anchored at the value M_o, and in the process causes R to accelerate (see §VI). Thus the stage in which m_{pl} drops faster than the universe expands involves important dynamics. Even without solving the equations of motion for specific cases, we can make some general comments about the kind of anchor needed.

We start with the equation of motion for the scale factor in eqn (12). We can solve the quadratic eqn (12) for H, with $\kappa = 0$ for simplicity, to find the two solutions

$$H + \frac{\mu}{2} = \pm\sqrt{H_R^2 + \frac{(1 + \frac{2\omega}{3})}{4}\mu^2} \quad . \tag{17}$$

Notice from the definition of H and μ that eqn (17) can be rewritten as

$$\frac{\dot{m}_{pl}}{m_{pl}} - \frac{\dot{T}}{T} = \pm\sqrt{H_R^2 + \frac{(1 + \frac{2\omega}{3})}{4}\mu^2} \quad . \tag{18}$$

In our solutions to the equations of motion, we implicitly chose the solution with the positive square root in (17) so that H was positive and the universe expands. As a result m_{pl}/T was always growing.

However, if the Planck mass is to drop faster than the temperature, then at some point during the expansion, the minus sign in (18) must be operative. Thus the square root must pass smoothly through zero, then becoming negative. We require that the Hubble parameter H continues to be both real and positive; then for the solution with negative square root in (18), we find that we need

$$0 < 1 + \frac{2\omega}{3} + \frac{4H_R^2}{\mu^2} < 1 \, . \tag{19}$$

The first inequality follows from requiring the quantity inside the square root in eqn (18) to be positive and the second inequality from requiring H to be positive. The difficulty in building an anchor which drives down the ratio of m_{pl}/T is meeting condition (19). The expression between inequalities in eqn (19) appears to be the sum of 1+ positive quantities and could therefore not be less than 1. For a successful anchor, there must be negative contributions to H^2. [This behavior of the equations of motion has also been found by Hu, Turner, and Weinberg[5]. Their results apply to the case of nonzero curvature as well.]

Eqn (19) constrains the specific model we chose. It implies that either the equations of motion must be modified, a different theory of gravity must be studied, or the assumption of adiabaticity must be removed. More generally, other theories of gravity might provide equations of motion which are not driven to this constraint.

To move beyond this obstacle for scalar gravity and anchor the Planck mass we could alter the equations of motion and look for negative contributions to H^2. For instance a negative ω could allow the Planck mass to drop faster than the temperature cools if $|\omega|\mu^2/6 > H_R^2$, and so the kinetic terms must become important at some point after T_c. Although we could easily write down such a scheme, negative kinetic energies are controversial. One would expected that with negative kinetic energies the cosmology is unstable to the runaway growth of anisotropies. Still, it has been argued by Linde that there is nothing wrong with changing the overall sign of the Lagrangian and allowing for particles with overall negative energy. As long as particles with positive energy and those with negative energy do not couple directly there is no difficulty. He used this argument to build a theory which has no net cosmological constant.[6] We take the attitude that negative energies ought to be approached with caution. Similarly, one could try a negative value of the Planck mass for some period of time; however, Starobinsky has shown that this leads to instability away from an isotropic universe.

Another possibility might be a negative potential for the Planck mass. This might be similar to a suggestion by A. Zee.[7] Zee tried to solve the horizon problem differently than we do; still we may be able to use his means to a different end. He commented that in a model of induced gravity, negative contributions to the effective potential for the field equivalent to Φ might be generated. Although he did not present his results in this framework, he did notice that this could drive a period where the Planck mass changed at least as rapidly as T, precisely the behavior needed here. He went further to argue that if this were so at $t = 0$, then the integral defining the horizon might diverge. Thus the particle horizons would be infinite and there would be no horizon problem. This is different from our approach to the horizon problem. We assume there are finite growing horizons and try to ensure that they grow large as the universe evolves. Lastly, a negative decaying vacuum energy density could satisfy eqn (19) (for a discussion of decaying vacuum energy density see Ref [8]).

This simplest case of the slow roll limit with $\Delta S = 0$ is pushed to an uncomfortable position. The possible mechanisms needed to anchor the Planck mass may be difficult to construct without introducing relatively ugly complications. However, if even moderate entropy is produced, the ratio m_{pl}/T does not need to drop and this constraint does not hold.

IIIC. Beyond the Slow Roll Limit with ΔS Still Zero

In the last section, we illustrated the causality condition in the slow roll limit (so that $H \gtrsim T^2/m_{pl}$) and discussed what would be needed to implement it. Whenever $H \gtrsim T^2/m_{pl}$, the Planck mass has to drop faster than T does subsequent to t_c. Above we mentioned the properties of a mechanism which might drive down the ratio m_{pl}/T and in the process accelerate $R(t)$, thereby completing the scenario.

However, such a mechanism could be exploited another way. If there are negative contributions to H^2, then H could be smaller than T^2/m_{pl}. If this is so, then the Planck mass does not need to drop as much as indicated in eqn (16). Instead one may be in an entirely different regime where the slow roll assumption $H^2 \geq 8\pi G\rho/3$ is unfounded. Then one obtains a different causality condition than eqn (16) and eqn (19) does not need to be satisfied. Below we discuss the fast roll case for scalar theories of gravity. In addition, in higher order theories of gravity, one may encounter $H^2 \leq T^2/m_{pl}$, and thus eqn (19) need not hold. For instance, the inclusion of a term proportional to $\mathcal{R}^2$ in the action leads to the additional term in H^2 of the form

$$\propto -H^2 \left[\dot{H} + 2\frac{\ddot{H}}{H} - \left(\frac{\dot{H}}{H}\right)^2 \right] \quad . \tag{20}$$

(This has been used in the context of Starobinsky inflation[9].) We point out here that such terms lead to a different causality condition than given in eqn (16) and thus avoid the constraint given in the previous section.

In the remainder of this section we focus on the fast roll limit of scalar theories of gravity. Without the introduction of any additional mechanism, we see that the kinetic energy terms of the Φ field make a negative contribution to H^2 in eqn (12). In particular, we found for this model of scalar gravity that $H \ll T/m_{pl}$ if there is a 'fast roll' period with large kinetic terms and with $\omega \ll 1$ (a large deviation from Einstein gravity).

To see this consider the expression $H^2 = H_R^2 - \mu H + \omega/6\mu^2$. We see that if $\omega \to 0$, and thus there is a large deviation from Einstein gravity, then $H^2 \approx H_R^2 - \mu H$. This can quickly be solved for H,

$$H = \frac{-\mu}{2}\left(1 \pm \sqrt{1 + \frac{4H_R^2}{\mu^2}}\right) \quad . \tag{21}$$

We focus on the early stages of the radiation-dominated era, when the Planck mass can change quickly before it has reached its asymptotic value. If m_{pl} is a growing function of time, then $\mu > 0$ and we take the $-$ sign to ensure the positivity of H. So, for $\mu > 0$,

$$H = \frac{-\mu}{2} + \frac{\mu}{2}\sqrt{1 + \frac{4H_R^2}{\mu^2}} \quad . \tag{22}$$

We can approximate H in the fast roll limit of $\mu \gg H_R$ to find

$$H \approx H_R^2/\mu \ . \tag{23}$$

In the slow roll limit of the previous section a large Planck mass was needed to slow the Hubble expansion; that is, a weak gravitational coupling strength. In the fast roll limit, for $\omega \to 0$, a large kinetic term can actually slow the cosmic expansion without demanding a large value of $m_{pl}(T)$. The appeal is that the evolution could be slowed at some early time while m_{pl} might be comparable to M_o. We could then avoid the complications introduced by the need for an additional mechanism to drive the Planck mass down by many orders of magnitude.

However, although the Hubble expansion can be slowed, it is not at all clear that this alone leads to a large causally connected region. In our previous example of the slow roll limit the two conditions $H_c^{-1}R_c^{-1} \gtrsim H_o^{-1}R_o^{-1}$ and $d_{\text{horiz}}(T_c)/R_c \gtrsim H_o^{-1}R_o^{-1}$ were nearly identical. In the limit of large kinetic behavior and $\omega \ll 1$ they are not identical. In fact, we found from the solutions in Ref. [2] that $d_{\text{horiz}} \ll H^{-1}$. Thus even if H is small so that the Hubble length H^{-1} is large, this does not tell us the distance a light pulse has traveled and so does not measure the size of a causally connected region.

The integral $\int_{t_1}^{t_2} dt/R(t)$ certainly is the maximum distance a light pulse could have traveled between t_1 and t_2 and therefore eqn (1) seems to be the more important condition to satisfy. However, if $\omega \to 0$ and $\omega(\Phi)$ changes rapidly, we found this cosmology can be sufficiently unusual that we will need to rethink the causality condition. Most notably, the cosmology may be nonsingular; that is, the spacetime may be geodesically complete. [For a hint that this may be the case, take the limit of $\omega \to 0$ in the solutions in Ref. [2]. To prove nonsingularity, a program such as that outlined in Ref [10] should be followed.] If the universe is nonsingular it is unclear what is meant by the beginning of the universe, the horizon etc. We will not venture too far down this path. We only conclude that it is unclear what the causality condition is exactly if the universe is nonsingular.

Instead of abandoning this case altogether, we think it is still interesting to point out that the Hubble expansion can be slowed if we consider both the fast roll and small ω limits. Although the relevance of the slowed Hubble expansion is not yet clear, we present this case for the record and the interested reader until a closer investigation can be carried out.

IV. MONOPOLES AND DENSITY PERTURBATIONS

If there is a grand unified epoch at temperature T_G, then magnetic monopoles are produced. In the standard cosmology, far too many are produced. If there is inflation at $T < T_G$, then the monopoles are inflated away. In our MAD model, the monopole problem can be resolved as well. In a first order phase transition approximately one monopole per horizon volume is produced. The number of monopoles in our observable universe today is then given by the number of comoving horizon volumes at T_G that would fit inside the comoving volume of our observable universe, $N = \left(\frac{1/(H_o R_o)}{1/(H_G R_G)}\right)^3$. In the slow roll limit, this becomes $N = \left(\frac{M_o}{m_{pl}(T_G)}\frac{T_G}{T_o}\right)^3$. Thus, in our model, for $\frac{1}{H_G R_G} \geq \frac{1}{H_o R_o}$ (similar to the causality condition), the number of monopoles in our observable universe can be very small. Also, there is an additional effect that can lower the monopole density relative to the standard model: if $T_G \geq T_c$, the slow Hubble expansion will keep the monopoles in equilibrium longer and therefore lead to more monopole/antimonopole annihilations and fewer monopoles. In fact, during the period of slowed expansion, any reaction in the early universe will have more time to go to completion.

As mentioned previously below eqn (2), density fluctuations that may be responsible for the formation of large-scale structure may be generated. It would be interesting to calculate these for the MAD model. In addition, since the horizon size at various phase transitions can be larger than in the ordinary universe, it may be possible to causally produce density perturbations relevant for large-scale structure.

V. FLATNESS

One way to state the flatness problem is to express surprise that Ω is near 1 today, although $\Omega = 1$ is an unstable value. Here $\Omega = \rho/\rho_{cr}$, $\rho_{cr} = 1.88 \times 10^{-29} h_o^2$ gm cm^{-3}, and $h_o = H_o/100$ km s^{-1} Mpc^{-1}. This is a problem if the age of the universe is much larger than the characteristic timescale t_* for Ω to deviate from 1. In the standard model, $t_* \sim \bar{S}^{2/3}/M_o$. So for moderate values of the entropy $\bar{S} \sim 1$, Ω deviates rapidly from 1 in a Planck time $\sim 10^{-45}$ sec. Thus it is surprising that the universe survived to its present age of 10^{10} years with Ω so near 1 today. In those variants of MAD where the initial value of the Planck mass is much larger than 10^{19} GeV, the timescale to deviate from 1 is even faster, $t_* \sim 1/\tilde{m}_{pl}$ for $\bar{S} \sim 1$. In both the standard and MAD models, the universe can be made flat with a large unexplained value of the entropy, $\bar{S} > 10^{90}$. To see this we write $\Omega - 1 = k/(R^2 H^2)$. Then Ω near 1 today requires $R_o/|\kappa|^{1/2} > H_o^{-1}$, or, equivalently, $\bar{S} > |\kappa|^{3/2}(M_o/T_o)^3 \sim 10^{90}$. Since the universe evolves adiabatically in the standard model and in a MAD cosmology, there is presently no explanation for the enormous value of the entropy.

MAD does make the universe flatter between t_c and today, but there is still a flatness problem at higher temperatures if adiabaticity is imposed. We define

$$\bar{\Omega} = \frac{H_R^2 - \mu H + \omega\mu^2/6}{H^2} \tag{24}$$

where $\rho_{cr} = 3H^2\Phi/(8\pi)$ is the critical energy density, including contributions from Φ, required to just close the universe. Notice that one could have $\bar{\Omega}(t_o) = 1$ while the contribution from matter $\Omega_m(t_o) \ll 1$, in agreement with observations. The kinetic energy density of the Planck field could account for the unseen energy density, i.e., the dark matter. With this definition, $\bar{\Omega} = \frac{1}{1-x}$, where $x = \frac{\kappa/R^2}{H^2+\kappa/R^2}$. Clearly, in a MAD model with a large ratio H_o/H_c, $\bar{\Omega}$ is driven towards 1. In the MAD slow roll limit, we find $x(T_o)/x(T_c) \sim (M_o/m_{pl}(T_c))^2(T_c/T_o)^2 \lesssim 1$. In the standard model, on the other hand, x would have grown by a factor $(T_c/T_o)^2$. Thus, MAD assists the approach to flatness in between the time t_c and t_o. However, because of the large early value of the Planck mass, with $\bar{S} \sim 1$ it is hard to see how the universe would be able to survive even to T_c; it is this flatness problem at high temperatures that MAD does not address.

Notice that inflation also has a residual flatness problem if $\kappa = +1$. In the case of a closed universe, the universe recollapses before the temperature ever has a chance to reach, say, a GUT temperature at which inflation begins, unless the entropy is $S \sim 10^{15}$.

There may be entirely different mechanisms to explain flatness.[11] We note that some alternate gravity models[12] claim to not have a flatness problem.

We treat next the issue of accelerations which has not yet been discussed in a MAD context.

VI. LOOKING FOR ACCELERATIONS

We argued in the introduction that the horizon condition can be expressed as $1/(H_cR_c) \gtrsim 1/(H_oR_o)$, at least for $d_{\rm horiz} \sim H^{-1}$. This condition is equivalent to the statement that $\dot{R}_o \gtrsim \dot{R}_c$. If $\dot{R}$ today is bigger than at some earlier time it must be that the scale factor accelerated during some era between the time t_c and today. In general,

$$\frac{\ddot{R}}{R} = -\frac{4\pi}{3\Phi}(\rho_{total} + 3p_{total}) \quad . \tag{25}$$

In the standard cosmology both ρ and p are always greater than or equal to zero so the universe always decelerates. During inflation on the other hand, the pressure associated with the potential is negative. In fact, since the equation of state is $p = -\rho = -V$, the negative pressure of the potential powers a period of acceleration for the scale factor (see eqn (4)).

In the MAD model we have so far focused on the era prior to t_c. Subsequent to t_c, we know the scale factor must accelerate according to the arguments given above. We know that for inflation a potential will power this period of acceleration. What will power the period of acceleration for a MAD model? We will try to answer that question in this section.

For instance, consider the slow roll limit solution in the absence of all entropy production. The Hubble expansion is slowed by a large Planck mass and thus weak gravitational coupling strength. In §IIIB we listed possible anchors to force the Planck mass down and complete the MAD solution. As the strength of gravity grows the Hubble expansion must also grow and the anchor must have driven a period of cosmic acceleration. By calculating the energy density and pressure in Φ, we will show here that each of the anchors suggested could in fact lead to a negative pressure. Expressing $\ddot{R}$ in terms of the energy density and pressure will reveal that the negative pressures could each in turn lead to an accelerating epoch.

Beyond the slow roll context, we find that in a scalar theory of gravity a changing $\omega(\Phi)$ can lead to a negative pressure. This negative pressure could then drive the scale factor to accelerate. The existence of an accelerating era does not ensure that the horizon problem is solved. The horizon problem requires that the causality condition be satisfied and that at the end of the day our universe is reproduced. An acceleration alone does not provide this insurance. The acceleration must be sown onto the property needed to meet the causality condition, say entropy production or a large ratio of H_o/H_c. The role of the acceleration is to connect the universe at t_c with the universe today. We have not yet integrated this gravity driven acceleration into a specific solution to the horizon problem, although we do make suggestions as to how to proceed.

We begin to look for negative pressures and accelerations in general. We have written the nonminimal coupling of the field to gravity as simple, $\Phi\mathcal{R}$, while we allowed the kinetic term to look complicated, leaving $\omega(\Phi)$ general. Equivalently, we could write the action in terms of the field ψ which has units of mass,

$$A = \int d^4x\sqrt{-g}\left[-\frac{f(\psi)}{16\pi}\mathcal{R} - \frac{\partial_\mu\psi\partial^\mu\psi}{2} - V(\psi) + \mathcal{L}_{\rm matter}\right] , \tag{26}$$

with the Brans-Dicke parameter $\omega(\Phi)$ defined by

$$\omega = 8\pi\frac{\Phi}{(\partial\Phi/\partial\psi)^2} \tag{27}$$

and $f(\psi) \equiv \Phi(\psi)$. This description, at least classically, is equivalent to that of eqn (11). The theory is then specified by identifying the functional form of the coupling $f(\psi) \equiv \Phi(\psi)$. In eqn (26), the coupling to $\mathcal{R}$ is general while the kinetic term looks simple. Although it appears as though ψ has the canonical kinetic term of a minimally coupled scalar field this is deceiving. The field ψ does not have a canonical kinetic term in the sense that the energy-momentum tensor does not have the canonical form. The energy-momentum tensor is defined by the variation of the action with respect to the metric

$$\delta A_\Phi = \frac{1}{2}\int d^4x\sqrt{-g}T^{\mu\nu}_\psi\delta g_{\mu\nu} . \tag{28}$$

Although the term $\partial_\mu\psi\partial^\mu\psi/2$ would give the same contribution to $T^{\mu\nu}_\psi$ that the kinetic energy of a minimally coupled scalar field would, there are additional kinetic terms from the variation of the $\Phi\mathcal{R}$ coupling. The nonminimal coupling inevitably generates unusual contributions to $T^{\mu\nu}_\psi$. (Of course, these additional contributes appear in $T^{\mu\nu}_\Phi$ if we work instead with Φ.) Both descriptions are useful and we shall use both to illustrate various points.

Since our action explicitly assumed no direct coupling of Φ to $\mathcal{L}_{\rm matter}$, the energy-momentum tensor for matter has the standard form. For instance, for a minimally coupled scalar field, η,

$$T^{\mu\nu}_{\rm matter} = \left[\partial^\mu\eta\partial^\nu\eta - \frac{1}{2}(\partial_\alpha\eta\partial^\alpha\eta)g^{\mu\nu}\right] - V(\eta)g^{\mu\nu} . \tag{29}$$

In comparison the energy-momentum tensor for Φ is given by

$$T^{\mu\nu}_\Phi = \frac{\omega}{8\pi\Phi}\left[\partial^\mu\Phi\partial^\nu\Phi - \frac{1}{2}(\partial_\alpha\Phi\partial^\alpha\Phi)g^{\mu\nu}\right] - V(\Phi)g^{\mu\nu} + \frac{1}{8\pi\Phi}\left(D^\mu D^\nu\Phi - g^{\mu\nu}\Box\Phi\right) \tag{30}$$

$\Box = g^{\mu\nu}D_\mu D_\nu$, and D_μ is the covariant derivative. Notice, as claimed, that additional kinetic terms appear in the Φ stress tensor, namely the last two terms in eqn (30). We will pay these particular attention when we look for accelerations of the scale factor. The variation of the action with respect to Φ gives

$$-\frac{2\omega}{\Phi}\Box\Phi + \mathcal{R} + \frac{\partial V}{\partial\Phi}16\pi + \frac{\partial_\mu\Phi\partial^\mu\Phi}{\Phi}\left[\frac{\omega}{\Phi} - \frac{\partial\omega}{\partial\Phi}\right] = 0 . \tag{31}$$

The energy density and pressure in the field Φ will be denoted by ρ_Φ and p_Φ, respectively, while the undecorated symbols ρ and p will refer to the energy density and the pressure in ordinary matter. We wrote down the energy-momentum tensor for a scalar field coupled nonminimally to gravity in eqn (30). We explicitly break up

$T_\Phi^{\mu\nu}$ here into the energy density in the Φ-field and the pressure in the Φ-field. The energy density is

$$\rho_\Phi \equiv T_\Phi^{00} = \frac{\omega}{16\pi}\frac{\dot{\Phi}^2}{\Phi} + V(\Phi) - \frac{3}{8\pi}H\dot{\Phi} \tag{32}$$

We can draw a connection with an ordinary scalar field by working in terms of a field ψ with units of mass. Given $\Phi(\psi)$, the Brans-Dicke parameter ω is defined as $\omega = 8\pi\Phi/(\frac{\partial\Phi}{\partial\psi})^2$. As described earlier, the field ψ has an almost canonical kinetic energy density term; that is the kinetic energy term in ρ_Φ can be rewritten as

$$\frac{\omega}{16\pi}\frac{\dot{\Phi}^2}{\Phi} = \frac{1}{2}\dot{\psi}^2 \quad . \tag{33}$$

This looks like the standard kinetic energy density for a scalar field. Still, under a variation of the action with respect to the metric, the coupling of $\Phi(\psi)$ to $\mathcal{R}$ generates additional kinetic terms for Φ and thus ψ. Since additional kinetic terms are generated, even the field ψ does not have a truly canonical kinetic energy density. The term proportional to H in (32) is precisely such an additional term and accounts for a kind of frictional drag on the Φ field from the expansion of the cosmology. For instance, consider the Brans-Dicke model which can be expressed by

$$\Phi(\psi) = \frac{2\pi}{\omega}\psi^2 \quad . \tag{34}$$

Then eqn (32) becomes

$$\rho_{\Phi(\psi)} = \frac{1}{2}\dot{\psi}^2 + V(\psi) - \frac{3}{2\omega}H\psi\dot{\psi} \quad . \tag{35}$$

Clearly, the last term appears in addition to the usual kinetic energy density, i.e. the usual first term. For comparison we write the kinetic energy density in a scalar field minimally coupled to gravity,

$$\rho = \frac{1}{2}\dot{\eta}^2 + V(\eta) \quad . \tag{36}$$

The pressure from the Φ-field is given by $p_\Phi\delta^{ij} = T^{ij}$,

$$\begin{aligned} p_\Phi = \frac{\omega}{16\pi}\frac{\dot{\Phi}^2}{\Phi} - \frac{1}{8\pi}H\dot{\Phi} - V(\Phi) + \frac{1}{(3+2\omega)}\left[4V - 2\Phi\frac{\partial V}{\partial\Phi}\right] + \frac{\rho - 3p}{(3+2\omega)} + \\ - \frac{\dot{\Phi}^2}{8\pi(3+2\omega)}\frac{\partial\omega}{\partial\Phi} \end{aligned} \tag{37}$$

while the pressure in a minimally coupled scalar field is much more simply

$$p = \frac{1}{2}\dot{\eta}^2 - V(\eta) \quad . \tag{38}$$

Although Φ does not couple directly to matter, notice the term in eqn (37) involving the energy density and pressure in ordinary matter. Because of the coupling of Φ to the Ricci scalar $\mathcal{R}$, ordinary matter contributes to the pressure of the Φ-field; there is a term in eqn (37) proportional to $[\mathrm{Tr}(T^{\mu\nu}_{\mathrm{matter}}) = \rho - 3p]$.

We put the results of eqns (32) and (37) to use in eqn (25) to find $\ddot{R}/R$. The lengthy expression for the acceleration is

$$\frac{\ddot{R}}{R} = -\frac{4\pi}{3\Phi}\left[\rho + 3p + \frac{(\rho - 3p)}{(1+2\omega/3)}\right] - \frac{\omega}{3}\left(\frac{\dot{\Phi}}{\Phi}\right)^2 + H\left(\frac{\dot{\Phi}}{\Phi}\right) + \frac{1}{2}\left(\frac{\dot{\Phi}}{\Phi}\right)^2 \frac{\Phi}{(3+2\omega)}\frac{\partial\omega}{\partial\Phi}$$
$$+ \frac{8\pi}{3\Phi}\left[V - \frac{1}{(1+2\omega/3)}\left\{2V - \Phi\frac{\partial V}{\partial\Phi}\right\}\right] . \tag{39}$$

To lend some context, imagine again the MAD solution in the slow roll limit. In §IIIB we enumerated possible mechanisms to add to the model in order to drive down the ratio of m_{pl}/T subsequent to t_c. Regardless of the varying degrees of attractiveness of these mechanisms, we note that they might in fact lead to a period of acceleration. For instance, a negative ω could render p_Φ negative and in turn $\ddot{R}$ positive. Or a negative potential, which could lead to m_{pl}/T dropping, could accelerate the scale factor, provided the last term in square brackets in eqn (39) is positive. For example, if ω is small and $\partial V/\partial\Phi \geq 0$, then the last term in square brackets in eqn (39) is positive and $\ddot{R} > 0$ is possible.

Remember that for inflation the potential domination provides two key ingredients; (1) a negative pressure which powers the acceleration and (2) entropy production. In the slow roll limit of MAD gravity, we demonstrated how a large ratio of Planck masses could replace the entropy production of inflation. Any mechanism which would anchor the Planck mass faster than the universe would cool will accelerate the scale factor. A completed picture would then provide two key ingredients; (1) acceleration and (2) a large ratio of Planck masses.

We went to the trouble of finding $\ddot{R}$ to check that as m_{pl}/T dropped the expansion of the universe would in fact accelerate as claimed. Now that we have found $\ddot{R}$, we can take this opportunity to move beyond the slow roll context for a minute. In particular the last term in (37) is of interest ,

$$p_{\Phi,\omega} = -\frac{\dot{\Phi}^2}{8\pi(3+2\omega)}\frac{\partial\omega}{\partial\Phi} . \tag{40}$$

Remarkably, if $\omega(\Phi)$ grows with Φ there is a negative pressure associated with the changing structure of gravity. To make the expression for $\ddot{R}$ manageable, we consider radiation domination so that $\rho = 3p$ and massless scalar theories so that $V(\Phi) = 0$. Then eqn (39) becomes

$$\frac{\ddot{R}}{R} = -H^2 - \frac{\omega}{6}\left(\frac{\dot{\Phi}}{\Phi}\right)^2 + \frac{1}{2}\left(\frac{\dot{\Phi}}{\Phi}\right)^2 \frac{\Phi}{(3+2\omega)}\frac{\partial\omega}{\partial\Phi} . \tag{41}$$

For Brans-Dicke gravity ω is a positive constant. As a result, the third term in eqn (41) vanishes. The remaining terms are negative. Therefore, without an additional ingredient, $\ddot{R} < 0$ and there is no opportunity to accelerate the expansion in Brans-Dicke gravity. Alternatively if $\omega(\Phi) \neq$constant, the last term in eqn (41) is not zero. In fact, if $\omega(\Phi)$ is a growing function of Φ, this last term will be positive and there is the opportunity to accelerate the scale factor.

In § (4.5) we make the last suggestion that a MAD acceleration plus entropy violation be considered. In this way, the insights of inflation could be integrated into this MAD model.

VII. ENTROPY PRODUCTION

We have set forth the suggestion that the structure of gravity alone, in the absence of potential domination, could lead to a resolution of the horizon problem. The action we have worked from was constructed precisely so that the energy-momentum tensor for the Planck mass was conserved independently of the energy-momentum tensor for matter. This leads to the adiabaticity of the cosmology. However, if a coupling of the Planck mass directly to matter is considered, then it could be that energy is transfered between the Planck sector and the matter sector. There is then the possibility that energy from the Planck mass is transfered into entropy and the assumption of adiabaticity on which this discussion is based should be dropped. In fact, as indicated implicitly in the introduction, both entropy production and a dynamical Planck mass combined could solve the horizon problem (see eqn (6)). If even moderate entropy is produced, then a smaller ratio of Planck masses would be needed to meet condition (2). Consequently the Planck mass would not have to drop as quickly as the temperature and the problems of the adiabatic slow roll limit could be circumvented.

In fact, in the spirit of inflation, we could now imagine gravity alone driving an accelerating phase at the end of which entropy is produced. At time t_c a small causally connected region grows large due to the negative pressure from the changing structure of gravity (eqn (40)). Perhaps at the end of the accelerating stage entropy could be generated as the Planck mass slows, transfering energy to ordinary matter.

VIII. CONCLUSIONS

If a comoving causally connected region at some high temperature is large enough to envelop everything we can see, there might be a causal explanation for the observed homogeneity and isotropy of our universe. This causality condition can be satisfied in different ways. Taking the two extreme cases, the causality condition can be met with either a large ratio of entropies or a large ratio of Hubble constants; that is, a slowed Hubble expansion during an early epoch relative to the Hubble expansion today. Inflation meets the causality condition with a large ratio of entropies. In an inflationary cosmology, there is an era of potential domination which powers an era of cosmic acceleration and subsequently generates entropy. In MAD models there is no period of potential domination and the entropy production of inflation can be replaced in part or in full by a change in the behavior of gravity. To illustrate the case most different from standard inflation, in this paper we first considered a universe where there is no entropy generation. More generally, we argue that a combination of entropy production and modified gravity could be used to meet the causality condition.

Our results for a scalar theory of gravity under the assumption of adiabaticity can be broken into the slow roll limit and the fast roll limit. In the slow roll limit a large Planck mass and thus weak gravitational coupling strength is needed to slow the Hubble expansion. Due to the slowed evolution, the universe grows old and a causally connected region grows large enough to envelop our observable universe. Subsequently the strength of gravity must grow and the cosmic evolution must speed up. In the absence of an additional ingredient, the temperature will cool faster than the strength of gravity grows and we are unable to reproduce our cosmology. An additional mechanism must be found which can sensibly drive the Planck mass down faster than the radiation bath cools. We found that the general properties of the anchor needed are unusual though not impossible to construct. This obstacle to completing the MAD model may be circumvented entirely in higher order theories of gravity or if even moderate entropy is produced in a scalar theory of gravity.

If in a scalar theory of gravity $\omega \ll 1$ initially, and thus there is a large deviation from Einstein gravity, then the fast roll limit hints of an alternative to the slow roll limit. The cosmic evolution is slowed due to the large deviation from Einstein gravity and could then speed up as ω grows and the deviation from Einstein gravity is diminished. Still, this case is not without its complications, as it is unclear how to state the correct causality condition. These issues need to be resolved before specific theories are pursued.

We have seen that any dynamical solution to the horizon problem requires a period of cosmic acceleration, $\ddot{R} > 0$. At the close of this paper, we showed that a negative pressure in a theory with a growing Brans-Dicke parameter may be able to drive an epoch of accelerated growth of the scale factor. Once we know that this negative pressure could in principle accelerate the scale factor, the possibility that gravity could address the horizon question seems promising.

ACKNOWLEDGEMENTS

We would like to thank F. Adams, A. Guth, I. Redmount, M. Turner, R. Watkins, E. Weinberg, and H. Zaglauer, for helpful conversations. KF and JJL thank each other for getting MAD. KF thanks the ITP at U.C. Santa Barbara and the Aspen Center for Physics, where part of this work was accomplished, for hospitality. We acknowledge support from NSF Grant No. NSF-PHY-92-96020, a Sloan Foundation fellowship, and a Presidential Young Investigator award.

REFERENCES

[1] A. H. Guth, *Phys. Rev.* D **23**, 347 (1981).
[2] J. J. Levin and K. Freese, *Phys. Rev.* D **47** (1993).
[3] G. W. Gibbons and S. W. Hawking, *Phys. Rev.* D **15**, 2738 (1977).
[4] C. Brans and C. H. Dicke, *Phys. Rev.* **24**, 925 (1961).
[5] Y. Hu, M. Turner, and E. Weinberg preprint also addresses some of these points, although we do not agree with their conclusions.
[6] A. D. Linde, *Phys. Lett.* **201 B**, 437 (1988); A. D. Linde, *Particle Physics and Inflationary Cosmology.* (New York: Hardwood Academic Publishers) (1990).
[7] A. Zee, *Phys. Rev. Lett.* **44**, 703 (1980).
[8] K. Freese, F. Adams, J. Frieman, and E. Mottola, *Nucl. Phys.* **287**, 797 (1987).
[9] A. D. Linde, *Particle Physics and Inflationary Cosmology.* (New York: Hardwood Academic Publishers) (1990); J. D. Barrow and A. Ottewill, *J. Phys.* **A16**, 2757 (1983); A. A. Starobinsky, *Sov. Astron. Lett.* **9**, 302 (1983).
[10] R. Brandenberger, V. Mukhanov, and A. Sornborger, preprint 1993.
[11] Note, e.g., that a large number of relativistic degrees of freedom early on, $g_*(t_{pl})/g_*(t_o) \gg 1$, could explain the large entropy of the universe. E.g., one

could imagine substructure of the fundamental particles as we know them now.
[12] P.D. Mannheim, *Ap. J.* **391**, 429 (1992).
[13] R. Brustein and G. Veneziano, preprint CERN-TH 7179/94.

AXITONS

Edward W. Kolb [1,2] and Igor I. Tkachev [1,3]

[1] NASA/Fermilab Astrophysics Center,
Fermi National Accelerator Laboratory, Batavia, IL 60510,
[2] Department of Astronomy and Astrophysics, Enrico Fermi Institute
The University of Chicago, Chicago, IL 60637
[3] Institute for Nuclear Research of the Academy of Sciences of Russia
Moscow 117312, Russia

INTRODUCTION

The invisible axion is one of the best motivated candidates for cosmic dark matter. The axion is the pseudo-Nambu–Goldstone boson resulting from the spontaneous breaking of a $U(1)$ global symmetry known as the Peccei–Quinn, or PQ, symmetry[1]. There are stringent astrophysical[2,3] and cosmological [4] constraints on the properties of the axion. In particular, the combination of cosmological and astrophysical considerations restrict[5] the axion decay constant f_a and zero temperature axion mass m_a to be in the narrow windows $10^{10}\,\mathrm{GeV} \lesssim f_a \lesssim 10^{12}\,\mathrm{GeV}$, and $10^{-5}\,\mathrm{eV} \lesssim m_a \lesssim 10^{-3}\,\mathrm{eV}$. The contribution to the mean density of the Universe from axions with mass in this window is guaranteed to be cosmologically significant.

After PQ symmetry breaking at $T \sim f_a$, but before QCD effects are important, the axion is massless. Axion mass arises at low temperatures because the PQ symmetry is anomalous, and it is broken explicitly by QCD instanton effects. The axion field is often represented in terms of an angular variable $\theta \equiv a/f_a$, and if θ is taken as the dynamical variable, its potential is

$$V(\theta) = m_a^2(T) f_a^2 (1 - \cos\theta) \equiv \Lambda_a^4(T)(1 - \cos\theta). \tag{1}$$

For $T \gg \Lambda_{\mathrm{QCD}}$, the temperature dependence of the axion mass scales as[6]

$$m_a^2(T) = m_a^2(T_*)(T/T_*)^{-n}; \qquad n = 7.4 \pm 0.2. \tag{2}$$

When the field $\theta(x)$ is created during the Peccei-Quinn symmetry breaking phase transition at $T \sim f_a$, it should be uncorrelated on scales larger than the Hubble radius at that time. Lack of coherense does not necessarily require the reheating temperature after inflation to be higher than f_a, since inflation itself can produce strong fluctuations in the axion field as discussed in Ref.[7]. As the temperature decreases and the Hubble radius grows (in a radiation-dominated Universe the Hubble radius grows as

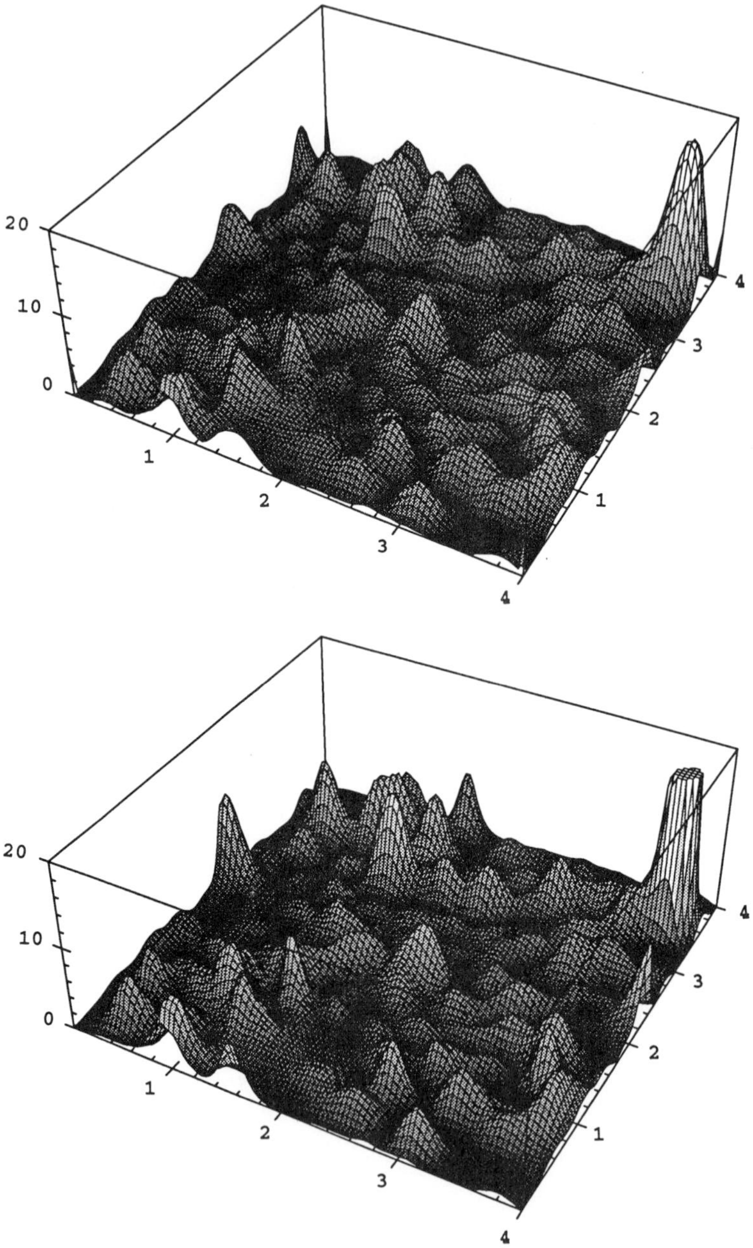

Figure 1. A 2-dimensional slice through the 3-dimensional box at time corresponding to $\eta = 2$ for the harmonic potential (top) and the axion potential (lower). The Hubble radius at this time is 2 units and the inverse of the axion mass is 0.038 units. The height of the figure corresponds to the energy density in the axion field normalized to the height for homogeneous field evolution: $\rho_a(\eta = 2)/\bar{\rho}_a(\eta = 2)$.

$H^{-1}(T) \propto T^{-2}$), the field becomes smooth on scales up to the Hubble radius just because subhorizon modes are redshifted away as any radiation. This continues until $T = T_1 \sim 1$ GeV when the axion mass "switches on," i.e., when $m_a(T_1) \approx 3H(T_1)$, and the axion mass begins to become important in the equations of motion. Coherent axion oscillations then transform fluctuations in the initial amplitude of the axion field into fluctuations in the axion density.

One expects that typical positive density fluctuations on Hubble scale at this time will satisfy $\rho_a \approx 2\bar{\rho}_a$, where $\bar{\rho}_a$ is mean cosmological density of axions[8]. At the temperature of equal matter and radiation energy density, $T_{\rm eq} = 5.5\,\Omega_a h^2\,{\rm eV}$ [9], these fluctuations will be already non-linear and will separate out as miniclusters[8] with $\rho_{\rm mc} \approx 10^{-14}\,{\rm g\,cm^{-3}}$. The minicluster mass will be of the order of the axion mass within the Hubble radius at temperature T_1, $M_{\rm mc} \sim 10^{-9}\,M_\odot$. The radius of the cluster is $R_{\rm mc} \sim 10^{13}$cm, and the gravitational binding energy will result in an escape velocity of $v_e/c \sim 10^{-8}$. Note that the mean phase-space density of axions in such a gravitational well is enormous: $\bar{n} \sim \rho_a m_a^{-4} v_e^{-3} \sim 10^{48} f_{12}^4$, where $f_{12} \equiv f_a/10^{12}\,{\rm GeV}$.

We will show below that non-linear effects of axion self-interaction lead to the formation of pseudo-soliton objects out of inhomogeneous axion field. Consequently, a substantial number of regions at $\Lambda_{\rm QCD} > T > T_{\rm eq}$ can have an axion density orders of magnitude larger than $2\,\bar{\rho}_a$. We call these objects axitons.

The axitons are not true solitons because the field coherently oscillates inside the axiton. In an expanding Universe the oscillating field will be red-shifted. Quantitatively, axitons resemble breathers of the (1 + 1)-dimensional sine-Gordon model[10]. Eventually the amplitude of axion field oscillations inside the axiton is red-shifted to sufficiently small values so that non-lineraities can be neglected, and the axiton configuration is frozen in the comoving volume as is any linear fluctuation. However the energy contrast relative to the homogeneous background will be large.

After that time and till $T > T_{\rm eq}$ the axiton energy density scales as T^3, so we can write $\rho_{\rm axiton} \equiv 3\,s\,(1+\Phi)T_{\rm eq}/4$, where Φ is the constant which depends upon the initial configuration of the axion field, i.e., the value of misalignment angle *and its gradients* at T_1 in the corresponding volume. Here, s is the entropy density, and in this normalization $\Phi = 0$ corresponds to the mean cosmological axion density, and $\Phi = 1$ corresponds to $2\bar{\rho}_a$. The energy density inside a given fluctuation is equal to the radiation energy density at $T = (1+\Phi)T_{\rm eq}$. At that time the self gravity of the fluctuation comes to dominate, and if $\Phi \gtrsim 1$ it separates out from the cosmological expansion, collapses, and forms a minicluster with density*

$$\rho_{\rm mc} \simeq 140\Phi^3(1+\Phi)\bar{\rho}_a(T_{\rm eq}) \approx 3\times 10^{-14}\Phi^3(1+\Phi)\left(\Omega_a h^2\right)^4 {\rm g\ cm^{-3}}, \qquad (3)$$

Even a relatively small increase in Φ is important because the final density depends upon Φ^4.

One reason why non-linear effects can lead to amplification of the axion density was recognized in Ref.
remt86. In that analysis it was proposed that some correlation regions can have values of Φ larger than one because the closer the initial value of θ is to the top of the axion potential, the later axion oscillations commence. However, this effect alone is not very significant and produces Φ significantly larger than 1 only for initial values very finely tuned to the top of the potential. Moreover, the axion field is not exactly coherent on scales of the Hubble radius, and even small fluctuations will spoil this simple picture.

Our scenario[11] for the generation of dense axion miniclusters mainly depends upon the interplay of the non-linear effects in the potential and gradients in the axion field.

*The factor of 140 results from a detailed calculation.

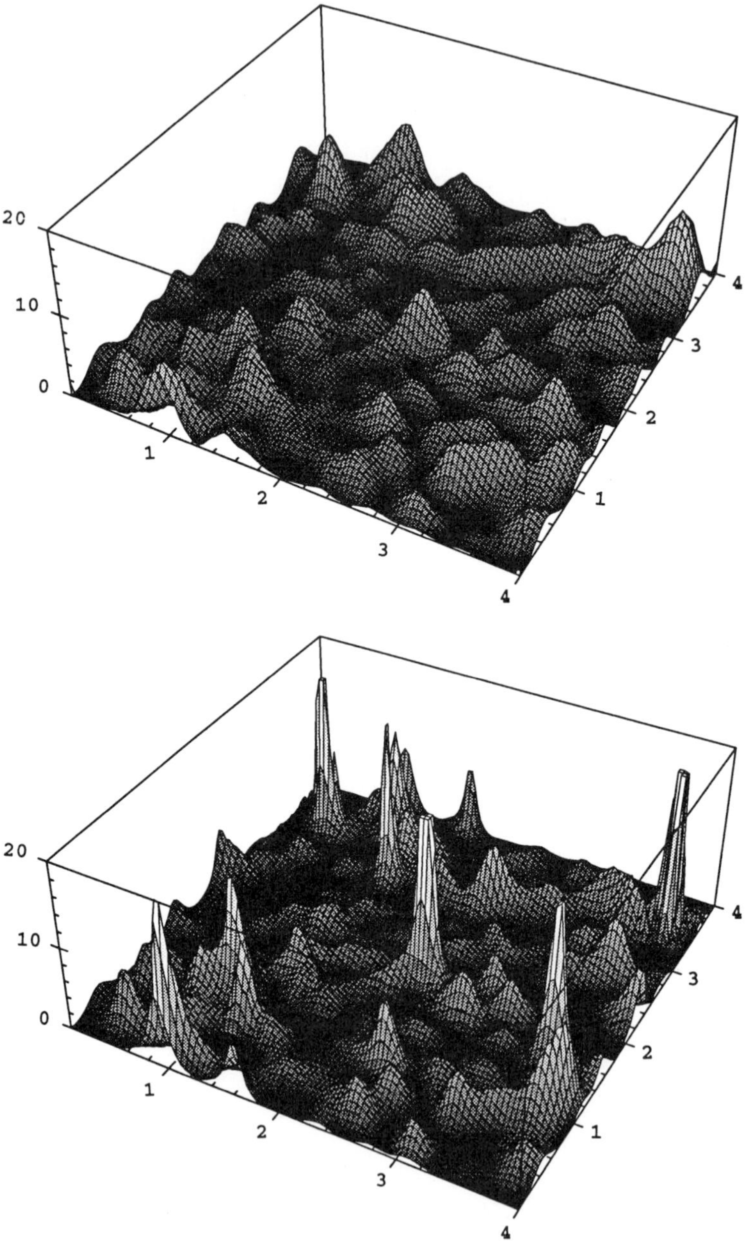

Figure 2. A 2-dimensional slice through the 3-dimensional box at time corresponding to $\eta = 3$ for the harmonic potential (top) and the axion potential (lower). The Hubble radius at this time is 3 units and the inverse of the axion mass is 0.005 units. The height of the figure corresponds to the energy density in the axion field normalized to the height for homogeneous field evolution: $\rho_a(\eta = 3)/\bar{\rho}_a(\eta = 3)$.

In the range of temperatures $T_1 \gg T \gg \Lambda_{QCD}$ the interplay of these two effects is important indeed. This can be seen in the following way. At temperatures $T \gg T_1$, the potential is negligible in the equations of motion compared to the gradient terms which force the field to be homogeneous on scales less than the Hubble radius. At $T \ll \Lambda_{QCD}$, on the contrary gradients can be neglected and one can treat the evolution of fluctuations as that of a pressureless gas. Clearly, around the QCD epoch when the potential just starts to become important in the equations of motion the gradient terms are still important, and since the initial amplitude can be close to π, the non-linear nature of the potential is also important. In order to find the energy density profile at freeze out one has to trace the non-linear inhomogeneous field evolution through the epoch $T_1 \gg T \gg \Lambda_{QCD}$.

Here, we present the results of a 3-dimensional numerical study of the evolution of the inhomogeneous axion field around the QCD epoch. We find that the resulting axion clumps are much denser than previously thought. The density in miniclusters exceeds at least by ten orders of magnitude the local dark matter density in the Solar neighborhood and even can reach the critical conditions for Bose star formation[12]. This might have a number of astrophysical consequences, as well as implications for laboratory axion searches[13].

AXION FIELD EVOLUTION

A. Equations of Motion

We start with deriving the equations of motion for the axion field in a form suitable for numerical calculations. In an expanding, spatially flat Universe with scale factor $R(t)$, the equation of motion for the axion field takes the familiar form

$$\ddot{\theta} + 3\frac{\dot{R}}{R}\dot{\theta} - \frac{\bar{\Delta}^2\theta}{R^2(t)} + m_a^2(t)\sin\theta = 0, \tag{4}$$

where dot denotes time derivative and $\bar{\Delta}$ is the Laplacian with respect to comoving coordinates $\bar{x}$.

Rather than cosmological time, it is convenient to work with a conformal-time coordinate which in a radiation-dominated Universe can be defined as $\eta \equiv R/R_1$, where R_1 is the scale factor at some arbitrary epoch. The equation of motion become

$$\theta'' + \frac{2}{\eta}\theta' - \frac{\bar{\Delta}^2\theta}{H^2(R_1)R_1^2} + \frac{\eta^2}{H^2(R_1)}m_a^2(R)\sin\theta = 0, \tag{5}$$

where prime denotes $d/d\eta$. We use Eq. (2) to find that in conformal time the mass evolves as $m_a^2(R) = m_a^2(R_1)\eta^n$. We will fix R_1 by making the choice $m_a^2(R_1) = H^2(R_1)$ (i.e., $\eta = 1$ corresponds to the instant of time when the inverse of the axion mass is equal to the Hubble radius) which simplifies the equation of motion:

$$\theta'' + \frac{2}{\eta}\theta' - \Delta^2\theta + \eta^{n+2}\sin\theta = 0, \tag{6}$$

where Δ is now the Laplacian with respect to rescaled comoving coordinates, $x \equiv H(R_1)R_1\bar{x}$. In other words, $x = 1$ corresponds to the Hubble distance at the epoch when the Hubble radius is the inverse of the axion mass.

The equation of motion can be written as a wave equation after redifinition of the field, $\psi \equiv \eta\theta$:

$$\ddot{\psi} - \Delta^2\psi + \eta^{n+3}\sin(\psi/\eta) = 0\ . \tag{7}$$

Table I. The scaling of physical quantities with conformal time η.

TIME	$t(\eta) = t(\eta = 1)\eta^2$
TEMPERATURE	$T(\eta) = T(\eta = 1)\eta^{-1}$
SCALE FACTOR	$R(\eta) = R(\eta = 1)\eta$
AXION MASS	$m_a(\eta) = m_a(\eta = 1)\eta^{n/2} \quad n = 7.4 \pm 0.2$
HUBBLE RADIUS	$R_H(\eta) \equiv H^{-1}(\eta) = R_H(\eta = 1)\eta^2$

The equation of motion is finally in a form convenient for the study of the evolution of the axion field during the epoch when the mass switches on. In Table I we give the scaling with η of several important physical length and mass scales. We next turn to the specification of the initial conditions.

B. Initial Conditions

At $\eta \ll 1$ the potential term in Eq. (7) can be neglected, and the solution of the wave equation can be expressed simply as a sum of Fourier harmonics. As usual, there will be two sums over frequency ω: one sum proportional to $\sin(\omega\eta)$ and one sum proportional to $\cos(\omega\eta)$. In the decomposition of the θ field, terms like $A(\omega)\sin(\omega\eta)/(\omega\eta)$ and $B(\omega)\cos(\omega\eta)/(\omega\eta)$ will appear. The terms proportional to $\cos(\omega\eta)$ correspond to decaying modes on all scales and can be neglected. Finally, assuming that on large scales the distribution for θ is a white noise, we obtain

$$\theta = A\pi \sum_{ijk} \frac{\sin(\omega\eta)}{\omega\eta} \sin(p_i x + \varphi_{1ijk}) \sin(p_j y + \varphi_{2ijk}) \sin(p_k z + \varphi_{3ijk}) \,, \tag{8}$$

where φ's are random phases and $\omega^2 = p_i^2 + p_j^2 + p_k^2$. On scales larger than the Hubble radius the resulting field distribution is frozen, while modes smaller than the Hubble radius are redshifted away, as expected.

We numerically evolved this distribution starting from initial time $\eta = 0.4$ in a comoving box of size $L = 4$ with periodic boundary conditions. There were 100^3 grid points in the box. Each of the momenta in the field decomposition took six discrete values, $p_n = 2\pi n/L$, with $n = 1, ..., 6$. So, in total there were 3×6^3 random phases, each with values in the interval $0 < \varphi < 2\pi$.

The final parameter to be chosen is the magnitude of A. Recall that, the axion potential is periodic with period 2π. We will consider $A = 1$ because in this case it is unlikely that domain walls will form in a box of the size we study and we are interested in the structure of density enhancements that are not associated with axion domain walls. At larger A domain walls are produced at about 1 per horizon.

C. Results of Numerical Calculations

1. (3 + 1)-*dimensional evolution*

In order to present the results of the (3+1)-dimensional calculations we will take a two-dimensional slice through the three-dimensional box, and plot the energy density as the height above the plane. We have analyzed the time evolution of the energy density in several different slices. All of the slices generally look alike. The most important (and generic) feature is the development of large-amplitude peaks. As the system evolves in

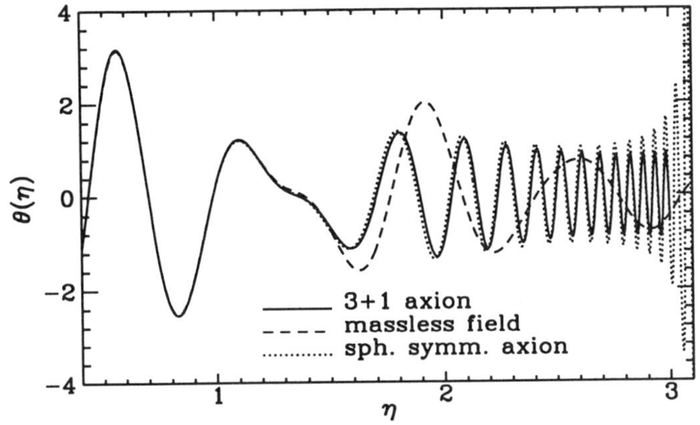

Figure 3. Time dependence of the axion field at the center of an axiton. The solid curve is the result of the full (3 + 1)-dimensional calculation. The dashed curve is an analytic calculation for a massless field with the same initial conditions. The dotted curve is the (1 + 1)-dimensional calculation assuming spherical symmetry.

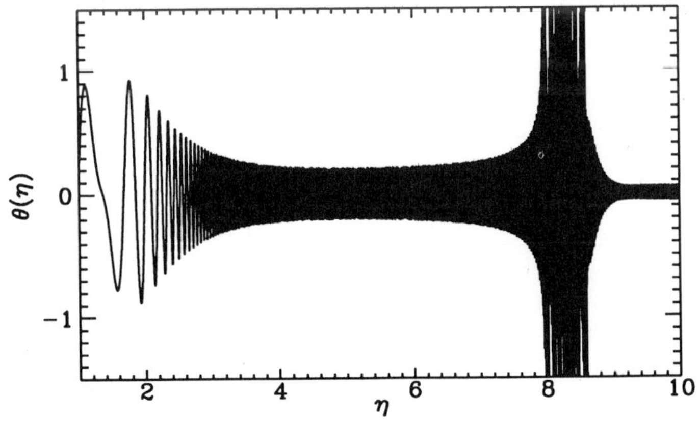

Figure 4. The time dependence of θ in the center of an axiton in the (1 + 1)-dimensional calculation. The axiton was generated by the choice $A = 0.73$.

time, the peaks in the energy density, the axitons, increase in magnitude and become more compact. We present the results in the $z = \text{const}$ plane, which intersects the point with the maximum energy density at late times. We emphasize that all slices through the box are quantitatively similar. We normalize the energy-density by comparing it to the energy density of a homogeneous axion field, $\bar{\rho}_a(\eta)$, with initial amplitude equal to the *rms* value of the misalignment angle, $\theta_{rms} = \pi/\sqrt{3}$.

In order to isolate the effect of the non-linearities in the axion potential, we also evolve the same initial conditions with a harmonic axion potential, $V(\theta) = m^2(T)f^2\theta^2/2$, and compare the evolution of the harmonic potential model to the axion model.

The distribution of the axion energy density in the reference plane at time corresponding to $\eta = 2$ is shown in Fig. 1a for the harmonic potential, and in Fig. 1b for the axion potential. The maximum energy density peak that picks the reference plane is clearly seen in Fig. 1b, its top portion is choped off to fit overall the scale of the figure.

The distribution of the axion energy density in the reference plane at time corresponding to $\eta = 3$ is shown in Fig. 2a for the harmonic potential, and in Fig. 2b for the axion potential. Again, the tops of the four peaks in Fig. 2b are chopped off; their heights are in excess of 100! Of course the peaks are only evident for the axion potential model.

Comparing Fig. 2b to Fig. 1b, we see that for the axion potential most of the high magnitude peaks grow considerably in height and became thinner, while most of the low amplitude peaks remain almost unchanged, i.e., they are in the linear regime and consequently are frozen by the cosmological expansion. There are some peaks (some even relatively high at $\eta = 2$) which decreased in amplitude. Those peaks represent the tales of the density clumps which reach their maximum at some other value of z. All high density peaks contract in the coordinate volume, those which decreased in height simply moved out of our reference plane. High density peaks do not develop in the evolution of the harmonic potential, and the evolution proceeds as was assumed in the linear analysis[6,8].

There is insufficient resolution on this grid to proceed further in time with the axion potential, nevertheless we had proceed up to $\eta = 4$. As expected the evolution of the field in harmonic case is frozen after $\eta = 3$ in comoving reference frame, while density contrasts continue to grow in the axion case.

There is a simple, heuristic explanation for the fact that non-linear effects lead to the formation of high density peaks. The average pressure over a period of homogeneous axion oscillations in the axion potential potential is negative,† and is equal to $\langle P \rangle \simeq -\Lambda_a^4(T)\theta_0^4/64$, where θ_0 is the amplitude of the oscillations[14] (this formula is valid for $\theta_0 \ll \pi$; as $\theta_0 \to \pi$, the field spends more and more time near the top of the potential, and $\langle P \rangle \to -2\Lambda_a^4$). In other words, the axion self-interaction is attractive. The larger the amplitude of oscillations inside the fluctuation, the more negative the pressure inside, and consequently, fluctuations with excess axions will contract in the comoving volume. In addition, matter with a smaller pressure suffers less redshift in cosmological expansion.

In order to learn the fate of the high density peaks, we have choosen one of them in Fig. 2 and generated the corresponding spherically symmetric initial conditions at $\eta = 0.4$ and evolved it in time. We now describe the results of this calculation.

2. $(1+1)$*-dimensional evolution*

†Of course the average pressure is dominated by relativistic species at this time. It is the pressure contributed by the axions that is negative.

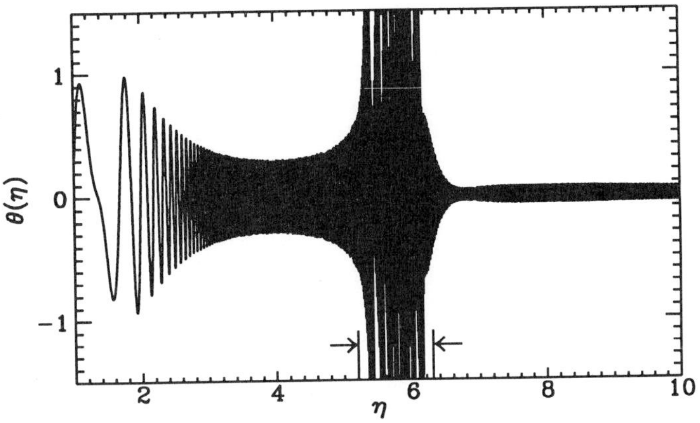

Figure 5. The time dependence of θ in the center of an axiton in the (1 + 1)-dimensional calculation. The axiton was generated by the choice $A = 0.77$.

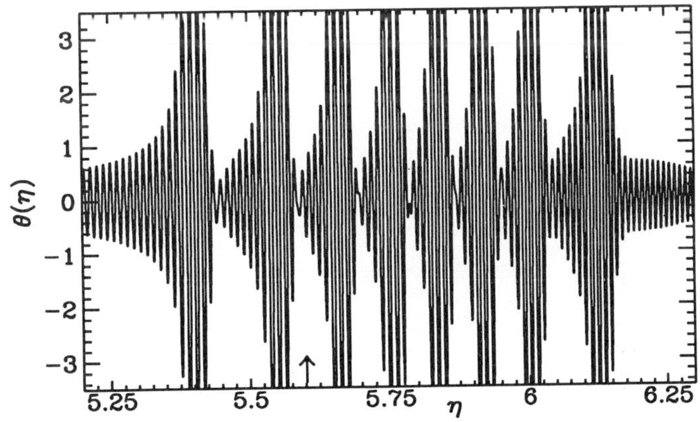

Figure 6. An exploded view of the large amplitude oscillations in the center of the axiton of Fig. 5. The region pictured here is indicated by the arrows in Fig. 5.

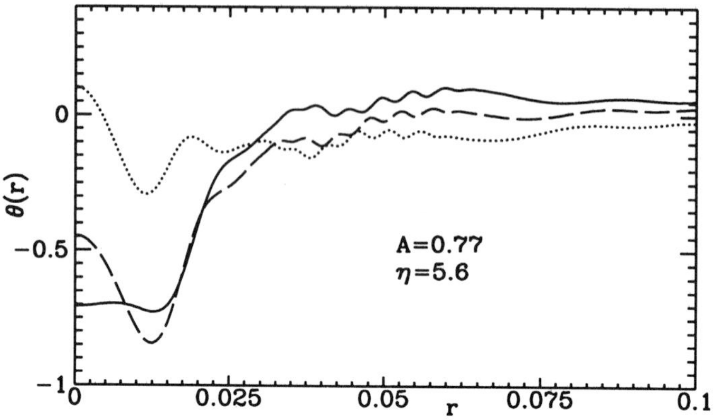

Figure 7. The radial profile of the axiton of Fig. 9 at three instants during one period of oscillation around time $\eta = 5.6$ indicated by the arrow in Fig. 6.

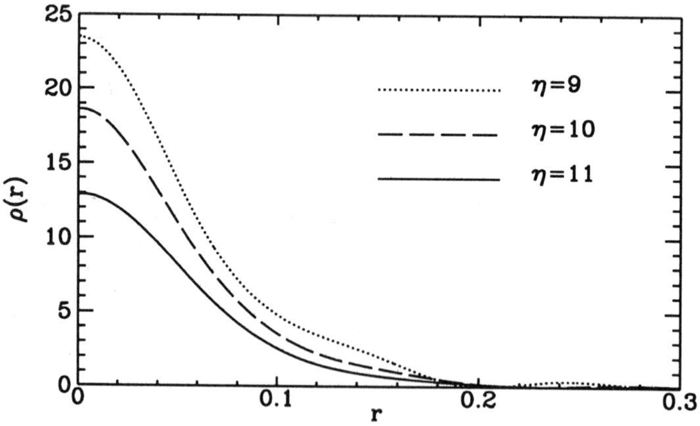

Figure 8. Energy density profiles of the axiton in the spherically symmetric calculation with $A = 0.77$ at three instants of time.

The axiton we choose to examine is the one near the center of the grid of Fig. 2, with grid coordinates $\{2.24, 1.92\}$ (the grid coordinate of the plane of Fig. 2 in the z-direction is 1.76—also near the center of the 3-dimensional box—and the axiton we chose is almost at its maximum in this plane, having an absolute maximum at $z = 1.80$).

The dependence of the axion field upon time at the reference point at the center of this peak in our $(3+1)$-dimensional numerical calculation is shown in Fig. 3 by the solid curve. We can compare this evolution to the evolution of a massless field which is still given by Eq. (8). The evolution of a massless field at the reference point is shown in the Fig. 3 by the dashed line.

It is then straightforward to construct a spherically symmetric solution to the massless counterpart of Eq. (7) which in the center of the configuration will have exactly the same time dependence as in Fig. 3 . We start with the field decomposition, Eq. (8), substitute the values of the coordinates of the given spatial point, and multiply each time harmonic by $\sin(\omega r)/\omega r$. So the dashed line in Fig. 3 represents the time dependence for a massless field both at the reference point of the $(3+1)$-dimensional calculation and at the center of a $(1+1)$-dimensional spherically symmetric configuration. We can use the resulting configuration $\theta = \theta(\eta, r)$ for generating spherically symmetric initial conditions which will approximate the peak of our choice for the runs with the actual axion potential. The result of such a calculation with $A = 1$ is shown in Fig. 3 by the dotted line.

From Fig. 3, we see that the axion mass effectively switches on at a time $\eta \sim 1.3$. By this time the amplitude of the massless field is greater than unity. This means that for the evolution of the field using the axion potential the oscillations start in the non-linear regime $\theta \gtrsim 1$ in the region that will develop into a axiton. We see also that the non-linearity is strong enough to force the density peak to collapse not only in coordinate space, but also in physical space as well, since by the time $\eta \sim 3$ the amplitude of axion field oscillations are in the non-linear regime and growing (this is somewhat difficult to see in the figure). The rate of growth in the non-spherical case is much slower (compare the solid and dotted lines). This makes sense because we expect a spherical collapse to lead to a denser central region.

So in general, there are two competing effects in the evolution of the axion field. 1) A contraction of the axiton due to the pressure difference, leading to an increase in amplitude in the center, and 2) a decrease in amplitude of the oscillations due to the expansion of the Universe. We found that with a sufficiently large initial amplitude at the start of oscillations, the first process wins, and the amplitude in the center of the axiton increases to values $\theta \gtrsim \pi$. In the opposite case, e.g., if the initial amplitude when oscillations commence is in the linear regime, the amplitude monotonically decreases in time.

In realistic axion models the axion mass does not continue to grow with η, but saturates at its zero-temperature value around $T \sim \Lambda_{QCD}$. For axion masses in the allowed window this corresponds to $\eta \gtrsim 6$. Due to the steep power-law dependence of the function $m(\eta)$ in Eq. (2), the period of the field oscillations becomes very small by $\eta \sim 6$, and direct numerical methods fail even in the case of spherical symmetry. In order to follow the evolution of the fluctuation up to freeze out we must assume that the mass saturates at a smaller value of η. This would correspond to a larger value of f_a. We have approximated the procces of axion mass saturation by the the simple formula

$$m^2(\eta) = m_a^2(\eta = 1)\eta^n/[1 + (\eta/\eta_c)^n], \tag{9}$$

taking $\eta_c = 3.5$. This value of η_c corresponds to too large value of $f_a \sim 4 \times 10^{13}$GeV, which would give $\Omega_a h^2$ in excess of one. However we expect that qualitatively the evolution of the axion field will have the same basic properties for larger values of η_c (smaller values of f_a).

We can vary the initial overall amplitude A of our spherically symmetric configurations. This has the effect of spanning different initial conditions of a well defined one-parameter family of axitons. Moreover, varing A is easier than choosing different peaks in Fig. 2.

The time dependence of the field in the center of the fluctuation that will develop into an axiton for $A = 0.73$ is presented in Fig. 4, and for $A = 0.77$ in Fig. 5. In both cases the configuration collapses and the amplitude of θ rapidly increases in the center, even exceeding the value of π. This is followed by a period of several rebounds. An expanded view of the rebounds is shown in Fig. 6. During each rebound (eight in total in both cases) relativistic axions are emitted. We can see the signature of axion emission by looking at the radial profile of an axiton. In Fig. 7 we show the profile of the axiton of Fig. 6 at three instants in time during one oscillation period. The emission of relativistic axions is seen in the outgoing waves of Fig. 7. The emission of relativistic axions reduces the energy of the central configuration below some critical value, at which point a pseudo-soliton, an axiton, is produced.

The final energy density profile of this configuration for the case $A = 0.77$ is shown in Fig. 8. At time $\eta = 9$ (dotted line), outgoing secondary waves are still seen in the tail of the configuration. By time $\eta = 11$ there is no evidence of outgoing radiation. The amplitude of the energy density at $r = 0$ is 23.5 at $\eta = 9$ and 12.9 at $\eta = 11$ (the energy density in this graph is not normalized to the homogeneous background). It is clear that the energy density in the center scales as η^{-3} (e.g., $23.5/12.9 = (11/9)^3$), confirming that the linear regime has been reached, the fluctuation is frozen, and the number of axions per comoving volume is conserved. The energy density of a homogenoeus background at $\eta = 10$ with $\eta_c = 3.5$ and initial amplitude equal to the *rms* value of θ is 0.85 in the units of the figure. Thus, the fluctuation of Fig. 8 has an energy density contrast of 20.

Not all fluctuations that pass through the non-linear regime contract in physical space. For example, a sample spherical fluctuation for $A = 0.70$ does not collapse. The corresponding energy density profiles of this fluctuation at two moments of time are presented in Fig. 9 by the dashed lines. This should be compared to the solid lines, which are the energy density profiles for a fluctuation with $A = 0.73$ which does undergo collapse. We see that the slope of the energy density in the non-linear tail tends to a power law $\rho \propto r^{-3}$ prior to the collapse. This leads to an increase in field amplitude in the center, while due to the overall expansion of the Universe, the amplitude decreases. For $A = 0.73$, the first process wins for some period of time, see Fig. 4, while for $A = 0.7$ the general expansion dominates at all times, and the amplitude of the oscillations decreases monotonically. However, the decrease in amplitude is much slower than it would be with the harmonic potential, and the final energy density contrast with $\eta_c = 3.5$ and $A = 0.7$ is 45.

For comparison we also present in Fig. 9 the energy density profile of the fluctuation with $A = 0.77$ at $\eta = 11$. Remarkably, it has the same power-law slope, $\rho \propto r^{-3}$, despite the fact that this profile represents a fluctuation that has undergone "violent oscillations" accompanied by axion emission (see Figs. 8, 10 and 11).

Since the axion interaction is attractive, one can expect that bound states of axions can form. One example of such a bound state is the well known "breather" solution in the $(1 + 1)$-dimensional sine-Gordon model. In $(3 + 1)$ dimensions this solution possesses planar symmetry and turns out to be unstable with respect to fragmentation. If a spherically symmetric counterpart of the "breather" would exist in Minkowski space-time, it would behave in an expanding Universe just as the fluctuation shown in Fig. 8. Thus the axiton is related to a spherically symmetric breather.

Suppose we can extrapolate these results to the range of realistic axion models, i.e., to larger values of η_c corresponding to smaller values of f_a. Then we must consider the possibility of producing enormous density contrasts. Indeed, both the increase in axion mass and the expansion of the Universe adiabatically decreases the amplitude of axion oscillations in the linear regime (or in the homogeneous state), so that at $T \lesssim \Lambda_{\rm QCD}$ the corresponding background energy density is about $\bar{\rho}_a \approx T_{\rm eq} T^3$, where $T_{\rm eq} \sim 5.5 \Omega_a h^2$eV is the temperature of equal radiation and axion energy density. In the case of a collapsing non-linear fluctuation, the final field configuration is the output of non-linear dynamics. Let θ_L be the amplitude of field oscillations in the axiton at the time when it enters the linear regime at $T_L = T_1/\eta_L$. Then the corresponding energy density in the fluctuation will be at this time about $\Lambda_a^4 \theta_L^2$. The ratio of the axiton energy density to the homogeneous background axion energy density will be

$$1 + \Phi \approx \Lambda_a^4 \theta_L^2 \eta_L^3 / T_{\rm eq} T_1^3. \tag{10}$$

Using the results from Figs. 7 and 8 ($\theta_L \sim 0.1$, and $\eta_L > 6$),[‡] we obtain $1 + \Phi \approx 10^4$ prior to gravitational decoupling of the fluctuations from the cosmological expansion.

Although this possibility is exciting, a word of caution is necessary. Non-linear dynamics is rather unpredictable, and one can not exclude the possibility that at $\eta_c > 6$ all collapsing non-linear fluctuations somehow dissipate, leaving very small θ_L. Note also that non-spherical configurations can evolve quite differently than the spherical configurations. Nevertheless, our point of view is that spectrum of energy density contrasts can span the entire range from order 1 up to of order 10^4 or even larger. However, at this time we have nothing to say in regard to the number density of peaks as a function of its amplitude.

DISCUSSION

In principle, all axion miniclusters could be relevant to laboratory axion search experiments, since even for Φ as small as 1, the density is 10^{10} times larger than the local galactic halo density [see Eq. (3)]. Moreover, as we have noted already, the energy density in an axiton after it separates out from the general expansion will be Φ^4 times larger than the energy density at $T_{\rm eq}$. For example, a rather moderate density contrast of $\Phi = 30$ at $\Lambda_{\rm QCD} > T > T_{\rm eq}$ will correspond to roughly an additional factor of 10^6 in the energy density of the axiton at $T \ll T_{\rm eq}$.

We can define the boundary of future miniclusters as a surface where the density of the axiton becomes equal to its mean cosmological value. It turns out that, in our calculation, 80 % of all axions belong to miniclusters at $\eta = 3$. See Fig. 3(b).

The probability of a direct encounter with a minicluster is small. Let's assume that all of the axions end up in miniclusters of mass $10^{-9} M_\odot$, density 10^{-14}g cm^{-3}, and radius 4×10^{12}cm. Using a local halo mass density of 5×10^{-25}g cm^{-3} would give a minicluster number density of 7,000,000 pc^{-3}. With a typical velocity of 250 km s^{-1} the encounter rate would be 1 per 30 million years with $\Phi = 1$. Although the signal in an axion detector [13] from a close encounter would be enormous, it might be a long wait with a weak signal between encounters if a major fraction of the axions are part of miniclusters.

‡In any case, η_L will be larger than η_c, the value of η where the axion mass saturates to its zero-temperature value [see Eq. (9)], and $\eta_L > 6$ seems a very conservative estimate.

There should be some miniclusters with Φ in the range $10^{-3} \lesssim \Phi \lesssim 1$. These collapse during the matter-dominated epoch and have a larger radii than those with $\Phi \gtrsim 1$ which collapsed in the radiation-dominared epoch, so the probability of an encounter with a clump with $\Phi \ll 1$ can be larger. From the point of view of direct searches, even miniclusters with density contrast of order two times the average with respect to the galactic halo density are important. Such miniclusters form just prior to the moment of galaxy formation and started with $\Phi \sim 10^{-3}$. For $\Phi < 1$ the expected time between encounters is given in terms of the number density of clumps, n, the geometric cross section of the clump $\sigma \sim R_{\rm mc}^2$, and the virial velocity v as

$$\tau = \frac{1}{n\sigma v} \simeq \frac{1}{\Phi}\frac{\rho_{\rm mc}}{\rho_H}\frac{R_{\rm mc}}{v}, \tag{11}$$

where ρ_H is the halo density, and $R_{\rm mc}/v$ is the time the Earth spends inside the minicluster. The factor of Φ^{-1} appears because the number density of miniclusters with $\Phi \ll 1$ is suppressed in our model. A minicluster with $\Phi \ll 1$ would require the initial misalignment angle θ (which is uniformally distributed in the range 0 to 2π) to be finely tuned to the mean misalignment angle to an accuracy $\delta\theta/\theta \simeq \Phi/2$. Using Eq. (3), we finally obtain

$$\tau = \Phi\tau_{\Phi=1} = \Phi \cdot 3 \times 10^7 \text{yr}. \tag{12}$$

Note that the miniclusters discussed so far appear if the axion field is uncorrelated on scales larger than the Hubble radius at $T \sim 1$ GeV and there is no other sources of density perturbations. However miniclusters with $\Phi \ll 1$ can appear from primordial density fluctuations generated by inflation without the suppression factor of Φ^{-1}. If this is the case, then $\tau \simeq \Phi^2 \cdot 3 \times 10^7$ yr. Since Φ would be small, this would give a reasonable encounter rate, and the question of formation and survival of small-scale clumps within the galaxy is worth further study in applications to direct dark matter searches.

The question arises, could there be any astrophysical consequenses of very dense axion clumps? One outcome sometimes mentioned in the literature is the possibility of gravitational microlensing. Two conditions must be satisfied for the clump to cause gravitational microlensing [15]. First, the mass of the clump has to be in a range near $0.1 M_\odot$. Second, the physical radius of the clump has to be smaller than the Einstein ring radius, $R_{\rm E} = 2\sqrt{GMd}$ where d is the effective distance to the lens (typically $d \sim 20$ kpc). The second condition restricts the density of the minicluster to be $\rho \gtrsim 10^7 \rho_{\rm eq}/\sqrt{M_{-1}}$, where $M_{-1} \equiv M/0.1 M_\odot$. If the lensing object is a clump of non-interacting cold dark matter, it has to be formed from a density fluctuation with $\Phi \gtrsim 20$.

Axion miniclusters can have $\Phi \gtrsim 20$; however, they are too light. While it is possible to invent models where both conditions are met for some of the clumps (one example could be an axion model with an extremely small, but non-zero, value for the u-quark mass), it is hardly likely that a substantial amount of the dark matter has evolved into clumps capable of lensing. On the other hand, anticipating significant numbers of microlensing events (for the first positive reports see Ref. [16]) in the future, it is not excluded that some of them could be caused by the clumps in such classes of models (especially if collisional relaxation is significant). The corresponding light curve will be different from the MACHO event since clumps are extended objects.

Another astrophysical outcome of very dense axion clumps can be the possibility of "Bose star" formation in axion miniclusters, which we will discuss below.

The physical radius of an axiton at $T_{\rm eq}$ is larger by many orders of magnitude than the de Broglie wavelength of an axion in the corresponding gravitational well. Consequently, the gravitational collapse of the axion clump and subsequent virialization

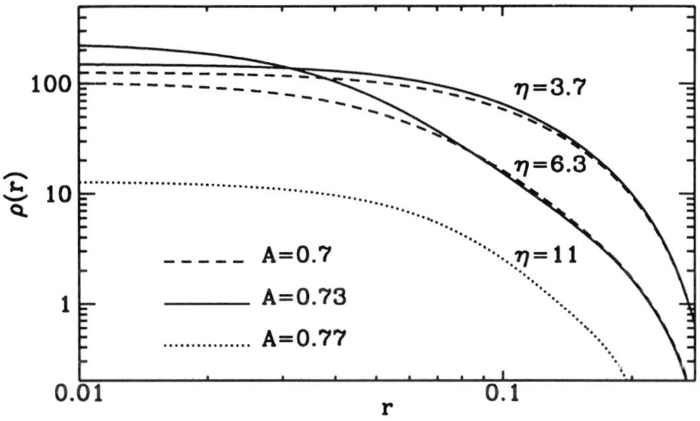

Figure 9. Energy density profiles of three axitons in the spherically symmetric calculation. The axitons were generated by the choices $A = 0.70$, $A = 0.73$, and $A = 0.77$. At $r \geq 0.1$ the curves separate into two pairs and a single configuration. The upper pair are the profiles at $\eta = 3.4$, the next pair are the profiles at $\eta = 6.3$, and the final dotted line is at $\eta = 11$.

can be described in the usual terms of cold dark matter particles. In a few crossing times some equilibrium distribution (presumably close to an isothermal distribution) of axions in phase space will be established. It is remarkable that in spite of the apparent smallness of axion quartic self-couplings, $|\lambda_a| = (f_\pi/f_a)^4 \sim 10^{-53} f_{12}^{-4}$, the subsequent relaxation in an axion minicluster due to $2a \to 2a$ scattering can be significant as a consequence of the huge mean phase-space density of axions [12]. In the case of Bose-Einstein statistics the inverse relaxation time is $(1+\bar{n})$ times the classical expression, or $\tau_R^{-1} \sim \bar{n}\, v_e \sigma \rho_a/m_a$, where σ is the corresponding cross section. For particles bound in a gravitational well, it is convenient to rewrite this expression in the form[12]

$$\tau_R \sim m_a^7 \lambda_a^{-2} \rho_a^{-2} v_e^2. \tag{13}$$

The shallower the gravitational well for a given density of axions, the larger the mean phase space density, and consequently the smaller the relaxation time due to the v_e^2 dependence in Eq. (13). Note also the dependence of the inverse relaxation time upon the square of the particle density.

The relaxation time (13) is smaller then the present age of the Universe if the energy density in the minicluster satisfies

$$\rho_{10} > 10^6 v_{-8} \sqrt{f_{12}}, \tag{14}$$

where $\rho_{10} \equiv \rho/(10\,\mathrm{eV})^4$ and $v_{-8} \equiv v_e/10^{-8}$. If this occurs, then an even denser core in the center of the axion cloud should start to form. An analogous process is the so-called gravithermal instability caused by gravitational scattering. This was studied in detail for star clusters, where the "particles" obey classical Maxwell–Boltzmann statistics. Axions will obey Bose–Einstein statistics, with equilibrium phase-space density $n(p) = n_{\mathrm{cond}} + [e^{\beta E} - 1]^{-1}$, containing a sum of two contributions, a Bose condensate and a thermal distribution. The maximum energy density that non-condensed axions can saturate is $\rho_{\mathrm{ther}} \sim m_a^4 v_e^3$, which corresponds to $\bar{n}_{\mathrm{ther}} \sim 1$. Consequently, given the initial condition $\bar{n} \gg 1$, one expects that eventually the number of particles in the condensate will be comparable to the total number of particles in the region if relaxation is efficient. Under the influence of self gravity, a Bose star [17,14] then forms[12].

Comparing Eqs. (3) and (14), we conclude that the relaxation time is smaller than the present age of the Universe and conditions for Bose star formation can be reached in miniclusters with a density contrast $\Phi \gtrsim 30$ at the QCD epoch.

Under appropriate conditions stimulated decays of axions to two photons in a dense axion Bose star are possible [14,18] (see also[19]), which can lead to the formation of unique radio sources—axionic masers.

Let some axion miniclusters start to shine indeed, and let and $n(\nu)$ is the number density of shining miniclusters divided by the mean number density of all miniclusters, $\bar{n} \sim 10^{-53} cm^{-3}$, assuming that all dark matter is segregated in miniclusters. We can estimate the averaged luminosity of this cosmic maser as $\bar{L} \sim M_{\mathrm{mc}}/\tau_R$, where τ_R is the relaxation time assumed to be equal to the age of the universe. We obtain

$$\bar{L} \sim 10^{-9} f_{12}^{-1/4}\, v_{-8}^{5/2}\, L_\odot, \tag{15}$$

which is quite high to be detected.

Axionic line in diffuse background radiation expected from uniform distribution of miniclusters is a superposition of the photon spectra generated in decays occurring at different times. We obtain at time t for intensity per unit frequency

$$I_\nu \sim (\nu/m_a)^{3/2}\, L(t_i)\, n(t_i)\, t/m_a, \tag{16}$$

where $n(t_i)$ is number density of shining miniclusters and $L(t_i)$ is cluster luminosity at the time of photon injection, $t_i = (\nu/m_a)^{3/2} t$. In units of (erg/Hz sec cm^2 sr) the intensity is

$$I_\nu \sim 10^{-13} (\nu/m_a)^{3/2} L(\nu) n(\nu) f_{12}^{3/4} v_{-8}^{5/2}, \tag{17}$$

where $L(\nu)$ is the luminosity normalized by the averaged luminosity (15), $L(\nu) \equiv L(t_i(\nu))/\bar{L}$. Already with $n(\nu) \sim 10^{-6}$ the estimate (17) is comparable to the intensity of CMBR on frequency $\sim 10^{-5}$eV. Note also that the intense maser line is expected when the shining minicluster passes near the solar system.

In view of the results of this paper we conclude that the questions of axion Bose star formation, structure, and possible astrophysical signatures deserve detailed study.

In conclusion, we have presented a 3-dimensional numerical study of the evolution of inhomogeneities in the axion field around the QCD epoch, including for the first time important non-linear effects. We found that the non-linear effects of the attractive self-interaction can lead to a much larger density of axions in miniclusters than previously estimated. Large amplitude density contrasts form solitons we call axitons, and resemble the bound-state "breather" solutions of the $(1+1)$-dimensional sine-Gordon model. The increase in the axion density may be sufficiently large that axion miniclusters formed by the fluctuations might exceed the critical density necessary for them to relax to form Bose stars.

ACKNOWLEDGMENTS

EWK and IIT are supported by the DOE and NASA under Grant NAGW–2381.

REFERENCES

1. R.D. Peccei and H. Quinn, *Phys. Rev. Lett.* 38:1440 (1977); *Phys. Rev.* D 16:1791 (1977); S. Weinberg, *Phys. Rev. Lett.* 40:223 (1978); F. Wilczek, *ibid.* 40:279 (1978).
2. D. Dicus, E. Kolb, V. Teplitz and R. Wagoner, *Phys. Rev. D* 18:1829 (1978) and *Phys. Rev. D* 22:839 (1980); M. Fukugita, S. Watamura and M. Yoshimura, *Phys. Rev. Lett.* 48:1522 (1982) and *Phys. Rev. D* 26:1840 (1982).
3. G. Raffelt and D. Seckel, *Phys. Rev. Lett.* 60:1793 (1988); M. S. Turner, *Phys. Rev. Lett.* 60:1797 (1988); R. Mayle et al., *Phys. Lett.* 203B:188 (1988); T. Hatsuda and M. Yoshimura, *Phys. Lett.* 203B:469 (1988).
4. J. Preskill, M. Wise and F. Wilczek, *Phys. Lett.* 120B:127 (1983); L.F. Abbott and P. Sikivie, *Phys. Lett.* 120B:133 (1983); M. Dine and W. Fischler, *Phys. Lett.* 120B:137 (1983).
5. For recent reviews, see M.S. Turner, *Phys. Rep.* C197:67 (1990); G.G. Raffelt, *Phys. Rep.* C198:1 (1990).
6. M.S. Turner, *Phys. Rev. D* 33:889 (1986).
7. A.D. Linde and D.H. Lyth, *Phys. Lett.* B246:353 (1990); D.H. Lyth and E.D. Stewart, *Phys. Rev. D* 46:532 (1992).
8. C.J. Hogan and M.J. Rees, *Phys. Lett.* B205:228 (1988).
9. E.W. Kolb and M.S. Turner, "The Early Universe", Addison-Wesley, Redwood City, Ca. (1990).
10. R. Rajaraman, "Solitons and Instantons", Elsevier Science Publishers, New York (1982).
11. E. Kolb and I.I. Tkachev, *Phys. Rev. Lett.* 71:3051 (1993).
12. I.I. Tkachev, *Phys. Lett.* B261:289 (1991).
13. P. Sikivie, *Phys. Rev. Lett.* 51:1415 (1983).
14. I.I. Tkachev, *Sov. Astron. Lett.* 12:305 (1986).

15. B. Paczynski, Astrophys. J. 304:1 (1986); K. Griest, Astrophys. J. 366:412 (1991).

16. E. Aubourg *et al.*, Nature 365:623 (1993); C. Alcock *et al.*, Nature 365:621 (1993).

17. R. Ruffini and S. Bonozzola, *Phys. Rev.* 187:1767 (1969); J.D. Breit, S. Gupta, and A. Zaks, *Phys. Lett.* B140:329 (1984); M. Gleiser, *Phys. Rev. D* 38:2376 (1988); E. Seidel and W.-M. Suen, *Phys. Rev. Lett.* 66:1659 (1991).

18. I.I. Tkachev, *Phys. Lett.* B191:41 (1987).

19. T.W. Kephart and T.J. Weiler, preprint VAND-TH-90-2, 1990 (unpublished).

POWER SPECTRUM OF COSMIC STRING PERTURBATIONS ON THE MICROWAVE BACKGROUND

LEANDROS PERIVOLAROPOULOS[1]

Division of Theoretical Astrophysics
Harvard-Smithsonian Center for Astrophysics
60 Garden Street
Cambridge, Mass. 02138, USA

Abstract

I review recent progress towards a detailed understanding of the power spectrum of cosmic string perturbations induced on the Microwave Background (CMB). The use of a simple analytic model allows to include important effects that have not been included in previous studies. These effects are potential fluctuations on the last scattering surface (the Sachs-Wolfe effect) and Doppler CMB perturbations. These previously neglected fluctuations are shown to dominate over post-recombination fluctuations (Gott-Kaiser-Stebbins effect) on angular scales less than about 2° assuming that no reionization occurs. The effective power spectrum index n_{eff} is calculated on COBE scales and is shown to be somewhat larger than 1 ($1.35 \lesssim n_{eff} \lesssim 1.5$). After normalizing the model on COBE data and string simulations, I derive the predicted values of $(\frac{\delta T}{T})_{rms}$ for several ongoing CMB experiments on medium and small angular scales. A comparison is also made with current observations and predictions of models based on inflation.

[1]E-mail address: leandros@cfata3.harvard.edu

1 Introduction

One of the most important characteristics of a model for large scale structure formation is the spectrum of primordial perturbations the model predicts. This spectrum is imprinted on the last scattering surface and parts of it show up on the cosmic microwave background (CMB) temperature maps of experiments at various angular scales.

The CMB power spectrum predicted by models based on inflation[2, 3] has been well studied analytically[4] on all scales larger than 1-2 arcmin. The main competition to these models comes from models where the primordial perturbations are generated by topological defects which produce seed-like primordial perturbations. These perturbations are imprinted on the CMB as temperature perturbations and can be detected by ongoing CMB experiments. A particularly interesting type of topological defect is the *cosmic string*.

The cosmic string theory [5, 6, 7] for structure formation is the oldest and (together with textures [8]) best studied theory of the topological defect class. By fixing its single free parameter $G\mu$ (μ is the *effective* mass per unit length of the wiggly string and G is Newtons constant) to a value consistent[9] with microphysical requirements coming from GUT's, the theory may automatically account for large scale filaments and sheets [10, 11, 12, 13, 14, 15], galaxy formation at epochs $z \sim 2-3$ [16, 17] and galactic magnetic fields [18]. It can also provide large scale peculiar velocities [19, 20] and is consistent with the amplitude, spectral index[21, 22, 23, 24, 25, 26] and the statistics[27, 28, 29, 30] of the cosmic microwave background (CMB) anisotropies measured by the COBE collaboration [31, 32] on large angular scales ($\theta \sim 10°$).

The CMB spectrum of the cosmic string model has been investigated using both simulations[22, 24] and analytical methods[25] but only on large angular scales (*i.e.* scales of a couple of degrees or larger). This paper shows the results of an attempt I have recently made[26] to extend the analysis of the CMB spectrum predicted by cosmic strings to arbitrarily small scales. Due to resolution limitations this extension can only be made using analytical methods.

In the next section I will briefly review the three basic mechanisms by which strings can induce perturbations on the CMB: post-recombination Gott-Kaiser-Stebbins[33, 34] (GKS) type perturbations, potential perturbations on the last scattering surface (LSS) and Doppler perturbations due to velocities of electron last scatterers[35]. Then I will describe an analytical method[19, 25, 28, 36, 20] used to derive the power spectrum of CMB fluctuations induced by strings, in terms of the single free parameter of the model ($G\mu$) and three evolution parameters that can be fixed from string simulations. In section 3, I will show how can these parameters be fixed by using the COBE detection for the free parameter $G\mu$ and string simulations for the evolution parameters. The scale invariance of the spectrum will also be demonstrated. Finally, in section 4, I will use the resulting normalized spectrum to make predictions for the rms temperature fluctuations detected by ongoing CMB experiments on medium and small angular scales.

I will assume $\Omega_0 = 1$ $h = 1/2$, Cold Dark Matter (CDM), $\Lambda = 0$ and standard recombination.

2 Types of CMB Fluctuations

The best studied mechanism for producing temperature fluctuations on the CMB by cosmic strings is the Gott-Kaiser-Stebbins (GKS) effect[33, 34]. According to this effect, moving long strings present between the time of recombination t_{rec} and today produce (due to their deficit angle[37]) discontinuities in the CMB temperature between photons reaching the observer through opposite sides of the string.

The second mechanism for producing CMB fluctuations by cosmic strings is based on potential fluctuations on the LSS. Long strings and loops present between the time of equal matter and radiation t_{eq} and the time of recombination t_{rec} induce density and velocity fluctuations to their surrounding matter. These fluctuations grow gravitationally and at t_{rec} they produce potential fluctuations on the LSS. Temperature fluctuations arise because photons have to climb out of a potential with spatially dependent depth.

The third mechanism for the production of temperature fluctuations is based on the Doppler effect. Moving long strings present on the LSS produce velocity fields to the surrounding plasma. Thus, photons scattered for last time on these perturbed last scatterers suffer temperature fluctuations due to the Doppler effect.

The total temperature perturbation may be obtained by superposing the effects of these three mechanisms at all times from t_{rec} to today. Clearly each mechanism involves the superposition of some type of seed present during a given period of time. Therefore, in order to construct the spectrum of the resulting perturbations we must address the following two questions: First, *how can we superpose the seeds in order to construct the resulting spectrum* and second *what is the type of seed corresponding to each mechanism for producing CMB fluctuations?*

I will first address the first question. Consider a great circle on the sky (*e.g.* a meridian) and a seed function $f_1^\Psi(\theta)$ of angular scale Ψ, superposed N times at random positions θ_n on the circle with variable magnitudes a_n. The resulting pattern will be

$$f(\theta) = \sum_{n=1}^{N} a_n f_1^\Psi(\theta - \theta_n) = \frac{1}{2\pi} \sum_{n=1}^{N} a_n \sum_{k=-\infty}^{+\infty} \tilde{f}_1^\Psi(k) e^{ik(\theta-\theta_n)} \tag{1}$$

where $\tilde{f}_1^\Psi(k)$ is the Fourier transform of $f_1^\Psi(\theta)$

$$\tilde{f}_1^\Psi(k) \equiv \int_{-\pi}^{+\pi} d\theta\, f_1^\Psi(\theta) e^{-ik\theta} \tag{2}$$

Therefore $\tilde{f}(k)$ (*i.e.* the Fourier transform of $f(\theta)$) can be expressed in terms of $\tilde{f}_1^\Psi(k)$

$$\tilde{f}(k) = \tilde{f}_1^\Psi(k) \sum_{n=1}^{N} a_n e^{ik\theta_n} \tag{3}$$

and the corresponding power spectrum $P_0(k)$ is easily obtained as the ensemble average of $|\tilde{f}(k)|^2$.

$$P_0(k) \equiv < |\tilde{f}(k)|^2 > = N |\tilde{f}_1^\Psi(k)|^2 < a_n^2 > \tag{4}$$

The 1σ error to $P_0(k)$ can also be obtained from the standard deviation of a_n^2. In realistic cases the superposed seed functions $f_1^{\Psi}(\theta)$ will be perturbations induced by topological defects which obey a *scaling solution* and therefore their size is a fixed fraction of the horizon at any given time. This implies that the size of the superposed seeds will be larger at later times due to the expansion of the comoving horizon. Thus when the comoving horizon grows by a factor α, the size of the superposed seeds will grow by the same factor while their total number N will be reduced by α since the total number of horizons included in the circle will be smaller at that time. By considering Q expansion steps for the comoving horizon we obtain

$$P_Q(k) = \sum_{q=0}^{Q} P_q(k) \equiv \sum_{q=0}^{Q} \frac{N}{\alpha^q} |\tilde{f}_1^{\alpha^q \Psi}(k)|^2 < a_n^2 > \quad (5)$$

where Q is determined by the maximum and minimum size of the superposed seeds (or the corresponding horizon scales) as

$$Q = \frac{\log(\frac{\Psi_{max}}{\Psi_{min}})}{\log \alpha} \quad (6)$$

while N is the product of the number of seeds per horizon M times the number of horizon scales included in the circle during the first expansion step when the horizon scale is $\Theta(t_i)$

$$N(t_i) = M \times \frac{2\pi}{\Theta(t_i)} \quad (7)$$

In order to obtain the power spectrum in the case of strings we must define the types of seed functions $f_1(\theta)$ that need to be superposed and the initial and final times of superposition for each mechanism generating perturbations.

A long string moving with velocity v_s between the LSS and an observer will induce, according to the GKS effect a temperature discontinuity of magnitude[33, 34, 38]

$$\frac{\delta T}{T} = 4\pi G\mu(v_s\gamma_s)_{rms}\hat{k} \cdot (\hat{v} \times \hat{s}) \quad (8)$$

where $\hat{k} \cdot (\hat{v} \times \hat{s})$ is a geometric factor which depends on the relative orientation between the string $\hat{s}$ and the photon unit wavevector $\hat{k}$. Therefore an observer scanning the sky along an *arc* that itersects the string will detect a temperature step function with amplitude $\frac{\delta T}{T}$ and angular scale 2Ψ depending on the string curvature radius ξ given as a fraction of the angular size Θ of the horizon at the time the photons interacted with the string ($\Psi(t) \equiv \xi\Theta(t)/2$).

Moving long strings present between t_{eq} and t_{rec} produce velocity perturbations towards the surface they sweep in space. Assuming CDM, these perturbations grow rapidly and form planar density enhancements called *wakes*. Let $\sigma(t_i, t_{rec})$ be the surface density of a wake present at t_{rec} and formed at an earlier time t_i. The wake, due to its surface density will induce potential fluctuations at a distance h from its surface.

The potential fluctuations in turn produce temperature fluctuations on the photons departing from the LSS, with magnitude

$$\frac{\delta T}{T} = \frac{1}{3}\Phi_w(x, t_{rec}) = \frac{1}{3}4\pi G\sigma(t_i, t_{rec})|h(x)| \quad (9)$$

and size Ψ determined by the scale of coherence of the long string that produced the wake. In (9) $h(x) = x \ \cos\phi$ is the *perpendicular* distance to the wake as a function of the angular distance x from the wake on the last scattering surface. The *angular* distance from the wake over which this perturbation persists is approximatelly $\Psi(t_i)$.Therefore, an observer scanning the sky along an *arc* that intersects a wake will detect a temperature fluctuation pattern described by equation (9) as a function of the angular scale x.

String simulations have shown that the string network contains in addition to long strings, a component of tiny loops with typical sizes about 10^{-4} the size of the horizon at any given time. A loop present on the LSS will also induce potential perturbations due both to its energy density and to the dark matter it has accreted. The corresponding temperature fluctuation may be approximated by a function that is 0 on scales much larger than the size of the loop and equal to $\frac{\beta G\mu}{3}(\frac{t}{t_i})^{2/3}$ on smaller scales *i.e.*

$$\frac{\delta T}{T} = \frac{1}{3}\Phi_l(x,t) \simeq \frac{\beta G\mu}{3}(\frac{t}{t_i})^{2/3} \qquad |x| \leq R \quad (10)$$

$$\frac{\delta T}{T} \simeq 0 \qquad |x| > R \quad (11)$$

The time dependent factor represents the gravitational growth of the loop induced perturbation on the CDM and β is a parameter of order smaller than 10 determining the length of the loop as a function of its radius R. Since the typical angular size of a loop present at t_{rec} is 1 arcsec or less, we expect the effects of loops to be negligible for all present experiments since their resolution is on much larger scales. This is verified from the results shown below.

The velocity perturbations induced by long strings present on the LSS produce velocity fields on the electrons last scatterers of CMB photons. These fields induce temperature fluctuations by the Doppler effect. The important perturbations are those produced on the plasma *at* t_{rec} since earlier perturbations are damped by photon drag and pressure effects. The magnitude of the produced temperature fluctuations is equal to the projection of the plasma velocity with respect to the observer on the photon unit wavevector

$$\frac{\delta T}{T} = \hat{k}\cdot\vec{v} = \lambda\pi G\mu v_s\gamma_s\hat{k}\cdot(\hat{v}\times\hat{s}) \quad (12)$$

where λ is defined as

$$\lambda = (1 + \frac{(1-\frac{T}{\mu})}{2(v_s\gamma_s)^2}) \quad (13)$$

and is different from 1 in the case of a wiggly string in which the tension T is not equal to the mass per unit length μ. It measures the Newtonian interaction with matter induced by the wiggles of the string. Simulations indicate $\lambda \simeq 6$. The scale of these Doppler perturbations is Ψ, determined by the coherence length of long strings present on the LSS.

Having specified the types of seeds that need to be superposed, it is now straightforward to use (5) to obtain the power spectrum component induced by each type of seed. The total spectrum is obtained by summing up the spectrum components.

$$P(k) = P_{GKS}(k) + P_W(k) + P_L(k) + P_D(k) \tag{14}$$

A typical spectrum component obtained for the GKS term is

$$P_{GKS}(k) = \frac{(4\,\pi\,G\,\mu)^2\ <(v_s\,\gamma_s)^2)>\ 200\ M}{3} \sum_{q=0}^{Q=23} \frac{16\ \sin^4(\xi\ 0.008\ \alpha^q\ k)}{\alpha^q k^2} \tag{15}$$

The part $(4\,\pi\,G\,\mu)^2\ <(v_s\,\gamma_s)^2)>$ comes from the amplitude of the GKS step function (8), the 1/3 comes by averaging over all string orientations, the $200M$ is the number of seeds included in the first expansion step occuring at t_{rec} and the sum is over the Fourier transform of the seed function scaled to expand with the comoving horizon at each expansion step. The assumption that each expansion step occurs when the physical horizon doubles in size implies that $\alpha \simeq 2^{1/3} \simeq 1.26$. Decreasing α has the effect of increasing the number of terms Q in the sum (15) but decreasing the value of each term. Thus we expect that (15) should not be very sensitive to the precise value of α.

The other spectrum components (Wakes, Loops and Doppler) have similar forms expressed in terms of sums[26]. They depend on four parameters: the only free parameter $G\mu$, the parameter $b \equiv M\ <(v_s\gamma_s)^2>$, the string coherence length (curvature radius) ξ as a fraction of the horizon scale $\Theta(t)$ ($\Psi(t) = \xi\Theta(t)/2$ and the parameter λ determined by the wiggliness of the string. The three evolution parameters b, ξ and λ may be fixed by comparing with numerical simulations while the free parameter $G\mu$ is fixed by comparing with observations (*e.g.* the COBE detection).

3 Fixing Parameters

In order to make predictions about ongoing CMB experiments we must determine the only free parameter $G\mu$ as well as the parameters b, λ and ξ. String simulations [41, 40] indicate that $M \simeq 10$ while $(v_s\gamma_s)_{rms} \simeq 0.15$ implying $b \simeq 0.24$ and $\lambda \simeq 6$. I will verify these values by directly fitting our spectrum with partial CMB spectra obtained by simulations on large angular scales. Bouchet, Bennett and Stebbins[22] (hereafter BBS) have used numerical simulations to calculate the term $P_{GKS}(k)$ for a single expansion step. Their result for the total power on angular scales smaller than θ_* with $t_i = t_{rec}$, $t_f \simeq 2\,t_{rec}$ is

$$\begin{aligned} P_{BBS}(\theta \le \theta_*, \Theta_i = \Theta_{rec}) &= \int_{2\pi/\theta_*}^{\infty} \frac{d^2k}{(2\pi)^2} P_{BBS}(k) \\ &= (6G\mu)^2 \left(\frac{\theta_*^{1.7}}{0.0012 + \theta_*^{1.7}}\right)^{0.7} \end{aligned} \tag{16}$$

The present analysis, focusing on a line across the sky rather than a patch predicts

$$P_{an}(\theta \leq \theta_*, \Theta_i = \Theta_{rec}) = 2 \int_{2\pi/\theta_*}^{\infty} \frac{dk}{(2\pi)} P_{KS}^{Q=0}(k) \tag{17}$$

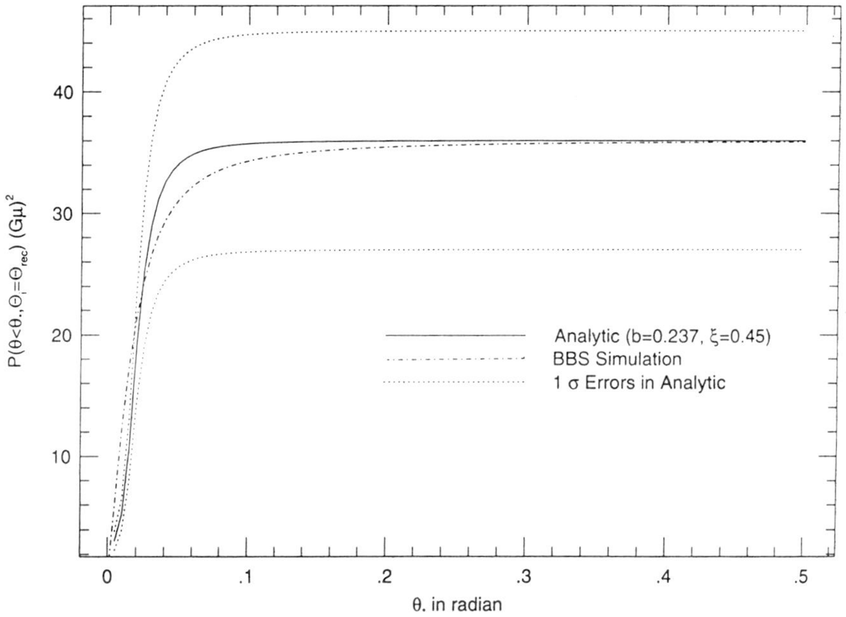

Figure 1. The total power on scale less than θ_* produced by cosmic strings during one expansion step starting at t_{rec}.

Figure 1 shows $P_{an}(\theta_*)$ for $b = 0.237$ and $\xi = 0.45$ (continous line) superimposed on $P_{BBS}(\theta_*)$ (dashed line). It also shows the 1σ errors to $P_{an}(\theta_*)$ obtained from the variance of a_n^2. The values $b = 0.237$ and $\xi = 0.45$ were chosen in order to obtain the best fit to P_{BBS} but they are in very good agreement with the expected values (obtained for $M \simeq 10$, $(v_s\gamma_s)_{rms} \simeq 0.15$ and string radius of curvature about half the horizon scale).

Figure 2a shows a superposition of the components of the spectrum (*e.g.* equation (15) for the GKS term) with the above choice of parameters. The sums were performed using *Mathematica*[42]. Clearly the GKS term (continous line) dominates on large angular scales ($\theta > 4°$) while the Doppler term (long dashed line) is dominant on smaller scales. The contribution of potential perturbations by wakes (dotted line) is less important but is clearly not negligible especially on scales of a few arcmin ($k \simeq 1500$). Finally, the contibution of loops (short dashed line) is negligible on all scales larger than 2-3 arcmin ($k \leq 8000$).

Figure 2b shows the product $kP(k)$ fot the total spectrum (equation (14)) with 1σ errors denoted by the dotted lines.

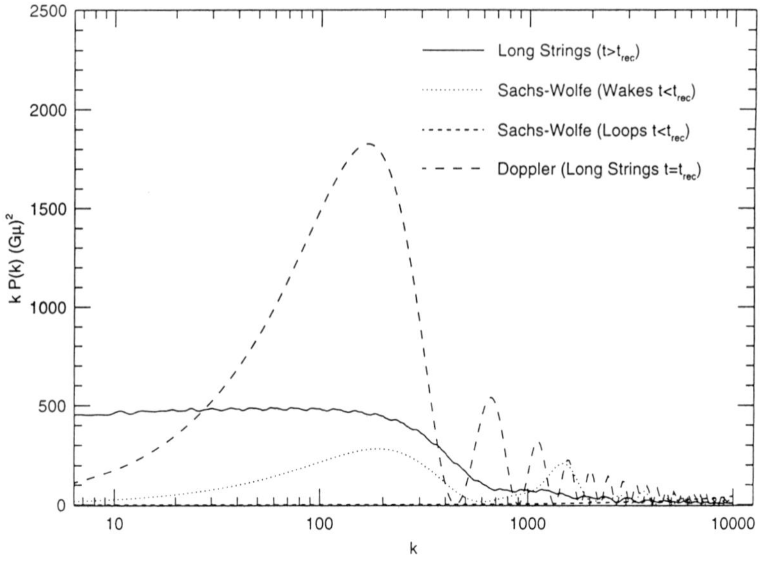

Figure 2a. The four components of the power spectrum of CMB perturbations induced by cosmic strings

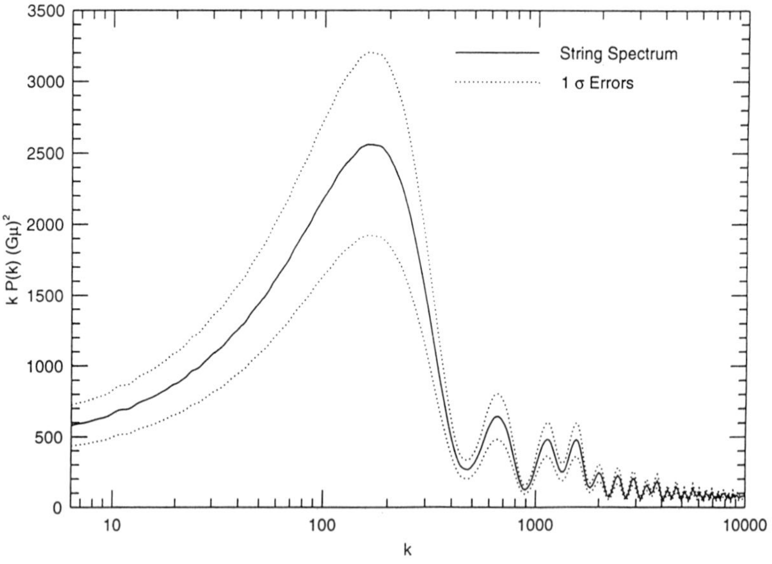

Figure 2b. The *total* power spectrum of string perturbations along a great circle on the sky.

One of the most interesting questions that may be addressed using the spectrum of Figure 2b is *'What is the effective power spectrum index n, predicted by cosmic strings on COBE angular scales?'*. Previous studies[24, 25] have addressed this question without taking into account the effects of potential and Doppler perturbations. The correlation function $C_1(\theta)$ for perturbations along a great circle is given in terms of $P(k)$ as

$$C_1(\theta) = <\frac{\delta T}{T}(\phi)\frac{\delta T}{T}(\theta+\phi)>_\phi = \frac{1}{(2\pi)^2}\sum_{k=-\infty}^{k=+\infty} P(k)e^{ik\theta} \tag{18}$$

For 2d maps the corresponding equation is ($l \gg 1$, $\theta \ll \pi$) [35]

$$C_2(\theta) \simeq \frac{1}{(2\pi)^2}\int d^2l\, C_l\, e^{i\vec{l}\cdot\vec{\theta}} \tag{19}$$

By isotropy we must have $C_1(\theta) = C_2(\theta)$. It may also be shown that $l^2 C_l \sim l^{n-1}$ where n is the power spectrum index. Since both k and l are Fourier conjugate of θ we have $k \simeq l$. Also (30) and (31) imply (with $\theta \simeq 0$) that $P(k) \simeq \pi l C_l$ and

$$kP(k) \sim k^{n-1} \tag{20}$$

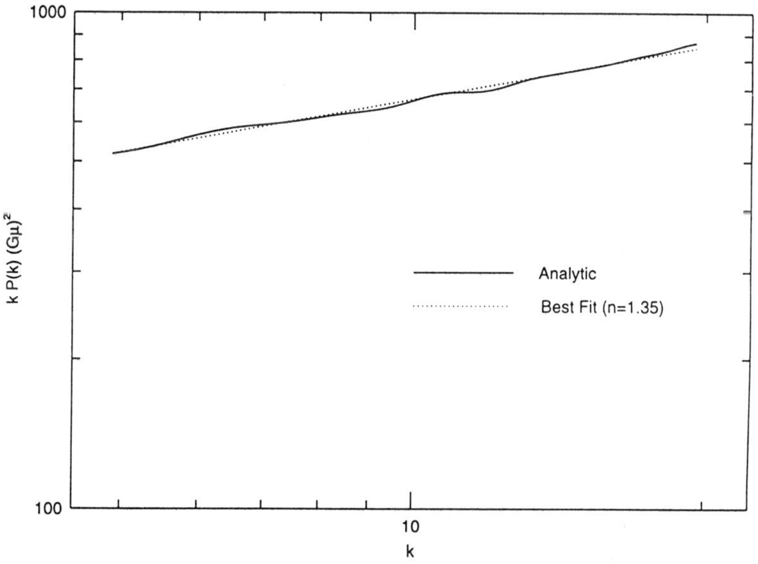

Figure 3a. The best linear fit to the total power spectrum for $5 \leq k \leq 20$.

Figures 3a and 3b show the best linear fit of the log-log plot $kP(k)$ vs k, for $5 \leq k \leq 20$ and $5 \leq k \leq 100$ respectively. The best fits give $n = 1.35$ (Figure 3a) and $n = 1.48$ (Figure 3b). This result indicates that cosmic strings favor values of n somewhat larger than 1 in agreement with recent indications from the Tenerife experiment and the second year data of COBE [43] which favor $n \simeq 1.5$. In contast it is much harder for inflationary models to explain such high values of n [44, 45].

There is a simple analytic way to show that in the sector of the power spectrum where the GKS effect dominates, a scale invariant ($n \simeq 1$) spectrum should be expected. For $\tilde{f}_1^\Psi(k) \sim \Psi \tilde{f}_1^{\Psi=1}(k\Psi)$ (as in the case of the GKS seed functions), (12) may be writen as

$$P(k) = \sum_{q=0}^{Q} P_q(k) = \sum_{q=0}^{Q} \alpha^q P_0(\alpha^q k) \simeq \alpha P(\alpha k), \quad \Psi_{max}^{-1} \leq k \leq \Psi_{min}^{-1} \tag{21}$$

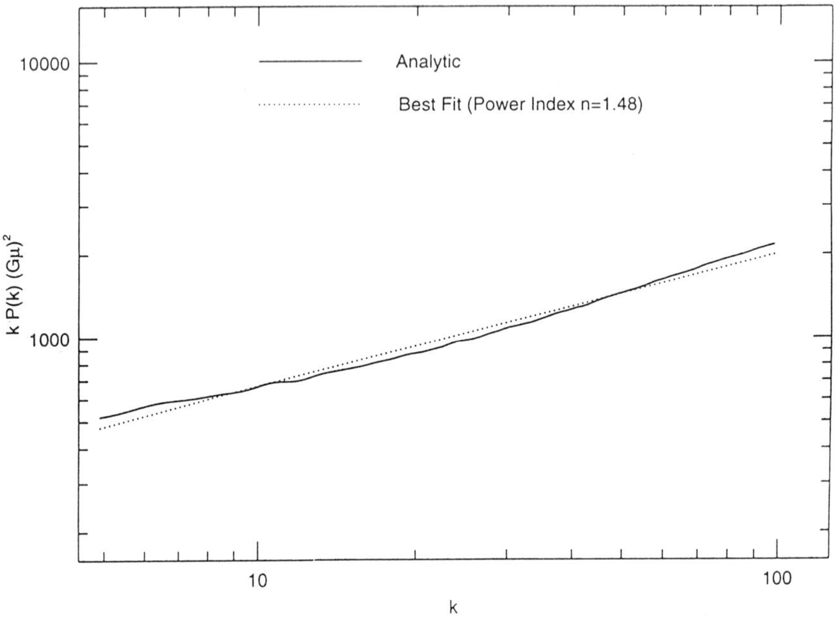

Figure 3b. The best linear fit to the total power spectrum for $5 \leq k \leq 100$.

Therefore

$$kP(k) \sim \text{const} \tag{22}$$

which indicates a scale invariant spectrum for the Kaiser-Stebbins term in the angular scale range $\theta \geq 2°$ ($k \leq \Psi_{min}^{-1}$). The Kaiser-Stebbins term plotted in Figure 2a (continous line) is in agreement with this result (the best fit for this component of the spectrum is obtained for $n = 1.12$).

4 Predictions

The predicted $\frac{\delta T}{T}_{rms}$ for ongoing experiments can only be obtained after the free parameter $G\mu$ is fixed. This may be achieved by using the COBE detection. Using the DMR filter function $W(k) \simeq e^{-k^2/18^2}$ and comparing the observed $\frac{\delta T}{T}_{rms} = 1.1 \pm 0.5$

with the predicted one

$$\frac{\delta T}{T}_{rms} = (C(0))^{1/2} = [\frac{1}{2\pi^2}\sum_{k=0}^{\infty} P(k)W(k)]^{1/2} \tag{23}$$

the single free parameter $G\mu$ is fixed to the value

$$G\mu \simeq 1.6 \pm 0.5 \tag{24}$$

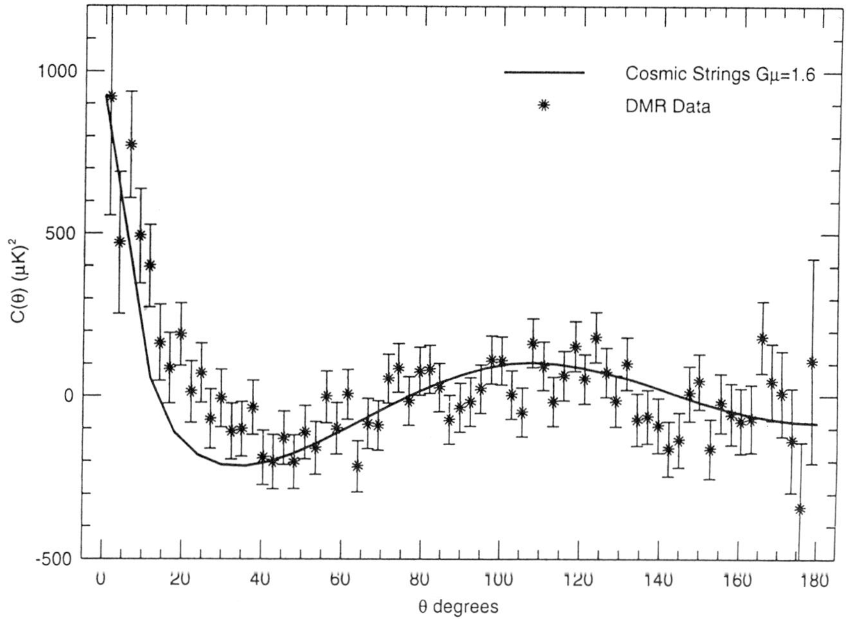

Figure 4. The cosmic string predicted correlation function smoothed on COBE scales. Superimposed are the first year COBE data.

Having completely fixed the spectrum we are now in position to use appropriate filter functions to make predictions for ongoing CMB experiments[46] on various scales. Figure 4 shows the correlation function obtained by Fourier transforming the power spectrum and smoothed by the COBE filter function. Superimposed are the COBE data[31, 32].

Table 1 shows the approximate location of the maxima (k_0) and the spreads (Δk) of the filter functions of several ongoing experiments. It also shows the coresponding detections or upper bounds. The cosmic string predictions were obtained by using (23) with gaussian filter functions of the form

$$W(k) = e^{-(k-k_0)^2/\Delta k^2} \tag{25}$$

and the power spectrum shown in Figure 3. The predictions of the standard model based on inflation ($0.8 \leq n \leq 1.0$, $\Lambda = 0$, CDM) have been derived in Ref. [4]. From

Table 1, it is clear that the predictions of both the standard CDM model and cosmic strings are consistent with detections at the 1σ level. However this may well change in the near future as the quality of observations improves.

Table 1. Detections of $\frac{\Delta T}{T}_{rms} \times 10^6$ and the corresponding predictions of the string and inflationary models ($0.8 \leq n \leq 1.0$, $\Lambda = 0$) normalized on COBE.

Experiment	k_0	Δk	Detection	Strings	Inflation
COBE	0	18	11 ± 2	11 ± 3	11 ± 2
TEN	20	16	≤ 17	13 ± 3	9 ± 1
SP91	80	70	11 ± 5	20 ± 5	12 ± 2
SK	85	60	14 ± 5	19 ± 4	12 ± 3
MAX	180	130	≤ 30 (μPeg)	21 ± 5.5	16 ± 5
MAX	180	130	49 ± 8 (GUM)	21 ± 5.5	16 ± 5
MSAM	300	200	16 ± 4	19 ± 4	24 ± 6
OVRO22	600	350	-	13 ± 4	17 ± 7
WD	550	400	≤ 12	17.5 ± 4.5	7 ± 2
OVRO	2000	1400	≤ 24	13.5 ± 3.5	7 ± 3

In conclusion, I have demonstrated, using a simple analytical method, that the CMB spectrum predicted by the cosmic string model can be calculated in a straightforward way including all the relevant sources of perturbations. I have also shown that the results are consistent with numerical simulations even though their validity extends beyond the resolution of present simulations. Finally I showed that the predicted power spectrum index is slightly larger than 1 ($n_{eff} \simeq 1.4$) and that the predicted rms temperature fluctuations $\frac{\Delta T}{T}_{rms}$ are consistent with detections to this date on all angular scales larger than 2-3 arminutes.

This analysis has assumed standard recombination and values of cosmological parameters ($\Omega_0 = 1$, $h = 1/2$, CDM, $\Lambda = 0$). It is important to extend these results to less standard cases including reionization or presence of Hot Dark Matter. Work in this direction is in progress.

Acknowledgements

This work was supported by a CfA Postdoctoral Fellowship.

References

[1] Guth A. 1981, Phys. Rev. **D23**, 327.

[2] Guth A. & Pi S. -Y. 1982, Phys.Rev.Lett.**49**, 110.

[3] Bardeen J., Steinhardt P. & M.Turner M. 1983, Phys.Rev. **D28**, 679.

[4] Bond R., Crittenden R., Davis R., Efstathiou G. & Steinhardt P. 1994, Phys. Rev. Lett. **72**, 13.

[5] Kibble T. W. B. 1976, J.Phys. **A9**, 1387.

[6] Vilenkin A. 1985, Phys.Rep. **121**, 263.

[7] Brandenberger R. 1992, 'Topological Defect Models of Structure Formation After the COBE Discovery of CMB Anisotropies', Brown preprint BROWN-HET-881 (1992), publ. in proc. of the International School of Astrophysics "D.Chalonge", 6-13 Sept.1992, Erice, Italy, ed. N.Sanchez (World Scientific, Singapore, 1993).

[8] Turok N. 1989, Phys. Rev. Lett. **63**, 2625.

[9] Turok N. and Brandenberger R. 1986, Phys. Rev. **D33**, 2175.

[10] Vachaspati T. 1986, Phys. Rev. Lett. **57**, 1655.

[11] Stebbins A. *et. al.* 1987, Ap. J. **322**, 1.

[12] Perivolaropoulos L., Brandenberger R. & Stebbins A. 1990, Phys.Rev. **D41**, 1764.

[13] Vachaspati T. & Vilenkin A. 1991. Phys. Rev. Lett. **67**, 1057-1061.

[14] Vollick D. N. 1992, Phys. Rev. D**45**, 1884.

[15] Hara T. & Miyoshi S. 1993, Ap. J. **405**, 419.

[16] Brandenberger R., Kaiser N., Shellard E. P. S., Turok N. 1987. Phys.Rev. **D36**, 335.

[17] Albrecht A. & Stebbins A. 1993. Phys. Rev. Lett. **69**, 2615.

[18] Vachaspati T. 1992b, Phys. Rev. D**45**, 3487.

[19] Vachaspati T. 1992a, Phys.Lett. **B282**, 305.

[20] Perivolaropoulos L. & Vachaspati T. 1994, Ap. J. Lett., **423**, L77, CfA-3590, ASTRO-PH/9303242.

[21] Brandenberger R. & Turok N. 1986, Phys. Rev. **D33**, 2182.

[22] Bouchet F. R., Bennett D. P. & Stebbins A. 1988, Nature **335**, 410.

[23] Veeraraghavan S. & Stebbins A. 1990, Ap.J. **365**, 37.

[24] Bennett D., Stebbins A. & Bouchet F. 1992, Ap.J. Lett. **399**, L5.

[25] Perivolaropoulos L. 1993a, Phys.Lett. **B298**, 305.

[26] Perivolaropoulos L. 1994, *Spectral Analysis of CMB Fluctuations Induced by Cosmic Strings*, Submitted to the Ap. J., CfA-3591, ASTRO-PH/9402024.

[27] Gott J. *et. al.* 1990, Ap.J. **352**, 1.

[28] Perivolaropoulos L. 1993b, Phys. Rev. **D48**, 1530.

[29] Moessner R., Perivolaropoulos L. & Brandenberger R. 1994, Ap. J., **425** 365, ASTRO-PH/9310001.

[30] Coulson D., Ferreira P., Graham P. & Turok N. 1993, Π *in the Sky? CMB Anisotrtopies from Cosmic Defects*, PUP-TH-93-1429, HEP-PH/9310322.

[31] Smoot G. *et. al.* 1992, (COBE), *Ap. J. Lett.* **396**, L1.

[32] Wright E. L. *et. al.* 1992, *Ap. J. Lett.* **396**, L5.

[33] Kaiser N. & Stebbins A. 1984, *Nature* **310**, 391.

[34] Gott R. 1985, Ap. J. **288**, 422.

[35] Efstathiou G. 1989, in 'Physics of the Early Universe', SUSSP 36, 1989, ed. J.Peacock, A.Heavens & A.Davies (IOP Publ., Bristol, 1990).

[36] Perivolaropoulos L. 1993c, *The Fourier Space Statistics of Seed-like Cosmological Perturbations*, M.N.R.A.S. in press, CfA-3591, ASTRO-PH/9309023.

[37] Vilenkin A. 1981, Phys.Rev. **D23**, 852.

[38] Stebbins A. 1988, Ap.J. **327**, 584.

[39] Bennett D. & Bouchet F. 1988, Phys.Rev.Lett. **60**, 257.

[40] Allen B. & Shellard E. P. S. 1990, Phys.Rev.Lett. **64**, 119.

[41] Bennett D. & Bouchet F. 1988, Phys.Rev.Lett. **60**, 257.

[42] Wolfram S. 1991, *Mathematica version 2.0*, Addison-Wesley.

[43] Bennett C. L. *et. al.* 1994, *Cosmic Temperature Fluctuations from Two Years of COBE DMR Observations*, submitted to Ap. J., (ASTROPH-9401012).

[44] Lyth D. & Liddle A. 1994,*Observational Constraints on the Spectral Index*, Contribution to the 1993 Capri CMB Workshop. SUSSEX-AST 93/12-1, ASTRO-PH/9401014.

[45] Steinhardt P. 1993, private communication.

[46] TEN: Watson R. A. *et. al.* 1992, Nature **357**, 660.
SP91: Gaier T. *et.al.* 1992, (SP91), Ap. J. Lett. **398**, L1.
SP91: Schuster J. *et. al.* 1993, (SP91), Ap. J. Lett. **412**, L47.
MSAM: Cheng E. S. *et. al.* 1993,*A Measurement of the Medium Scale Anisotropy in the CMB*, preprint MSAM-93A.
SK: Wollack E. J. *et. al.* 1993, Ap. J. Lett. **419**, L49.
MAX1: Meinhold P. *et. al.* 1993, Ap. J. Lett. **409**, L1.
MAX2: Gunderson J. *et. al.* 1993, Ap. J. Lett. **413**, L1.
WD: Tucker G. S. *et. al.* 1993, (WD), Princeton preprint.
OVRO: Readhead A.C.S. *et.al.* 1989, (OVRO), Ap. J. **346**, 556.

SECTION III

PROGRESS IN NEW AND OLD IDEAS

TIME REVERSAL FOR SPACETIME AND INTERNAL SYMMETRY

E.C.G. Sudarshan

Center for Particle Physics
Department of Physics
University of Texas
Austin, TX 78712

ABSTRACT

The standard time reversal transformations have to be generalized when we deal with systems with internal symmetry. This generalization is formulated and adapted to higher symmetries involving the angular momentum.

Time reversal in quantum mechanics is a discrete transformation which associates with each motion a "time reversed motion" determined *kinematically*. When the reversed motion obeys the *same* equations of motion as the direct equations we state that there is *time reversal invariance*. Since the choice of the time origin is arbitrary, time reversal invariance automatically implies *time translation invariance*, whether it is reversible or irreversible.

THE WIGNER TIME REVERSAL

In quantum mechanics the time reversal transformation is realized anti-unitarily:

$$\psi(t) \longrightarrow e^{i\Theta}\,\psi^*(-t) \tag{1}$$

where the phase Θ maybe chosen at our convenience. We recognize that if we make a global phase change

$$\psi \longrightarrow e^{i\alpha}\,\psi \tag{2}$$

for all wavefunctions the time reversal transformation changes; but then α can be exploited to restore the time reversal to the standard form. On the other hand

$$C_1\,\psi_1(t) + C_2\,\psi_2(t) \longrightarrow C_1^*\,\psi_1^*(-t) + C_2^*\,\psi_2^*(-t)\,. \tag{3}$$

When the particles have spin this is no longer sufficient. Since all $SU(2)$ representations are either real (bosons) or pseudo real (fermions)[2] we have to define

$$\psi(t) \longrightarrow \exp(i\pi J_2)\psi^*(-t)\,. \tag{4}$$

Unified Symmetry: In the Small and in the Large,
Edited by B.N. Kursunoglu *et al.*, Plenum Press, New York, 1995

Rotations are unaffected by this transformation but the generators of rotation change sign:

$$\boldsymbol{J} \longrightarrow \exp(i\pi J_2)\boldsymbol{J}^* \exp(-i\pi J_2) = -\boldsymbol{J} \tag{5}$$

in accordance with the behaviour expected of angular momentum under time reversal. All this is well known and introduced here for establishing the formalism.

SEQUENCES OF ROTATIONS UNDER TIME REVERSAL

We have seen that angular momentum $\boldsymbol{J}$ undergoes a change of sign under time reversal. What happens to a finite rotation? If

$$\psi_2 = e^{i\boldsymbol{J}\cdot\boldsymbol{\theta}}\psi_1 \tag{6}$$

we find

$$\tilde{\psi}_2 = e^{i\pi J_2}\, e^{-i\boldsymbol{J}^*\cdot\boldsymbol{\theta}}\, e^{-i\pi J_2}\,\tilde{\psi}_1 = e^{i\boldsymbol{J}\cdot\boldsymbol{\theta}}\,\tilde{\psi}_1\,. \tag{7}$$

That is, the time reversed states are related by the *same* rotation.

This is to be contrasted with a *sequence* of rotations at various time intervals. For example if we write

$$\psi(t_3) = e^{iH(t_3-t_2)}\, e^{i\boldsymbol{J}\cdot\boldsymbol{\theta}}\, e^{-iH_2(t_2-t_1)}\, e^{i\boldsymbol{J}\cdot\boldsymbol{\theta}'}\, e^{-iH(t_1-t_0}\psi(t_0) \tag{8}$$

the time reversed states will satisfy

$$\tilde{\psi}\,(-t_3) = e^{+iH(t_3-t_2)}\, e^{i\boldsymbol{J}\cdot\boldsymbol{\theta}} e^{+iH(t_2-t_1)} e^{i\boldsymbol{J}\cdot\boldsymbol{\theta}'}\, e^{+iH(t_1-t_0)}\,\tilde{\psi}(-t_0)\,. \tag{9}$$

But if we choose

$$t_3 > t_2 > t_1 > t_0\,, \tag{10}$$

the second equation is better re-expressed in the form

$$\tilde{\psi}(-t_0) = e^{-iH(-t_0+t_1)} e^{-i\boldsymbol{J}\cdot\boldsymbol{\theta}'} e^{iH(-t_1+t_2)} e^{-i\boldsymbol{J}\cdot\boldsymbol{\theta}}\, e^{-iH(-t_2+t_3)}\tilde{\psi}(-t_3) \tag{11}$$

with

$$-t_0 > -t_1 > -t_2 > -t_3\,. \tag{12}$$

So when the time reversed evolution unfolds the rotations are *inverses* of the original rotations in the *reverse* order. We recognize that these also furnish a representation of the rotation group:

$$\begin{aligned} R_1\,R_2 \quad \longrightarrow \quad & \mathcal{D}^{-1}(\mathcal{R}_2)\,\mathcal{D}^{-1}(\mathcal{R}_1) = (\mathcal{D}(\mathcal{R}_1)\,\mathcal{D}(\mathcal{R}_2))^{-1} \\ & = \mathcal{D}^{-1}(\mathcal{R}_1\,\mathcal{R}_2)\,. \end{aligned} \tag{13}$$

All the rotations that happened are reversing themselves in the time reversed picture.

TIME REVERSAL FOR SYSTEMS WITH INTERNAL SYMMETRIES

We now seek the behaviour of particles transforming as *isospin* (flavor $SU(2)$) multiplets. In this case neither the simple conjugation

$$\psi(t) \longrightarrow \psi^*(-t) \tag{14}$$

nor the augmented conjugation

$$\psi(t) \longrightarrow \exp(i\pi I_2)\psi^*(-t) \tag{15}$$

would be appropriate. The first one would not commute with the $SU(2)$ group, while in the second one the isospin $\boldsymbol{I}$ would be reversed and along with it the electric change, say of the pion triplet. We need a new kind of time reversal transformation.

Since we want the time reversed pion of a definite charge to be a pion state of the same charge, we write:

$$\begin{pmatrix} \psi_+(t) \\ \psi_0(t) \\ \psi_-(t) \end{pmatrix} \longrightarrow \begin{pmatrix} \psi_+^*(-t) \\ \psi_0^*(-t) \\ \psi_-^*(-t) \end{pmatrix} . \tag{16}$$

This means that the internal symmetry labels are unaffected by time reversal. For the nucleon doublet we have

$$\begin{pmatrix} p(t) \\ n(t) \end{pmatrix} \longrightarrow \exp\left(\frac{i\pi}{2}\sigma_2\right) \begin{pmatrix} p^*(-t) \\ n^*(-t) \end{pmatrix} . \tag{17}$$

These transformations point out that the time reversal transformation is not a straight forward antiunitary transformation but requires some delicate handling.

For a system with internal symmetry we proceed as follows: Choose a complete set of commuting symmetry operators $\{H_\ell\}$ and their normalized eigenfunctions $\eta_{\lambda(\ell)}$. Consider these quantitites as "real," that is unchanged under time reversal. The generic state may be written in the form

$$\psi(t) = \sum_{\{\lambda(\ell)\}} \psi_{\lambda(\ell)}(t) \cdot \eta_{\lambda(\ell)} . \tag{18}$$

The time reversal transformation is defined as

$$\psi(t) \longrightarrow \tilde{\psi}(-t) = \sum_{\{\lambda(\ell)\}} \exp(i\pi J_2)\psi^*_{\lambda(\ell)}(-t) \cdot \eta_{\lambda(\ell)} \tag{19}$$

that is,

$$\begin{aligned} \eta_{\lambda(\ell)} &\longrightarrow \eta_{\lambda(\ell)} \\ \psi_{\lambda(\ell)} &\longrightarrow \exp(i\pi J_2)\psi^*_{\lambda(\ell)}(-t) . \end{aligned} \tag{20}$$

In the Cartan notation[4] $E_\alpha \longrightarrow E_\alpha$, $H_\ell \longrightarrow H_\ell$. It follows that internal quantum numbers like hypercharge, baryon number or isospin are *not* changed so that for example the $SU(3)$ decuplet goes into the decuplet (and note the $\overline{10}$); but the angular momentum changes to its negative.

SEQUENCE OF INTERNAL SYMMETRY TRANSFORMSTIONS

For time reversal the Cartan-Weyl basis of the Lie algebra $(\{H_\ell\}, \{E_\alpha\})$ is considered invariant. Hence an arbitrary element of the Lie algebra undergoes the transformation

$$A \longrightarrow \tilde{A} \tag{21}$$

with

$$A = \sum C_\ell H_\ell + \sum C_\alpha E_\alpha \ , \ \tilde{A} = \sum C_\ell H_\ell + \sum C_\alpha^* E_\alpha = \sum C_\ell H_\ell + \sum C_{-\alpha} E_\alpha \tag{22}$$

with the understanding that

$$C_\ell^* = C_\ell \ , \ C_{-\alpha} = C_\alpha^* . \tag{23}$$

For a finite transformation

$$\exp(iA) \longrightarrow \exp(-i\tilde{A}) . \tag{24}$$

The finite transformations generated by the any linear combination of H_ℓ, $E_\alpha + E_{-\alpha}$ goes into its inverse, while any transformation generated by $i(E_\alpha - E_{-\alpha})$ remains unaffected. (For example the isospin rotation around the second axis is a real orthogonal transformation between the pions and is invariant.)

When we consider a sequence of such transformations

$$\psi(t_3) = e^{-iH(t_3-t_2)} e^{iA_2} e^{-iH(t_2-t_1)} e^{iA_1} e^{-iH(t_1-t_0)} \psi(t_0) \tag{25}$$

with

$$t_3 > t_2 > t_1 > t_0 , \tag{26}$$

the time reversed states obey

$$\tilde{\psi}(-t_3) = e^{iH(t_3-t_2)} e^{-i\tilde{A}_2} e^{iH(t_2-t_1)} e^{-i\tilde{A}_1} e^{iH(t_1-t_0)} \tilde{\psi}(-t_0) . \tag{27}$$

In the natural temporal sequence

$$-t_0 > t_1 > t_2 > t_3 \tag{28}$$

we may write

$$\tilde{\psi}(-t_0) = e^{-iH(-t_0+t_1)} e^{i\tilde{A}_1} e^{-iH(-t_1+t_2)} e^{i\tilde{A}_2} e^{-iH(-t_2+t_3)} \tilde{\psi}(-t_3) . \tag{29}$$

Therefore for the "real" generators (that is real combinations of H_ℓ, $E_\alpha + E_{-\alpha}$) the corresponding group elements are *not* inverted but the exponentials of the multiples of $E_\alpha - E_{-\alpha}$ are inverted. This behaviour is distinctly different from that of angular momentum and finite rotations.

With the standard phase conventions for representations with $\{H_\ell\}$ diagonal and the $\{E_\alpha\}$ having only real matrix elements

$$\tilde{A} = A^T \quad , \quad e^{i\tilde{A}} = (e^{iA})^T \, . \tag{30}$$

so that

$$\tilde{\psi}(-t_0) = e^{-iH(-t_0+t_1)}\, e^{iA_1^T} e^{-iH(-t_1+t_2)} e^{iA_2^T} e^{-iH(-t_2+t_3)} \tilde{\psi}(-t_3) \, . \tag{31}$$

Thus the *transposed* internal symmetries are unfolding in the *reverse* order. These furnish a representation of the internal symmetry transformation group:

$$\begin{aligned} \mathcal{S}_1 \cdot \mathcal{S}_2 \quad \longrightarrow \quad & \mathcal{D}^T(\mathcal{S}_2)\mathcal{D}^T(\mathcal{S}_1) = (\mathcal{D}(\mathcal{S}_1)\mathcal{D}(\mathcal{S}_2))^T \\ & = \mathcal{D}^T(\mathcal{S}_1 \mathcal{S}_2) \, . \end{aligned} \tag{32}$$

DISCUSSIONS

The time reversal transformation has a "split" action. The spacetime groups and internal symmetries are not treated in the same fashion. The internal symmetry algebra undergoes a transformation which maybe realized as a transposition in the standard phase conventions where the Cartan-Weyl basis is realized by real matrices. In contrast for rotations we have to make the additional unitary transformation of rotation through 180° about the second axis. Consequently the isospin transformations and the spin transformations behave differently under time reversal.

Two further remarks are in order. First: if an $SL(3,R)$ or $SU(3)$ spectrum generating group is generated the quadrupole generators must be odd under time reversal rather than even. So we cannot use the moment of inertia[5] but may use its time derivative[6]. Second the nuclear p shell $SU(3)$ as formulated[7] seems to violate time reversal invariance. As far as CPT invariance of local field theory is concerned both C and CPT would affect the internal symmetry labels the same way.

The aim of this work has been to complete the proposal by Wigner of the time reversal transformation in quantum mechanics.

The extension of time reversal for the Galitli/Poincaré groups in addition to the internal symmetry groups is straightforward; and so is the discussion for CPT.

REFERENCES

1. E.P. Wigner, Gott. Nachr. Math. Nr. **31**, 546 (1932).

2. E. Cartan, Lesscons sur la Spineurs I, Herman et Cie, Paris (1938).

3. E.C.G. Sudarshan and L.C. Biedenharn, UT preprints DOE-ER 40757-034 and DOE-ER 40757-035.

4. R.E. Behrends et al. Rev. Mod. Phys. **34**, 1 (1962).

5. A.P. Balachandran, P. Nair, S.G. Rajeev, and A. Stern, Phys. Rev. Lett. **49**, 1124 (1982); Phys. Rev. **D27**, 1153 (1983).

6. Y. Dothan, M. Gell-Mann, and Y. Néeman, Phys. Lett. **17**, 148 (1965).

7. R.J. Elliot, Proc. Roy. Soc. **245A**, 128 (1958).

SUPERSTRING FERMION VERTEX AND GAUGE SYMMETRY IN FOUR DIMENSIONS

L. Dolan

Department of Physics, University of North Carolina
Chapel Hill, North Carolina 27599-3255, USA

INTRODUCTION

In superstring theory, the conformal fields which correspond to the ground states in the Ramond sector are the spin fields. Our objective is to discuss the properties of these operators, and the four-dimensional nature of the fermion wavefunction which labels the states. We show how the choice of the supercurrent determines the gauge symmetry charges of these massless Ramond ground states. In sect. 2, we describe the BRST expressions of their vertex operators, including the ghost fields in bosonized form; and relate BRST invariance for states in the canonical ghost picture to the physical state gauge conditions of the old covariant formalism. In sect. 3, we compute examples of tree amplitudes. We present a non-standard choice of BRST vertex operators which allow the massless Ramond states to carry non-zero charge under a Yang-Mills gauge symmetry. The BRST invariance of these operators is shown to depend on the ability to define an appropriate supercurrent. In sect. 4, we present alternative constructions of the supercurrent which satisfy the super Virasoro world sheet symmetry with shifted central charge.

BRST GHOST SUPERCONFORMAL FIELD THEORY

Non-unitary representations of the superVirasoro algebra are provided by the BRST superconformal ghost system[1−3]. The ghost superfields are $B(z) = \beta(z) + \theta b(z)$ and $C(z) = c(z) + \theta\gamma(z)$ with conformal spin $h_\beta = \frac{3}{2}$, $h_c = -1$, etc. The commutation relations on the Ramond sector are $\{b_n, c_m\} = \delta_{n,-m}$, $[\beta_n, \gamma_m] = -\delta_{n,-m}$. Normal ordering is defined by putting the annihilation operators b_n for $n \geq -1$, c_n for $n \geq 2$ to the right of the creation operators b_n for $n \leq -2$, c_n for $n \leq 1$, then

$$b(z)c(\zeta) = {}^{\times}_{\times} b(z)c(\zeta) {}^{\times}_{\times} + \frac{1}{z-\zeta} \quad ; \quad c(z)b(\zeta) = {}^{\times}_{\times} c(z)b(\zeta) {}^{\times}_{\times} + \frac{1}{z-\zeta} \qquad (1)$$

This is a natural definition for normal ordering as the "vacuum" expectation value of this normal ordered product $\langle 3|\, {}^{\times}_{\times} c(z)b(\zeta) {}^{\times}_{\times} |0\rangle = 0$. In the bosonized form it will

Unified Symmetry: In the Small and in the Large,
Edited by B.N. Kursunoglu *et al.*, Plenum Press, New York, 1995

be natural to define a set of momentum states $|q\rangle_{bc}$, where $b_n|q\rangle = 0\,, n > -q-2$, $c_n|q\rangle = 0\,, n \geq q+2$. The annihilation operators used to define the normal ordering are those which annihilate the bc vacuum $|0\rangle_{bc}$. For the superconformal ghosts, normal ordering is defined similarly by

$$\beta(z)\gamma(\zeta) = {}^{\times}_{\times}\beta(z)\gamma(\zeta){}^{\times}_{\times} - \frac{1}{z-\zeta} \quad ; \quad \gamma(z)\beta(\zeta) = {}^{\times}_{\times}\gamma(z)\beta(\zeta){}^{\times}_{\times} + \frac{1}{z-\zeta} \tag{2}$$

The superVirasoro representation has $c = -15$:

$$L(z) = -2\,{}^{\times}_{\times} b\partial c {}^{\times}_{\times} - {}^{\times}_{\times}(\partial b)c{}^{\times}_{\times} - \tfrac{3}{2}\,{}^{\times}_{\times}\beta\partial\gamma{}^{\times}_{\times} - \tfrac{1}{2}\,{}^{\times}_{\times}(\partial\beta)\gamma{}^{\times}_{\times} \tag{3a}$$
$$F(z) = b\gamma - 3\beta\partial c - 2(\partial\beta)c \tag{3b}$$

In the BRST formulation, physical states are BRST singlets. Furthermore, the vanishing of the BRST charge Q on the states of the canonical ghost picture implies the physical state conditions for the matter states of the old covariant formalism. For the N=1 worldsheet supersymmetry system, $Q \equiv \frac{1}{2\pi i}\oint dz Q(z)$ is given from the general form

$$Q(z) \sim c(L^{\text{matter}} + \tfrac{1}{2}L^{\text{ghost}}) - \gamma\tfrac{1}{2}(F^{\text{matter}} + \tfrac{1}{2}F^{\text{ghost}}) \tag{4a}$$

by the BRST current

$$\begin{aligned} Q(z) &= Q_0(z) + Q_1(z) + Q_2(z) \\ Q_0(z) &= cL^{X,\psi} - {}^{\times}_{\times} cb\partial c {}^{\times}_{\times} + \tfrac{3}{2}\partial^2 c + cL^{\beta\gamma} + \partial(\tfrac{3}{4}\,{}^{\times}_{\times}\gamma\beta{}^{\times}_{\times}) \\ Q_1(z) &= -\gamma\,\tfrac{1}{2}F^{X,\psi} \\ Q_2(z) &= -\tfrac{1}{4}\gamma b\gamma \end{aligned} \tag{4b}$$

Here the matter fields $L^{X,\psi}(z)$ and $F^{X,\psi}(z)$ close the superconformal algebra (30) with $c = 15$; and L^{ghost} and F^{ghost} denoted in (4a) are given in (3). Due to the underlying $N = 2$ symmetry of the b, c, β, γ ghost system, we can in fact define a general $F^{\text{ghost}} \sim AF^{+} + BF^{-}$ which closes on L^{ghost} given in (3a) by the following:

$$F(z) = Ab\gamma + B(-3\beta\partial c - 2(\partial\beta)c) \tag{4c}$$

where $AB = 1$. However, the expression for $F^{\text{ghost}}(z)$ given in (3b) should be used in defining the BRST current via (4a), since this is the ghost supercurrent used to define the upper and lower components of the ghost superfields $B(z)$ and $C(z)$ in terms of $b(z), c(z), \beta(z), \gamma(z)$. The BRST current operator product is:

$$\begin{aligned} Q(z)Q(\zeta) = {} & \frac{d}{d\zeta}\Big[\frac{1}{(z-\zeta)^2}(-\frac{7}{16})\frac{dc(\zeta)}{d\zeta}c(\zeta)\Big] \\ & + \frac{1}{(z-\zeta)}\frac{d}{d\zeta}\Big[-\frac{3}{4}\frac{d^2c(\zeta)}{d\zeta^2}c(\zeta) - \frac{1}{8}c(\zeta)\gamma(\zeta)F^{X,\psi}(\zeta) \\ & \qquad + \frac{1}{16}\,{}^{\times}_{\times}\beta(\zeta)\gamma^3(\zeta){}^{\times}_{\times} + \frac{1}{8}\,{}^{\times}_{\times} b(\zeta)c(\zeta){}^{\times}_{\times}\gamma^2(\zeta) - \frac{3}{8}\frac{d\gamma(\zeta)}{d\zeta}\gamma(\zeta)\Big] \end{aligned} \tag{4d}$$

so that

$$Q^2 = \tfrac{1}{2}\{Q,Q\} = 0\,. \tag{4e}$$

The derivation of (4d) showing the relationship of the components is as follows:

$$\begin{aligned}
Q_0(z)Q_0(\zeta) &= \frac{d}{d\zeta}\Big[\frac{1}{(z-\zeta)^2}\Big(-\frac{7}{16}\Big)\frac{dc(\zeta)}{d\zeta}c(\zeta)\Big] \\
&\quad + \frac{1}{(z-\zeta)}\frac{d}{d\zeta}\Big[-\frac{3}{4}\frac{d^2c(\zeta)}{d\zeta^2}c(\zeta)\Big] \\
Q_0(z)Q_1(\zeta)+Q_1(z)Q_0(\zeta) &= \frac{1}{(z-\zeta)}\frac{d}{d\zeta}\Big[-\frac{1}{8}c(\zeta)\gamma(\zeta)F^{X,\psi}(\zeta)\Big] \\
Q_0(z)Q_2(\zeta)+Q_2(z)Q_0(\zeta)+Q_1(z)Q_1(\zeta) &= \frac{1}{(z-\zeta)}\frac{d}{d\zeta}\Big[\frac{1}{16}\,{}^{\times}_{\times}\beta(\zeta)\gamma^3(\zeta){}^{\times}_{\times} \\
&\quad + \frac{1}{8}\,{}^{\times}_{\times}b(\zeta)c(\zeta){}^{\times}_{\times}\gamma^2(\zeta) - \frac{3}{8}\frac{d\gamma(\zeta)}{d\zeta}\gamma(\zeta)\Big]
\end{aligned} \tag{4f}$$

It is useful to consider the 'bosonized' conformal field theories of these fermionic $b(z), c(z)$ and bosonic $\beta(z), \gamma(z)$ systems. A 'bosonized' theory is one in which the operator fields are associated with a lattice of momentum vectors. In analogy with the usual space-time coordinates

$$X^\mu(z)X^\nu(\zeta) =: X^\mu(z)X^\nu(\zeta) : -\,\eta^{\mu\nu}\ln(z-\zeta) \tag{5}$$

we define for the fermionic b, c system, the boson fields

$$\phi(z)\phi(\zeta) =: \phi(z)\phi(\zeta) : +\,\eta^{\mu\nu}\ln(z-\zeta) \tag{6}$$

so that

$$: e^{\phi(z)} :: e^{\phi(\zeta)} : =: e^{\phi(z)}\, e^{\phi(\zeta)} : (z-\zeta)\,. \tag{7}$$

We can now make the identifications

$$\begin{aligned}
c(z) &=: e^{\phi(z)}: \quad ; \qquad J(z) = -\,{}^{\times}_{\times}bc{}^{\times}_{\times} = \partial\phi \\
b(z) &=: e^{-\phi(z)} :
\end{aligned} \tag{8}$$

An arbitrary state in the basis space of the bosonized theory is given by

$$\psi = \prod_{a=1}^{M} J_{-m_a}|q\rangle \quad ; \qquad \text{where} \quad |q\rangle \equiv : e^{q\phi(0)} : |0\rangle \quad \text{for} \quad q \in \mathcal{Z}\,. \tag{9}$$

The conformal dimension of $: e^{q\phi(z)} :$ is $\frac{1}{2}q(q-3)$, so that the spectrum generated by $\partial\phi(z)$, and $|q\rangle$ on $\mathcal{Z}$ has the partition function

$$\begin{aligned}
PF_{\partial\phi,|q\rangle} &= \prod_{n=1}^{\infty}(1-\omega^n)^{-1}\sum_{m}\omega^{\frac{1}{2}(m^2-3m)} \\
&= \prod_{n=-1}^{\infty}(1+\omega^n)\prod_{m=2}^{\infty}(1+\omega^m)
\end{aligned} \tag{10}$$

The equality follows from the generalized Jacobi theta function identity:

$$\prod_{r=\lambda}^{\infty}(1+\omega^{r}) \prod_{s=1-\lambda}^{\infty}(1+\omega^{s}) = \omega^{-\frac{1}{2}\nu^{2}}[f(\omega)]^{-1}\sum_{n\in\mathbf{Z}}\omega^{\frac{1}{2}(n+\nu)^{2}}$$
$$= \omega^{-\frac{1}{2}\nu^{2}}[f(\omega)]^{-1}\Theta\left[{}^{2\nu}_{0}\right](0|\tau) \tag{11}$$

where $f(\omega)=\prod_{n=1}^{\infty}(1-\omega^{n})$ and $\omega=e^{2\pi i\tau}$. The partition function for the b,c system is

$$PF_{b,c} = \prod_{n=-1}^{\infty}(1+\omega^{n})\prod_{m=2}^{\infty}(1+\omega^{m}) \tag{12}$$

which is equal to (10), the partition function of the 'bosonized' conformal field theory.

Similarly, for the bosonic β,γ conformal field theory, we define the boson fields

$$\Phi(z)\Phi(\zeta) =: \Phi(z)\Phi(\zeta) : -\eta^{\mu\nu}\ln(z-\zeta) \tag{13}$$

so that

$$: e^{\Phi(z)} :: e^{\Phi(\zeta)} : = : e^{\Phi(z)}\, e^{\Phi(\zeta)} : (z-\zeta)^{-1}\,. \tag{14}$$

Because $: e^{\Phi(z)} :$ is a fermion field and $\beta(z)$,$\gamma(z)$ are bosons, an additional bosonic field $\chi(z)$ is introduced and

$$\gamma(z) =: e^{\Phi(z)} :: e^{-\chi(z)} : \quad ; \qquad J(z) = -\,{}^{\times}_{\times}\beta\gamma{}^{\times}_{\times} = \partial\Phi$$
$$\beta(z) =: e^{-\Phi(z)} :: \partial e^{\chi(z)} : \tag{15}$$

Here

$$\chi(z)\chi(\zeta) =: \chi(z)\chi(\zeta) : +\eta^{\mu\nu}\ln(z-\zeta) \tag{16}$$

Because the $\beta\gamma$ spectrum is unbounded from below, it is useful to define an infinite number of $\beta\gamma$ 'vacua' $|q\rangle_{\beta\gamma}$ where $\beta_n|q\rangle=0\,, n>-q-\frac{3}{2}$, $\gamma_n|q\rangle=0\,, n\geq q+\frac{3}{2}$, where $|q\rangle_{\beta\gamma}=e^{q\Phi(0)}|0\rangle_{\beta\gamma}$ and $L_0^{\beta\gamma}|q\rangle=-\frac{1}{2}q(q+2)|q\rangle$. The bosonic β,γ ghost system has two sectors: one is Neveu-Schwarz where $q\in\mathcal{Z}$, and the fields $\beta(z)$, $\gamma(z)$, and $: e^{\Phi(z)} :$ are periodic, i.e. half-integrally moded; the other sector is Ramond, where $q\in\mathcal{Z}+\frac{1}{2}$, and the fields $\beta(z)$, $\gamma(z)$, and $: e^{\Phi(z)} :$ are anti-periodic, i.e. integrally moded. We note that the conformal fields $: e^{q\Phi(z)} :$ for q odd have the same periodicity and statistics as the supercurrent. Their conformal dimensions given by $L_0^{\beta\gamma}|q\rangle=-\frac{1}{2}q(q+2)|q\rangle$ are $\frac{1}{2}$ for $q=-1$; $-\frac{3}{2}$ for $q=1,-3$; and $-\frac{15}{2}$ for $q=3,-5$; etc.

An arbitrary open string state for the combined matter ghost system in the q ghost picture is given by

$$|\Psi\rangle = |\phi\rangle_{X,\psi}\otimes c_1|0\rangle_{bc}\otimes|q\rangle_{\beta\gamma} = V_q(\phi,0)|0\rangle \tag{17}$$

In the Neveu-Schwarz sector, vertex operators in the canonical ghost picture have $q=-1$. In the Ramond sector, the canonical ghost picture is given by $q=-\frac{1}{2}$. BRST invariance of the state $|\Psi\rangle$, *i.e.* $Q|\Psi\rangle=0$ implies the old covariant physical state conditions when $|\Psi\rangle$ in the canonical ghost picture. For eg., for Ramond states with $q=-\frac{1}{2}$ in $|\Psi\rangle$ given by (17), we have

$$Q_0|\Psi\rangle = (c_0(L_0^{X,\psi}-\tfrac{5}{8}))|\Psi\rangle + \sum_{n>0}c_{-n}L_n^{X,\psi}|\Psi\rangle = 0$$
$$Q_1|\Psi\rangle = (\gamma_0F_0^{X,\psi} + \sum_{n>0}\gamma_{-n}F_n^{X,\psi})|\Psi\rangle = 0$$
$$Q_2|\Psi\rangle = 0\,. \tag{18}$$

The BRST vertex operators involve both superconformal ghost and matter fields. Four-dimensional superstring constructions are given by conformal field theories which describe the internal string degrees of freedom and the four-dimensional space-time degrees of freedom. The massless space-time fermion states are described by Ramond sector ground states. In type II superstring theories, the additional feature of space-time bosons formed from Ramond-Ramond states occurs. In order to construct general scattering amplitudes, it is useful to know the possible forms of the boson and fermion vertex operators for the four-dimensional states. In particular, any gauge symmetry carried by the fermion vertex operators could provide a new source of gauge bosons in type II theories. As in ten dimensions, any state has a vertex operator in each different ghost picture. Also, states in a given cohomology class have the same ghost number: these states are equivalent, ie. they differ from each other by a null vector, and matrix elements do not depend on the choice of representative of a cohomology class.

For *e.g*, the open string vertex operator for the state ϕ in the q ghost picture is given by $V_q(\phi, z)$, so for the massless vector Neveu-Schwarz state $\epsilon\cdot b_{-\frac{1}{2}}|k\rangle$, the vertex operator in the canonical $q=-1$ ghost picture is

$$V_{-1}(k,\epsilon,z) = \epsilon\cdot h(z)e^{ik\cdot X(z)}c(z)e^{-\Phi(z)} \tag{19a}$$

Since $k^2=0$ and the conformal dimension of c is -1 and of $:e^{-\Phi}:$ is $\frac{1}{2}$, the vertex operator $V_{-1}(k,\epsilon,z)$ has zero conformal weight. It's copy in the $q=0$ picture is

$$V_0(k,\epsilon,\zeta) = \lim_{z\to\zeta} e^{\Phi(z)}F(z)V_{-1}(k,\epsilon,\zeta) = (k\cdot h(\zeta)\epsilon\cdot h(\zeta) + \epsilon\cdot a(\zeta))e^{ik\cdot X(\zeta)}c(\zeta) \tag{19b}$$

where

$$F(z) = F^{X,h} = a_\mu(z)h^\mu(z) + \bar{F}(z). \tag{20}$$

Here $0\le\mu\le3$ and ϵ_μ is a four-dimensional polarization wavefunction in the Lorentz gauge: $k\cdot\epsilon=0$. In models where the internal degrees of freedom are described by free world sheet fermions, similarly for the scalar Neveu-Schwarz state $b^a_{-\frac{1}{2}}|k\rangle$, the vertex operator in the canonical $q=-1$ ghost picture is

$$V^a_{-1}(k,z) = h^a(z)e^{ik\cdot X(z)}c(z)e^{-\Phi(z)} \tag{21a}$$

It's copy in the $q=0$ picture depends on the choice of the internal supercurrent $\bar{F}(z)$ which is determined by the gauge symmetry of the theory:

$$\begin{aligned} V^a_0(k,\zeta) &= \lim_{z\to\zeta} e^{\Phi(z)}F(z)V^a_{-1}(k,\zeta) \\ &= \lim_{z\to\zeta}[\,k\cdot h(\zeta)\,h^a(\zeta) + (z-\zeta)\bar{F}(z)h^a(\zeta)\,]e^{ik\cdot X(\zeta)}c(\zeta) \end{aligned} \tag{21b}$$

As examples of different choices of the supercurrent, we consider

$$\tilde{V}^a_0(k,\zeta) = [\,k\cdot h(\zeta)\,h^a(\zeta) + T^a(\zeta)\,]e^{ik\cdot X(\zeta)}c(\zeta) \tag{21c}$$

$$V'^a_0(k,\zeta) = [\,k\cdot h(\zeta)\,h^a(\zeta) - \frac{i}{2\sqrt{\frac{c_\psi}{2}}}f_{abc}h^b(z)h^c(z)\,]e^{ik\cdot X(\zeta)}c(\zeta) \tag{21d}$$

where $T^a(\zeta)$ are the currents of an abelian Kac-Moody algebra, and f_{abc} are the structure constants of some non-abelian Lie algebra.

The matter spin fields are the conformal fields which correspond to the ground states $|\mathcal{A};k\rangle$ in the Ramond sector:

$$|\mathcal{A}\rangle = S_{\mathcal{A}}(0)|0\rangle \qquad ; \qquad |\mathcal{A};k\rangle = S_{\mathcal{A}}(0)e^{ik\cdot X(0)}|0\rangle \tag{22}$$

In ten dimensions, a degenerate ground state is denoted by $|\mathcal{A};k\rangle\chi^{\mathcal{A}}(k)$; here there is a sum over $1 \le \mathcal{A} \le 32$, and the ground state is a thirty-two fold degenerate fermion state. The spinor wavefunction $\chi^{\mathcal{A}}(k)$ is restricted to be Majorana-Weyl, ie that $\Gamma^{11}\chi = \chi$; and in a Majorana representation that χ is real: $\chi^* = \chi$. The vertex operator for the massless Ramond state $|\mathcal{A};k\rangle\chi^{\mathcal{A}}(k)$ in the canonical $q = -\frac{1}{2}$ picture is given for $k\cdot\Gamma\chi = 0$ by

$$V_{-\frac{1}{2}}(k,z) = \chi^{\mathcal{A}}(k)S_{\mathcal{A}}(z)e^{ik\cdot X(z)}c(z)e^{-\frac{1}{2}\Phi(z)}\,. \tag{23}$$

Eq. (23) can be derived from the spin field in the $q = -\frac{3}{2}$ picture given by:

$$V_{-\frac{3}{2}}(k,z) = -v_{\mathcal{A}}(k)S^{\mathcal{A}}(z)e^{ik\cdot X(z)}c(z)e^{-\frac{3}{2}\Phi(z)} \tag{24}$$

with use of the supercurrent leading to (21c) and $k\cdot\Gamma v = \chi$.

It will be interesting also to consider if there is a choice of supercurrent in analogy with (21d) which leads to

$$\begin{aligned} &V'_{-\frac{1}{2}}(k,\zeta)\\ &= \lim_{z\to\zeta} e^{\Phi(z)}F(z)V_{-\frac{3}{2}}(k,\zeta)\\ &= e^{-\frac{1}{2}\Phi(\zeta)}[\chi^{\mathcal{A}}(k) - \frac{i}{6\sqrt{\frac{c_\psi}{2}}}\frac{1}{2\sqrt{2}}f_{abc}v^{\mathcal{B}}(k)(\Gamma^a\Gamma^b\Gamma^c)_{\mathcal{B}}^{\ \mathcal{A}}\,]\,S_{\mathcal{A}}(\zeta)e^{ik\cdot X(\zeta)}\,c(\zeta) \end{aligned} \tag{25}$$

Such Ramond states would carry a non-zero charge under a Yang-Mills gauge symmetry. The BRST invariance of the vertex operator (25), i.e. that $[Q\,,\,V(k,\zeta)]$ vanishes up to a total derivative, would also follow from the existence of the supercurrent.

Furthermore, if there is a choice of supercurrent which when substituted into the BRST current (4b) leads to the BRST invariance of both $\tilde{V}_0^a(k,\zeta)$ in (21d) and $V_{-\frac{1}{2}}(k,z)$ in (23), then we can show that we can derive a consistent non-abelian three gluon coupling for the emission of the massless Neveu-Schwarz vector $\epsilon\cdot b^{\,L}_{\frac{1}{2}}b^{aR}_{-\frac{1}{2}}|k\rangle$ from Ramond-Ramond vector bosons via the amplitude of a closed superstring given by $A_3^L A_3^R$ where

$$\begin{aligned} A_3^L &= \bar{\Psi}_G(-k_1)\langle -k_1;G|\,V_0(k_2,\epsilon_2,1)\,|\,A;k_3\rangle\Psi_A(k_3)\\ &= \langle 0|V_{-\frac{1}{2}}(k_1,z_1)V_{-1}(k_2,\epsilon,z_2)V_{-\frac{1}{2}}(k_3,z_3)|0\rangle\\ &= \langle 0|c(z_1)c(z_2)c(z_3)|0\rangle\langle 0|\,:e^{-\frac{1}{2}\phi(z_1)}::e^{-\phi(z_2)}::e^{-\frac{1}{2}\phi(z_3)}:\,|0\rangle\\ &\quad\cdot\chi^{\mathcal{G}}(k_1)\chi^{\mathcal{A}}(k_3)\epsilon_{2\mu}\,\langle 0|S_{\mathcal{G}}(z_1)\psi^\mu(z_2)S_{\mathcal{A}}(z_3)|0\rangle\\ &= [(z_1-z_2)(z_2-z_3)(z_1-z_3)]\,[(z_1-z_2)^{-\frac{1}{2}}(z_2-z_3)^{-\frac{1}{2}}(z_1-z_3)^{-\frac{1}{4}}\,]\\ &\quad\cdot\chi^{\mathcal{G}}(k_1)\chi^{\mathcal{A}}(k_3)\epsilon_{2\mu}\,\Gamma^\mu_{\mathcal{GA}}\,[(z_1-z_2)^{-\frac{1}{2}}(z_2-z_3)^{-\frac{1}{2}}(z_1-z_3)^{-\frac{3}{4}}\,]\\ &= \bar{\chi}_{\mathcal{G}}(k_1)\Gamma^{\mu\mathcal{G}}{}_{\mathcal{A}}\chi^{\mathcal{A}}(k_3)\,\epsilon_{2\mu} \end{aligned} \tag{26a}$$

and

$$
\begin{aligned}
&A_3^R \\
&= \bar{\Psi}_D^\dagger(-k_1)\langle -k_1; D|V_0^a(k_2,1)\bar{F}_0^1|B;k_3\rangle(\Gamma^0\frac{1}{\sqrt{2}\,k_3^0}\Psi(k_3))_B \\
&= \langle 0|V_{-\frac{1}{2}}(k_1,z_1)V'^a_0(k_2,z_2)V_{-\frac{3}{2}}(k_3,z_3)|0\rangle \\
&= -\langle 0|c(z_1)c(z_2)c(z_3)|0\rangle\langle 0| : e^{-\frac{1}{2}\phi(z_1)} :: e^{-\frac{3}{2}\phi(z_3)} : |0\rangle \\
&\quad \cdot\chi^{\mathcal{A}}(k_1)v_{\mathcal{G}}(k_3)\,\langle 0|S_{\mathcal{A}}(z_1)[\,k_2\cdot h(z_2)h^a(z_2) - \frac{i}{2\sqrt{\frac{c_\psi}{2}}}f_{abc}h^b(z_2)h^c(z_2)\,]S^{\mathcal{G}}(z_3)|0\rangle \\
&= -[(z_1-z_2)(z_2-z_3)(z_1-z_3)]\,[(z_1-z_3)^{-\frac{3}{4}}] \\
&\quad \cdot\chi^{\mathcal{A}}(k_1)v_{\mathcal{G}}(k_3)\,[k_{2\mu}(\Gamma^\mu\Gamma^a)_{\mathcal{A}}{}^{\mathcal{G}} - \frac{i}{2\sqrt{\frac{c_\psi}{2}}}f_{abc}(\Gamma^b\Gamma^c)_{\mathcal{A}}{}^{\mathcal{G}}\,] \\
&\quad \cdot[(z_1-z_2)^{-1}(z_2-z_3)^{-1}(z_1-z_3)^{-\frac{1}{4}}\,] \\
&= -k_{2\mu}\chi_{\mathcal{A}}(k_1)(\Gamma^\mu\Gamma^a)^{\mathcal{A}}{}_{\mathcal{G}}v^{\mathcal{G}}(k_3) + \frac{i}{2\sqrt{\frac{c_\psi}{2}}}f_{abc}\chi_{\mathcal{A}}(k_1)(\Gamma^b\Gamma^c)^{\mathcal{A}}{}_{\mathcal{G}}\,v^{\mathcal{G}}(k_3) \\
&= k_{2\mu}\bar{\chi}_{\mathcal{A}}(k_1)(\Gamma^\mu\Gamma^a)^{\mathcal{A}}{}_{\mathcal{G}}v^{\mathcal{G}}(k_3) - \frac{i}{2\sqrt{\frac{c_\psi}{2}}}f_{abc}\bar{\chi}_{\mathcal{A}}(k_1)(\Gamma^b\Gamma^c)^{\mathcal{A}}{}_{\mathcal{G}}\,v^{\mathcal{G}}(k_3)\,. \qquad (26b)
\end{aligned}
$$

The first lines in (26 a,b) correspond to a modification of the standard vertex operator given in old covariant gauge notation in [4]. Eq. (26) transcribes these amplitudes into BRST form. The brackets in (26) denote the respective z_i dependence of the various correlation functions.

In order to describe four-dimensional states in the above analysis of Ramond ground states, we chose a particular ansatz for the ten-dimensional wave function χ. Also, because in ten dimensions there are six 'internal' Γ matrices we let the structure constants (21d) correspond to the symmetry group $SU(2)\times SU(2)$. In particular (25) is then given by

$$
\begin{aligned}
&V'_{-\frac{1}{2}}(k,\zeta) \\
&= \lim_{z\to\zeta} e^{\Phi(z)}F(z)V_{-\frac{3}{2}}(k,\zeta) \\
&= e^{-\frac{1}{2}\Phi(\zeta)}[\chi^{\mathcal{A}}(k) - \frac{i}{6\sqrt{\frac{c_\psi}{2}}}\frac{1}{2\sqrt{2}}f_{abc}v^{\mathcal{B}}(k)(\Gamma^a\Gamma^b\Gamma^c)_{\mathcal{B}}^{\mathcal{A}}\,]\,S_{\mathcal{A}}(\zeta)e^{ik\cdot X(\zeta)}\,c(\zeta) \\
&= e^{-\frac{1}{2}\Phi(\zeta)}[\chi^{\mathcal{A}}(k) - \frac{i}{2\sqrt{2}}v^{\mathcal{B}}(k)(\Gamma^4\Gamma^5\Gamma^6)_{\mathcal{B}}^{\mathcal{A}} - \frac{i}{2\sqrt{2}}v^{\mathcal{B}}(k)(\Gamma^7\Gamma^8\Gamma^9)_{\mathcal{B}}^{\mathcal{A}}\,]\,S_{\mathcal{A}}(\zeta)e^{ik\cdot X(\zeta)}\,c(\zeta) \\
&= e^{-\frac{1}{2}\Phi(\zeta)}[\chi^{i\alpha(k)} + \frac{i}{2\sqrt{2}}((1-\tilde{\gamma}^5)v^i(k))^\alpha]\,S_\alpha(\zeta)e^{ik\cdot X(\zeta)}\,c(\zeta)\,. \qquad (27)
\end{aligned}
$$

An internal supercurrent $\bar{F}(z)$ used to derive (27) from (24) is

$$
\bar{F}(z) = -\frac{i}{6\sqrt{\frac{c_\psi}{2}}}f_{abc}h^a(z)h^b(z)h^c(z) \qquad (28)
$$

but (28) closes a super Virasoro algebra with anomaly $c = 3$, not the value $c = 9$ required for the complete description of the internal conformal field theory. We note that what is needed is an additional matter system with $c = 6$. Since $c = 6$ is the critical dimension of a conformal field theory with $N = 2$ superconformal world sheet invariance[5–8], an $N = 1$ supercurrent might be constructed as $F = F^+ + F^-$. It may also be that the recent novel constructions[9] of the matter supercurrent involving the introduction of ghost states with wrong sign statistics could be of use here. In addition, we have investigated expressions where the total matter - ghost supercurrent with $c = 0$ has the ghost system mixed with the matter system, so that $F^{\text{total}} \neq F^{\text{matter}} + F^{\text{ghost}}$.

An alternative possibility is to consider the spin fields directly in four dimensions, without starting from a ten-dimensional Marjorana-Weyl wave function as in (23) and from Ramond ground states corresponding to the spinor space of ten-dimensional Γ matrices. Then, some of the restrictions on four-dimensional fermion ground states derived from ten-dimensional Γ matrix space might be absent. Instead of (23), we can then use

$$V_{-\frac{1}{2}}(k, z) = u^{i\alpha}(k) S_\alpha(z) \Sigma_i(z) e^{ik \cdot X(z)} c(z) e^{-\frac{1}{2}\Phi(z)} \tag{29}$$

where the fields $\Sigma_i(z)$ have conformal dimension equal to $\frac{3}{8}$.

Alot appears to depend on the ability to construct a supercurrent for the four-dimensional system. In the next section we present a shifted supercurrent construction of the super Virasoro algebra which results in modified central charge. Since the mixing of these new super Virasoro generators with the vertex operators is now different from the conventional case given in (32), the role of this construction in string theory may be relevant for symmetry breaking.

SVA CONSTRUCTION WITH SHIFTED CENTRAL CHARGE

Matter superconformal fields of conformal weight one-half close to form a super Kac-Moody algebra (SKMA). The mixing between the super Virasoro algebra (SVA) and the SKMA differs depending on the particular construction of the supercurrent. Given a construction with the conventional mixing of (32), we can shift those generators to define a new construction with an altered conformal anonomaly. The $N = 1$ super Virasoro algebra is given by operator products where the right hand side holds for $|z| > |\zeta|$ up to terms regular as $z \to \zeta$:

$$\begin{aligned} L(z)L(\zeta) &= \frac{\frac{c}{2}}{(z-\zeta)^4} + \frac{2L(\zeta)}{(z-\zeta)^2} + \frac{\frac{dL(\zeta)}{d\zeta}}{(z-\zeta)} \\ L(z)F(\zeta) &= \frac{\frac{3}{2}F(\zeta)}{(z-\zeta)^2} + \frac{\frac{dF(\zeta)}{d\zeta}}{(z-\zeta)} \\ F(z)F(\zeta) &= \frac{\frac{2c}{3}}{(z-\zeta)^3} + \frac{2L(\zeta)}{(z-\zeta)} \end{aligned} \tag{30}$$

The super Kac-Moody algebra is

$$\begin{aligned} T^a(z)T^b(\zeta) &= \frac{k\delta_{ab}}{(z-\zeta)^2} + \frac{if_{abc}T^c(\zeta)}{(z-\zeta)} \\ T^a(z)h^a(\zeta) &= \frac{if_{abc}T^c(\zeta)}{(z-\zeta)} \\ h^a(z)h^b(\zeta) &= \frac{\delta_{ab}}{(z-\zeta)}. \end{aligned} \tag{31}$$

Here $f_{abc}f_{abe} = c_\psi \delta_{ce}$; the level of the KMA is $x = \frac{2k}{\psi^2} = \frac{2k}{c_\psi}\tilde{h}$, where $\tilde{h}$ is the dual Coxeter number of the compact Lie algebra with structure constants f_{abc}. Conventional free fermion representations of the internal space SVA and SKMA algebras have the following Kac-Todorov[10] mixing characteristic of a weight one-half superfield.

$$
\begin{aligned}
L(z)T^a(\zeta) &= \frac{T^a(\zeta)}{(z-\zeta)^2} + \frac{\frac{dT^a(\zeta)}{d\zeta}}{(z-\zeta)} \\
L(z)h^a(\zeta) &= \frac{\frac{1}{2}h^a(\zeta)}{(z-\zeta)^2} + \frac{\frac{dh^a(\zeta)}{d\zeta}}{(z-\zeta)} \\
F(z)T^a(\zeta) &= \sqrt{k}[\frac{h^a(\zeta)}{(z-\zeta)^2} + \frac{\frac{dh^a(\zeta)}{d\zeta}}{(z-\zeta)}] \\
F(z)h^a(\zeta) &= \frac{1}{\sqrt{k}}\frac{T^a(\zeta)}{(z-\zeta)}
\end{aligned}
\tag{32}
$$

We now define a new set of SVA generators in terms two arbitrary c-number vectors ρ^a and $\tilde{\rho}^a$ and any set of generators $L(z), F(z), T^a(z), h^a(z)$ and c satisfying (30-32). The new SVA anomaly is $c^{\text{new}} = c - 3\rho^2$ and the new SVA generators have standard Kac-Todorov mixing only with the 'unbroken' SKMA:

$$
\begin{aligned}
L^{\text{new}}(z) &= L(z) + \frac{1}{2\sqrt{k}}\rho^a\partial T^a(z) + \frac{1}{\sqrt{k}z}\tilde{\rho}^a T^a(z) + \frac{1}{2z^2}(\tilde{\rho}^2 - \rho{\cdot}\tilde{\rho}) \\
F^{\text{new}}(z) &= F(z) + \rho^a\partial h^a(z) + \frac{1}{z}\tilde{\rho}^a h^a(z) \\
c^{\text{new}} &= c - 3\rho^2
\end{aligned}
\tag{33}
$$

In component form, we have

$$
\begin{aligned}
L_n^{\text{new}} &= L_n - n\frac{1}{2\sqrt{k}}\rho^a T_n^a + \frac{1}{\sqrt{k}z}(\tilde{\rho}^a - \tfrac{1}{2}\rho^a)T_n^a + \delta_{n,0}\frac{1}{2}(\tilde{\rho}^2 - \rho{\cdot}\tilde{\rho}) \\
F_n^{\text{new}} &= F_n - n\rho^a h_n^a + (\tilde{\rho}^a - \tfrac{1}{2}\rho^a)h_n^a \\
c^{\text{new}} &= c - 3\rho^2
\end{aligned}
\tag{34}
$$

We note here a possible origin of the free vectors, i.e. if ρ^a is the half sum of positive roots of some Lie algebra g, i.e. if $\rho^a \equiv \frac{1}{2}\sum_{\alpha>0}\alpha^a$, then the Freudenthal-de Vries strange formula implies $\frac{\rho^2}{c_\psi} = \frac{\dim \mathrm{g}}{24}$. Clearly, $\rho^a \neq 0$ shifts the central charge, ρ^a real or imaginary decreases or increases the central charge, $\rho^a = \frac{1}{2}\tilde{\rho}^a$ shifts L_0, and $\rho^a \neq \frac{1}{2}\tilde{\rho}^a$ shifts F_0.

CONCLUSIONS

The origin of Yang-Mills symmetry in four-dimensional superstring theory is conventionally described in massless Neveu-Schwarz sectors[11,12]. The discussion presented in this lecture is intended to investigate further aspects of superstring theory relevant to its description of four dimensions, including the role of BRST ghosts, new constructions of the supercurrent, and the nature of the fermion vertex and the massless states of the Ramond sector. To extend this analysis, we describe the internal gauge symmetry properties of the states associated with the spin fields. Arguments prohibiting such states from carrying non-trivial charge under such symmetry have been given in the literature[13–15], but they require certain assumptions. Possible counter examples are shown to arise naturally in a covariant BRST formalism, and it will be interesting to see if their consistency can be established.

REFERENCES

1. D. Friedan, E. Martinec, and S. Shenker, Nucl. Phys. **B271** (1986) 93.
2. D. Lust and S. Theisen, *Lectures on String Theory.* New York: Springer-Verlag, 1989.
3. J. Cohn, D. Friedan, Z. Qui, and S. Shenker, Nucl. Phys. **B278** (1986) 577.
4. L. Dolan and S. Horvath, Nuclear Physics **B416** (1994) 87.
5. G. Waterson, Phys. Lett. **B171B** (1986) 77.
6. P. Di Vecchia, J. Petersen and H. Zheng. Phys. Lett. **162** (1985) 327.
7. M. Bershadsky, W. Lerche, D. Nemeschansky and N. Warner, preprint hep-th/9211040 (November 1992).
8. T. Banks, L. Dixon, D. Friedan and E. Martinec, Nucl. Phys. **B299** (1988) 613.
9. C. Vafa and N. Berkovits, preprint hep-th/9310170 (October 1993).
10. V. Kac and I. Todorov, Comm. Math. Phys. **102** (1985) 337.
11. R. Bluhm, L. Dolan and P. Goddard, Nucl. Phys. **B289** (1987) 364.
12. R. Bluhm, L. Dolan and P. Goddard, Nucl. Phys. **B309** (1988) 330.
13. I. Antoniadis, C Bachas, C. Kounnas, and P. Windey, Phys. Lett. **B171** (1986) 51.
14. L. Dixon, V. Kaplunovsky, and C. Vava, Nucl. Phys. **B355** (1991) 649.
15. M. Chen and W. Yeung, Phys. Rev. **D40** (1989) 1120, 1129.

MASSIVE STRING STATES AS EXTREME BLACK HOLES

M. J. Duff† and J. Rahmfeld

Center for Theoretical Physics
Physics Department
Texas A & M University
College Station, Texas 77843

ABSTRACT

We consider the Schwarz-Sen spectrum of elementary electrically charged massive $N_R = 1/2$ states of the four-dimensional heterotic string and show the maximum spin 1 supermultiplets to correspond to extreme black hole solutions. The $N_L = 1$ states and $N_L > 1$ states (with vanishing left-moving internal momentum) admit a single scalar-Maxwell description with parameters $a = \sqrt{3}$ or $a = 1$, respectively. The corresponding solitonic magnetically charged spectrum conjectured by Schwarz and Sen on the basis of S-duality is also described by extreme black holes.

The idea that elementary particles might behave like black holes is not a new one [1]. Intuitively, one might expect that a pointlike object whose mass exceeds the Planck mass, and whose Compton wavelength is therefore less than its Schwarzschild radius,

Talk presented by M.J. Duff

† Research supported in part by NSF Grant PHY-9106593.

would exhibit an event horizon. In the absence of a consistent quantum theory of gravity, however, such notions would always remain rather vague. Superstring theory, on the other hand, not only predicts such massive states but may provide us with a consistent framework in which to discuss them. The purpose of the present paper is to confirm the claim [2] that certain massive excitations of four-dimensional superstrings are indeed black holes. Of course, non-extreme black holes would be unstable due to the Hawking effect. To describe stable elementary particles, therefore, we must focus on extreme black holes whose masses saturate a Bogomol'nyi bound [1]. The present paper therefore remains agnostic concerning the stronger claims[3, 4] that *all* black holes are single string states or, conversely, that all massive string states are black holes.

Specifically, we shall consider the four-dimensional heterotic string obtained by toroidal compactification. At a generic point in the moduli space of vacuum configurations the unbroken gauge symmetry is $U(1)^{28}$ and the low energy effective field theory is described by $N = 4$ supergravity coupled to 22 abelian vector multiplets. A recent paper [2] showed that this theory exhibits both electrically and magnetically charged black hole solutions corresponding to scalar-Maxwell parameter $a = 0, 1, \sqrt{3}$. In other words, by choosing appropriate combinations of dilaton and moduli fields to be the scalar field ϕ and appropriate combinations of the field strengths and their duals to be the Maxwell field F, the field equations can be consistently truncated to a form given by the Lagrangian

$$\mathcal{L} = \frac{1}{32\pi}\sqrt{-g}\left[R - \frac{1}{2}(\partial\phi)^2 - \frac{1}{4}e^{-a\phi}F^2\right] \tag{1}$$

for these three values of a. (A *consistent* truncation is defined to be one for which all solutions of the truncated theory are solutions of the original theory). In the case of zero angular momentum, the bound between the black hole ADM mass m, and the electric charge $Q = \int e^{-a\phi}\tilde{F}/8\pi$, where a tilda denotes the dual, is given by

$$m^2 \geq Q^2/4(1 + a^2) \tag{2}$$

where, for simplicity, we have set the asymptotic value of ϕ to zero. The $a = 0$ case yields the Reissner-Nordstrom solution which, notwithstanding contrary claims in the literature, does solve the low-energy string equations. The $a = 1$ case yields the dilaton black hole [7, 8]. The $a = \sqrt{3}$ case corresponds to the Kaluza-Klein black hole and the "winding" black hole [2] which are related to each other by T-duality. The Kaluza-Klein solution has been known for some time [7] but only recently recognized [2] as a heterotic string solution.

Let us denote by N_L and N_R the number of left and right oscillators respectively. We shall consider the Schwarz-Sen [9] $O(6, 22; Z)$ invariant spectrum of elementary electrically charged massive $N_R = 1/2$ states of this four-dimensional heterotic string, and show that the spin zero states correspond to extreme limits of black hole solutions which preserve 1/2 of the spacetime supersymmetries. By supersymmetry, the black hole interpretation then applies to all members of the $N = 4$ supermultiplet [10, 11], which has $s_{max} = 1$. For a subset of states the low-energy string action can be truncated to (1). The scalar-Maxwell parameter is given by $a = \sqrt{3}$ for $N_L = 1$ and $a = 1$ for $N_L > 1$ (and vanishing left-moving internal momenta). The other states with $N_L > 1$ are extreme black holes too, but are not described by a single scalar truncation of the

---

[1] The relationship between extremal black holes and the gravitational field around some of the elementary string states has also been discussed in [5] and [6].

type (1). The $N = 4$ supersymmetry algebra possesses two central charges Z_1 and Z_2. The $N_R = 1/2$ states correspond to that subset of the full spectrum that belong to the 16 complex dimensional ($s_{max} \geq 1$) representation of the $N = 4$ supersymmetry algebra, are annihilated by half of the supersymmetry generators and saturate the strong Bogomol'nyi bound $m = |Z_1| = |Z_2|$. As discussed in [12, 9], the reasons for focussing on this N=4 theory, aside from its simplicity, is that one expects that the allowed spectrum of electric and magnetic charges is not renormalized by quantum corrections, and that the allowed mass spectrum of particles saturating the Bogomol'nyi bound is not renormalized either.

Following [14] (see also [13]), Schwarz and Sen have also conjectured [9] on the basis of string/fivebrane duality [16] that, when the solitonic excitations are included, the full string spectrum is invariant not only under the target space $O(6, 22; Z)$ (T-duality) but also under the strong/weak coupling $SL(2, Z)$ (S-duality). The importance of S-duality in the context of black holes in string theory has also been stressed in [15]. Schwarz and Sen have constructed a manifestly S and T duality invariant mass spectrum. T-duality transforms electrically charged winding states into electrically charged Kaluza-Klein states, but S-duality transforms elementary electrically charged string states into solitonic monopole and dyon states. We shall show that these states are also described by the extreme magnetically charged black hole solutions. Indeed, although the results of the present paper may be understood without resorting to string/fivebrane duality, it nevertheless provided the motivation. After compactification from $D = 10$ dimensions to $D = 4$, the solitonic fivebrane solution of $D = 10$ supergravity [17] appears as a magnetic monopole [18] or a string [20] according as it wraps around 5 or 4 of the compactified directions [2]. Regarding this dual string as fundamental in its own right interchanges the roles of T-duality and S-duality. The solitonic monopole states obtained in this way thus play the same role for the dual string as the elementary electric winding states play for the fundamental string. The Kaluza-Klein states are common to both. Since these solitons are extreme ($a = \sqrt{3}$) black holes [2], however, it follows by S-duality that the elementary Kaluza Klein states should be black holes too! By T-duality, the same holds true of the elementary winding states. Rather than invoke S-duality, however, we shall proceed directly to establish that the elementary states described above are in one-to-one correspondence with the extreme electric black holes[3]. Now this leaves open the possibility that they have the same masses and quantum numbers but different interactions. Although we regard this possibility as unlikely given the restrictions of $N = 4$ supersymmetry, the indirect argument may be more compelling in this respect (even though it suffers from the drawback that S-duality has not yet been rigorously established). Of course, elementary states are supposed to be singular and solitonic states non-singular. How then can we interchange their roles? The way the theory accommodates this requirement is that when expressed in terms of the fundamental metric $e^{a\phi}g_{\mu\nu}$ that couples to the worldline of the superparticle the elementary solutions are singular and the solitonic solutions are non-singular, but when expressed in terms of the dual metric $e^{-a\phi}g_{\mu\nu}$, it is the other way around [21, 2].

Let us begin by recalling the bosonic sector of the four dimensional action for the massless fields obtained by dimensional reduction from the usual (2-form) version of

[2]It could in principle also appear as a membrane by wrapping around 3 of the compactified directions, but the $N = 4$ supergravity theory (3) obtained by naive dimensional reduction does not admit the membrane solution [20].

[3]The idea that there may be a dual theory which interchanges Kaluza-Klein states and Kaluza-Klein monopoles was previously discussed in the context of $N = 8$ supergravity by Gibbons and Perry [19]

$D = 10$ supergravity:

$$S = \frac{1}{32\pi}\int d^4x\sqrt{-G}e^{-\Phi}[R_G + G^{\mu\nu}\partial_\mu\Phi\partial_\nu\Phi - \frac{1}{12}G^{\mu\lambda}G^{\nu\tau}G^{\rho\sigma}H_{\mu\nu\rho}H_{\lambda\tau\sigma}$$
$$-\frac{1}{4}G^{\mu\lambda}G^{\nu\tau}F_{\mu\nu}{}^a(LML)_{ab}F_{\lambda\tau}{}^b + \frac{1}{8}G^{\mu\nu}Tr(\partial_\mu ML\partial_\nu ML)] \tag{3}$$

where $F_{\mu\nu}{}^a = \partial_\mu A_\nu{}^a - \partial_\nu A_\mu{}^a$ and $H_{\mu\nu\rho} = (\partial_\mu B_{\nu\rho} + 2A_\mu{}^a L_{ab}F_{\nu\rho}{}^b) +$ permutations. Here Φ is the $D = 4$ dilaton, R_G is the scalar curvature formed from the string metric $G_{\mu\nu}$, related to the canonical metric $g_{\mu\nu}$ by $G_{\mu\nu} \equiv e^\Phi g_{\mu\nu}$. $B_{\mu\nu}$ is the 2-form which couples to the string worldsheet and $A_\mu{}^a$ ($a = 1,...,28$) are the abelian gauge fields. M is a symmetric 28×28 dimensional matrix of scalar fields satisfying $MLM = L$ where L is the invariant metric on $O(6,22)$:

$$L = \begin{pmatrix} 0 & I_6 & 0 \\ I_6 & 0 & 0 \\ 0 & 0 & -I_{16} \end{pmatrix}. \tag{4}$$

The action is invariant under the $O(6,22)$ transformations $M \to \Omega M\Omega^T$, $A_\mu{}^a \to \Omega^a{}_b A_\mu{}^b$, $G_{\mu\nu} \to G_{\mu\nu}$, $B_{\mu\nu} \to B_{\mu\nu}$, $\Phi \to \Phi$, where Ω is an $O(6,22)$ matrix satisfying $\Omega^T L\Omega = L$. T-duality corresponds to the $O(6,22;Z)$ subgroup and is known to be an exact symmetry of the full string theory. The equations of motion, though not the action, are also invariant under the $SL(2,R)$ transformations: $\mathcal{M} \to \omega\mathcal{M}\omega^T, \mathcal{F}_{\mu\nu}{}^{a\alpha} \to \omega^\alpha{}_\beta\mathcal{F}_{\mu\nu}{}^{a\beta}, g_{\mu\nu} \to g_{\mu\nu}$, $M \to M$ where $\alpha = 1,2$ with $\mathcal{F}_{\mu\nu}{}^{a1} = F_{\mu\nu}{}^a$ and $\mathcal{F}_{\mu\nu}{}^{a2} = \left(\lambda_2(ML)^a{}_b\tilde{F}_{\mu\nu}{}^b + \lambda_1 F_{\mu\nu}{}^a\right)$, where ω is an $SL(2,R)$ matrix satisfying $\omega^T\mathcal{L}\omega = \mathcal{L}$ and where

$$\mathcal{M} = \frac{1}{\lambda_2}\begin{pmatrix} 1 & \lambda_1 \\ \lambda_1 & |\lambda|^2 \end{pmatrix}, \quad \mathcal{L} = \begin{pmatrix} 0 & 1 \\ -1 & 0 \end{pmatrix}. \tag{5}$$

λ is given by $\lambda = \Psi + ie^{-\Phi} \equiv \lambda_1 + i\lambda_2$. The axion Ψ is defined through the relation $\sqrt{-g}H^{\mu\nu\rho} = -e^{2\Phi}\epsilon^{\mu\nu\rho\sigma}\partial_\sigma\Psi$. S-duality corresponds to the $SL(2,Z)$ subgroup and there is now a good deal of evidence [9] in favor of its also being an exact symmetry of the full string theory. For the restricted class of configurations obtained by setting to zero the 16 gauge fields $F^{13\to 28}$ originating from the ten-dimensional gauge fields, it is possible to define a dual action [9] which has manifest $SL(2,R)$ symmetry. The field strengths $F^{1\to 6}$, whose origin resides in the $D = 10$ metric, remain the same but the $F^{7\to 12}$, whose origin resides in the $D = 10$ 2-form, are replaced by their duals. The equations of motion are also invariant under $O(6,6)$; the action is not except for the $SL(6,R)$ subgroup which acts trivially. This action is precisely the one obtained by dimensional reduction from the dual (6-form) version of $D = 10$ supergravity which couples to the worldvolume of the fivebrane [16] and for which the axion is just the 6-form component lying in the extra 6 dimensions. This provides another reason for believing that the roles of S and T duality are interchanged in going from string to fivebrane [9, 22, 20] and is entirely consistent with an earlier observation that the dual theory interchanges the worldsheet and spacetime loop expansions [25]. In this light, the need to treat the above 16 gauge fields on a different footing is only to be expected since in the dual formulation their kinetic terms are 1-loop effects [25].

We now turn to the electric and magnetic charge spectrum. Schwarz and Sen [9] present an $O(6,22;Z)$ and $SL(2,Z)$ invariant expression for the mass of particles saturating the strong Bogomol'nyi bound $m = |Z_1| = |Z_2|$:

$$m^2 = \frac{1}{16}(\alpha^a \;\; \beta^a)\mathcal{M}^0(M^0 + L)_{ab}\begin{pmatrix} \alpha^b \\ \beta^b \end{pmatrix} \tag{6}$$

where a superscript 0 denotes the constant asymptotic values of the fields. Here α^a and β^a $(a = 1, ..., 28)$ each belong to an even self-dual Lorentzian lattice Λ with metric given by L and are related to the electric and magnetic charge vectors (Q^a, P^a) by $(Q^a, P^a) = \left(M_{ab}{}^0(\alpha^b + \lambda_1{}^0\beta^b)/\lambda_2{}^0, L_{ab}\beta^b\right)$. As discussed in [9] only a subset of the conjectured spectrum corresponds to elementary string states. First of all these states will be only electrically charged, i.e. $\beta = 0$, but there will be restrictions on α too. Without loss of generality let us focus on a compactification with $M^0 = I$ and $\lambda_2{}^0 = 1$. Any other toroidal compactifications can be brought into this form by $O(6,22)$ transformations and a constant shift of the dilaton. The mass formula (6) now becomes

$$m^2 = \frac{1}{16}\alpha^a(I+L)_{ab}\alpha^b = \frac{1}{8}\left(\alpha_R\right)^2 \tag{7}$$

with $\alpha_R = \frac{1}{2}(I+L)\alpha$ and $\alpha_L = \frac{1}{2}(I-L)\alpha$. In the string language $\alpha_{R(L)}$ are the right(left)-moving internal momenta. The mass of a generic string state in the Neveu-Schwarz sector (which is degenerate with the Ramond sector) is given by

$$m^2 = \frac{1}{8\lambda_2{}^0}\left\{(\alpha_R)^2 + 2N_R - 1\right\} = \frac{1}{8\lambda_2{}^0}\left\{(\alpha_L)^2 + 2N_L - 2\right\}. \tag{8}$$

A comparison of (7) and (8) shows that the string states satisfying the Bogomol'nyi bound all have $N_R = 1/2$. One then finds

$$N_L - 1 = \frac{1}{2}\left((\alpha_R)^2 - (\alpha_L)^2\right) = \frac{1}{2}\alpha^T L\alpha, \tag{9}$$

leading to $\alpha^T L\alpha \geq -2$. We shall now show that extreme black holes with $a = \sqrt{3}$ are string states with $\alpha^T L\alpha$ null $(N_L = 1)$ and those with $a = 1$ are string states with $\alpha^T L\alpha$ spacelike $(N_L > 1)$. We have been unable to identify solutions of the low-energy field equations (3) corresponding to states with $\alpha^T L\alpha$ timelike $(N_L < 1)$. [4]

Let us first focus on the $a = \sqrt{3}$ black hole. To identify it as a state in the spectrum we have to find the corresponding charge vector α and to verify that the masses calculated by the formulas (2) and (6) are identical. The action (3) can be consistently truncated by keeping the metric $g_{\mu\nu}$, just one field strength ($F = F^1$, say), and one scalar field ϕ via the ansatz $\Phi = \phi/\sqrt{3}$ and $M_{11} = e^{2\phi/\sqrt{3}} = M_{77}^{-1}$. All other diagonal components of M are set equal to unity and all non-diagonal components to zero. Now (3) reduces to (1) with $a = \sqrt{3}$. (This yields the electric and magnetic Kaluza-Klein (or "F") monopoles. This is not quite the truncation chosen in [2], where just F^7 was retained and $M_{11} = e^{-2\phi/\sqrt{3}} = M_{77}^{-1}$. This yields the the electric and magnetic winding (or "H") monopoles. However, the two are related by T-duality). We shall restrict ourselves to the purely electrically charged solution with charge $Q = 1$, since this one is expected to correspond to an elementary string excitation. The charge vector α for this solution is obviously given by $\alpha^a = \delta^{a,1}$ with $\alpha^T L\alpha = 0$. Applying (6) for the mass of the state we find $m^2 = 1/16 = Q^2/16$, which coincides with (2) in the extreme limit. This agreement confirms the claim that this extreme $a = \sqrt{3}$ black hole is a state in the Sen-Schwarz spectrum and preserves 2 supersymmetries.

Next we turn to the $a = 1$ black hole. The theory is consistently truncated by keeping the metric, $F = F^1 = F^7$ and setting $M = I$. The only non-vanishing scalar is the dilaton $\Phi \equiv \phi$. Now (3) reduces to (1) with $a = 1$ but $Q^2 = 2$. An extreme $a = 1$ black hole with electric charge Q is then represented by the charge configuration $\alpha^a = \delta^{a,1} + \delta^{a,7}$. Applying (6) we find $m^2 = 1/4 = Q^2/8$ which coincides with (2) in the

[4] In the *non-abelian* theory Sen [9] identifies these states with the electric analogues of BPS monopoles.

extreme limit. Therefore the $a = 1$ extreme solution is also in the spectrum, and has $\alpha^T L \alpha = 2$ or $N_L = 2$.

Although physically very different, we can see with hindsight that both the $a = \sqrt{3}$ and $a = 1$ black holes permit a uniform mathematical treatment by noting that both may be obtained from the Schwarzschild solution by performing an $[O(6,1) \times O(22,1)]/[O(6) \times O(22)$ transformation [6]. The 28 parameters of this transformation correspond to the 28 $U(1)$ charges. If γ and u correspond to the boost angle and a 22 dimensional unit vector respectively, associated with O(22,1)/O(22) transformations, δ and v denote the boost angle and the 6 dimensional unit vector respectively, associated with the O(6,1)/O(6) transformations, and m_0 is the mass of the original Schwarzschild black hole, then the mass and charges of the new black hole solution are given by [6]:

$$m = \frac{1}{2} m_0 (1 + \cosh\gamma \cosh\delta)$$

$$\alpha_L = \sqrt{2} m_0 \cosh\delta \sinh\gamma \, u$$
$$\alpha_R = \sqrt{2} m_0 \cosh\gamma \sinh\delta \, v \qquad (10)$$

(Note that the convention about R and L of [6] is opposite to the one used in the present paper). Black holes with $\alpha^T L \alpha = 0$ are generated by setting $\gamma = \delta$, whereas black holes with $\alpha^T L \alpha > 0$ are generated by setting $\gamma < \delta$. The Bogomol'nyi bound given in (7) corresponds to $m^2 = (\alpha_R)^2/8$. This bound is saturated by taking the limit where the mass m_0 of the original Schwarzschild black hole approaches 0 and the parameter δ approaches ∞, keeping the product $m_0 \sinh\delta$ fixed. As discussed in [6], this is precisely the extremal limit. Thus we see that extremal black holes satisfy the Bogomol'nyi relation, both for $\alpha^T L \alpha = 0$ and $\alpha^T L \alpha > 0$.

From the above $a = \sqrt{3}$ solution we can generate the whole set of supersymmetric black hole solutions with $\alpha^T L \alpha = 0$ in the following way: first we note that we are interested in constructing black hole solutions with different charges but with fixed asymptotic values of M (which here has been set to the identity). Thus we are not allowed to make $O(6,22)$ transformations that change the asymptotic value of M. This leaves us with only an $O(6) \times O(22)$ group of transformations. The effect of these transformations acting on the parameters given in (10) above is to transform the vectors u and v by $O(22)$ and $O(6)$ transformations respectively without changing the parameters γ and δ. Now, the original $a = \sqrt{3}$ solution corresponds to a choice of parameters $\gamma = \delta$, $u^m = \delta_{m1}$ and $v^m = \delta_{k1}$. It is clear that an $O(6) \times O(22)$ transformation can rotate u and v to arbitrary 22 and 6 dimensional unit vectors respectively, without changing γ and δ. Since this corresponds to the most general charge vector satisfying $\alpha^T L \alpha = 0$, we see that the $O(6) \times O(22)$ transformation can indeed generate an arbitrary black hole solution with $\alpha^T L \alpha = 0$ starting from the original $a = \sqrt{3}$ solution. This clearly leaves the mass invariant, but the new charge vector α' will in general not be located on the lattice. To find a state in the allowed charge spectrum we have to rescale α' by a constant k so that $\alpha'' = k\alpha'$ is a lattice vector. Clearly the masses calculated by (2) and (6) still agree (this is obvious by reversing the steps of rotation and rescaling), leading to the conclusion that all states obtained in this way preserve 1/2 of the supersymmetries. Therefore all states in the spectrum belonging to $s_{max} = 1$ supermultiplets for which $N_R = 1/2, N_L = 1$ are extreme $a = \sqrt{3}$ black holes.

Let us now turn to the case of the $a = 1$ solution. In this case the original solution corresponds to the choice of parameters $\gamma = 0$, $v^m = \delta_{m1}$. (For $\gamma = 0$, the parameter u is irrelevant). An $O(6) \times O(22)$ transformation can rotate v to any other 6 dimensional unit vector, but it cannot change the parameters δ and γ. As a result, the final solution will continue to have $\gamma = 0$ and hence $\alpha_L = 0$. Since this does not represent the most general charge vector α, with $\alpha^T L \alpha > 0$, we see that the most general black hole representing states with $\alpha^T L\alpha > 0$ is not obtained in this way even after rescaling. The missing states with $\alpha_L \neq 0$ are constructed by choosing γ so that $\tanh^2\gamma = \alpha_L^2/\alpha_R^2$, and u, v as for the $a = \sqrt{3}$ case, followed by a suitable $O(6)\times O(22)$ rotation. Clearly, those solutions are extreme black holes too. However, for these solutions a truncation to an effective action of the form (1) is not possible. The following picture arises: for a fixed value of α_R^2, α_L^2 can vary in the range $\alpha_R^2 \geq \alpha_L^2 \geq 0$. The boundary states are described by the well-known $a = \sqrt{3}$ ($\alpha_R^2 = \alpha_L^2$) and $a = 1$ ($\alpha_L^2 = 0$) black holes, whereas the states in between cannot be related to a single scalar-Maxwell parameter a. But all solutions preserve 1/2 of the supersymmetries.

It should also be clear that the purely magnetic extreme black hole solutions [2] obtained from the above by the replacements $\phi \to -\phi, \alpha \to \beta$ will also belong to the Schwarz-Sen spectrum of solitonic states. Starting from either the purely electric or purely magnetic solutions, dyonic states in the spectrum which involve non-vanishing axion field Ψ can then be obtained by $SL(2,Z)$ transformations. Specifically, a black hole with charge vector $(\alpha, 0)$ will be mapped into ones with charges $(a\alpha, c\alpha)$ with the integers a and c relatively prime [9].

Not all black hole solutions of (3) belong to the Sen-Schwarz spectrum, however. Let us first consider the Reissner-Nordstrom solution. Since this black hole solves the equations of $N = 2$ supergravity, whose bosonic sector is pure Einstein-Maxwell, it solves (3) as well. The required consistent truncation is obtained by keeping $g_{\mu\nu}$, $F = F^1 = F^7 = \tilde{F}^2 = \tilde{F}^8$ and setting $\Phi = 0$, $M = 1$. Now (3) effectively reduces to (1) with $a = 0$ but $Q^2 = 4$. On the other hand, if it were in the Schwarz-Sen spectrum its charge vectors would be given by $\alpha^a = \delta^{a,1} + \delta^{a,7}$ with $\alpha^T L\alpha = 2$ and $\beta^a = \delta^{a,2} + \delta^{a,8}$ with $\beta^T L\beta = 2$. Applying (6) for the mass of the state we find $m^2 = 1/2$, which disagrees with the result $m^2 = 1$ obtained from the extreme limit of (2). So the test fails and the $a = 0$ black hole does not belong to the Schwarz-Sen spectrum. This was only to be expected since it breaks 3/4 of the supersymmetries and hence saturates the weaker Bogomol'nyi bound $m = |Z_1|, |Z_2| = 0$ [23]. Such black holes belong to the 32 complex dimensional ($s_{max} = 3/2$) supermultiplet. We see no reason to exclude these states from the full string spectrum, however. Another example of a black hole solution not in the Schwarz-Sen spectrum is the $a = 1$ dilaton black hole of [8] where the only non-vanishing gauge field is F^{13}. This has mass $m^2 = Q^2/8$ but according to (6) its mass would vanish. Again, this contradiction is only to be expected since this solution breaks all the supersymmetries, in contrast with the $F = F^1 = F^7$ embedding discussed above. We do not know whether such black holes saturating no Bogomol'nyi bound ($m > |Z_1|, |Z_2|$), which include the neutral Schwarzschild black holes ($Z_1 = Z_2 = 0$), are also in the string spectrum. States with these quantum numbers would belong to the 256 dimensional ($s_{max} \geq 2$) supermultiplets. According to [10], however, black holes breaking all the supersymmetries do not themselves form supermultiplets. This would appear to contradict the claim that *all* black holes are string states.

In the supersymmetric case, all values of a lead to extreme black holes with zero entropy but their temperature is zero, finite or infinite according as $a < 1$, $a = 1$ or $a > 1$, and so in [24] the question was posed: can only $a > 1$ scalar black holes describe elementary particles? We have not definitively answered this question but a tentative response would be as follows. First we note that the masses and charge vectors are such that the lightest $a = 0$ black hole may be regarded as a bound state (with zero binding energy) of two lightest $a = 1$ black holes which in turn can each be regarded as bound states (again with zero binding energy) of two lightest $a = \sqrt{3}$ black holes. Thus if by elementary particle one means an object which cannot be regarded as a bound state, then indeed extreme scalar black holes with $a > 1$ are the only possibility, but if one merely means a state in the string spectrum then $a \leq 1$ extreme scalar black holes are also permitted.

We have limited ourselves to $N_R = 1/2$ supermultiplets with $s_{min} = 0$. Having established that the $s = 0$ member of the multiplet is an extreme black hole, one may then use the fermionic zero modes to perform supersymmetry transformations to generate the whole supermultiplet of black holes [10, 11] with the same mass and charges. Of course, there are $N_R = 1/2$ multiplets with $s_{min} > 0$ coming from oscillators with higher spin and our arguments have nothing to say about whether these are also extreme black holes. They could be naked singularities. Indeed, although in this paper we have focussed primarily on identifying certain massive heterotic string states with extreme black holes, perhaps equally remarkable is that these elementary string states can be described at all by solutions of the supergravity theory. In a *field* theory, as opposed to a *string* theory, one is used to having as elementary massive states only the Kaluza-Klein modes with $s_{max} = 2$. However, as we have already seen, the winding states (usually thought of as intrinsically stringy) are on the same footing as Kaluza-Klein states as far as solutions are concerned, so perhaps the same is true for the $s > 2$ states.

None of the spinning $N_R = 1/2$ states is described by extreme *rotating* black hole metrics because they obey the same Bogomol'nyi bound as the $s_{min} = 0$ states, whereas the mass formula for an extreme *rotating* black hole depends on the angular momentum J. Moreover, it is the fermion fields which carry the spin in the $s_{min} = 0$ supermultiplet. (For the $a = 0$ black hole, they yield a gyromagnetic ratio $g = 2$ [11]; the $a = \sqrt{3}$ and $a = 1$ superpartner g-factors are unknown to us). It may be that there are states in the string spectrum described by the extreme rotating black hole metrics but if so they will belong to the $N_R \neq 1/2$ sector[5]. Since, whether rotating or not, the black hole solutions are still independent of the azimuthal angle and independent of time, the supergravity theory is effectively *two-dimensional* and therefore possibly integrable. This suggests that the spectrum should be invariant under the larger duality $O(8, 24; Z)$ [13], which combines S and T. The corresponding Kac-Moody extension would then play the role of the spectrum generating symmetry [26].

Conversations with G. Gibbons, R. Khuri and J. Liu are gratefully acknowledged.

References

[1] S. W. Hawking, Monthly Notices Roy. Astron. Soc. **152** (1971) 75; Abdus Salam in *Quantum Gravity: an Oxford Symposium* (Eds. Isham, Penrose and Sciama, O.U.P. 1975); G. 't Hooft, Nucl. Phys. **B335** (1990) 138.

[2] M. J. Duff, R. R. Khuri, R. Minasian and J. Rahmfeld, Nucl. Phys. **B418** (1994) 195.

[3] J. Ellis, N. E. Mavromatos and D. V. Nanopoulos, Phys. Lett. **B278** (1992) 246.

[5]The gyromagnetic and gyroelectric ratios of the states in the heterotic string spectrum would then have to agree with those of charged rotating black hole solutions of the heterotic string. This is indeed the case: the $N_L = 1$ states [27] and the rotating $a = \sqrt{3}$ black holes [28] both have $g = 1$ whereas the $N_L > 1$ states [4] and the rotating $a = 1$ [29] (and $a = 0$ [30]) black holes both have $g = 2$. In fact, it was the observation that the Regge formula $J \sim m^2$ also describes the mass/angular momentum relation of an extreme rotating black hole which first led Salam [1] to imagine that elementary particles might behave like black holes!

[4] L. Susskind, RU-93-44, hep-th/9309145; J. G. Russo and L. Susskind, UTTG-9-94,hep-th/9405117.

[5] M. Fabbrichesi, R. Jengo and K. Roland, Nucl. Phys. **B402** (1993) 360.

[6] A. Sen, Nucl. Phys. **B404** (1993) 109.

[7] G. W. Gibbons, Nucl. Phys. **B207** (1982) 337.

[8] D. Garfinkle, G. Horowitz and A. Strominger, Phys. Rev. **D43** (1991) 3140; S. B. Giddings, J. Polchinski and A. Strominger, NSF-ITP-93-62.

[9] J. H. Schwarz and A. Sen, Phys. Lett. **B312** (1993) 105; A. Sen, TIFR/TH/94-03, hep-th/9402002.

[10] G. W. Gibbons, in *Supersymmetry, Supergravity and Related Topics*, eds. F. del Aguila, J. A. Azcarraga and L. E. Ibanez (World Scientific, 1985).

[11] P. Aichelburg and F. Embacher, Phys. Rev. **D37** (1986) 3006.

[12] E. Witten and D. Olive, Phys. Lett. **B78** (1978) 97; H. Osborn, Phys. Lett. **B83** (1979) 321.

[13] M. J. Duff and J. X. Lu, Nucl. Phys. **B347** (1990) 394.

[14] A. Font, L. Ibanez, D. Lust and F. Quevedo, Phys. Lett. **B249** (1990) 35.

[15] S. Kalara and D. V. Nanopoulos, Phys. Lett. **B267** (1991) 343.

[16] M. J. Duff, Class. Quantum Grav. **5** (1988) 189; A. Strominger, Nucl. Phys. **B343** (1990) 167.

[17] M. J. Duff and J.X. Lu, Nucl. Phys. **B354** (1991) 141.

[18] R. R. Khuri, Phys. Lett. **B259** (1991) 261; Nucl. Phys. **B387** (1992) 315; J. P. Gauntlett, J. A. Harvey and J. T. Liu, Nucl. Phys. **B409** (1993) 363.

[19] G. W. Gibbons and M. J. Perry, Nucl. Phys. **B248** (1984) 629.

[20] M. J. Duff and R. R. Khuri, Nucl. Phys. **B411** (1994) 473.

[21] M. J. Duff and J. X. Lu, Nucl. Phys. **B416** (1994) 301.

[22] P. Binetruy, Phys. Lett. **B315** (1993) 80.

[23] R. Kallosh, A. Linde, T. Ortin, A. Peet and A. Van Proeyen, Phys. Rev. **D46** (1992) 5278.

[24] C. F. E. Holzhey and F. Wilczek, Nucl. Phys. **B380** (1992) 447.

[25] M. J. Duff and J. X. Lu, Nucl. Phys. **B357** (1991) 534.

[26] R. Geroch, J. Math. Phys. **13** (1972) 394.

[27] A. Hosoya, K. Ishikawa, Y. Ohkuwa and K. Yamagishi, Phys. Lett. **B134** (1984) 44.

[28] G. W. Gibbons and D. L. Wiltshire, Ann. of Phys. **167** (1986) 201.

[29] A. Sen, Phys. Rev. Lett. **69** (1992) 1006.

[30] G. C. Debney, R. P. Kerr and A. Schild, J. Math. Phys. **10** (1969) 1842.

Z' DIAGNOSTICS AT FUTURE COLLIDERS

M. CVETIČ
Department of Physics
University of Pennsylvania
Philadelphia, PA 19104–6396, USA

Abstract

We summarize the status of heavy gauge boson (Z') diagnostics at future hadron and e^+e^- colliders. The emphasis is on the model independent determination of gauge couplings of Z' to quarks and leptons at the pp (CERN LHC) and e^+e^- (NLC) colliders. For $M_{Z'} \sim 1$ TeV, the NLC would have a capability to probe *all* the quark and lepton charges to around $10 - 20\%$, provided heavy flavor tagging and longitudinal polarization of the electron beam is available. On the other hand, at the LHC primarily the magnitude of normalized couplings can be determined, with typical uncertainties by about a factor of 2 smaller, however, still comparable to the ones at the NLC. The complementarity between the diagnostic power of the two types of machines is emphasized.

Introduction

Current bounds on the mass of heavy neutral (Z') and charged (W') gauge bosons are weak. Bounds on the Z' mass are around $150 - 300$ GeV, and somehow higher for models with the specified Higgs sector. Bounds on the W' mass are around 300 GeV, and around 1.4 TeV for specific left-right symmetric models. Near-future e^+e^- (LEPII) and hadron (upgraded Tevatron) colliders will improve such bounds. *E.g.*, the upgraded Tevatron with integrated luminosity of 100 pb^-1 would improve bounds on a Z' mass to around 600 GeV.[1]

On the other hand, future hadron colliders, *e.g.*, the large hadron collider (LHC) at CERN, would be ideal place to discover heavy gauge bosons Z''s with a mass up to around 5 TeV. An immediate need after Z' discovery would be to learn more about its properties. In particular, a determination of Z' couplings to quarks and leptons is useful. By now a series of such probes was proposed, allowing for a model independent determination of quark and lepton charges provided $M_{Z'} \lesssim 2$ TeV.

[1]for a review on $M_{z'}$ bounds see for example Ref.[1].

Unified Symmetry: In the Small and in the Large,
Edited by B.N. Kursunoglu *et al.*, Plenum Press, New York, 1995

On the other hand, future e^+e^- colliders with large enough c.m. energy, *e.g.*, $\sqrt{s} = 2$ TeV, could provide a clean way to discover and study the properties of Z''s. A more likely possibility, however, is the next linear collider (NLC) with $\sqrt{s} = 500$ GeV. There, due to the interference effects of the Z' propagator with the photon and Z propagator, the probes with the two–fermion final states yield a complementary information on the existence of a Z'. An extensive study [2, 3] showed that effects of Z''s would be observable at the NLC for a large class of models with $M_{Z'}$ up to $1-3$ TeV. In particular, sensitivity of the NLC to discriminate [2] between specific classes of extended electroweak models, *e.g.*, different E_6 motivated models described by a parameter $\cos\beta$ (the mixing between the Z_χ and the Z_ψ defined below) or left-right symmetric models parameterized by the ratio $\kappa = g_R/g_L$ for the $SU(2)_{L,R}$ gauge coupling constants $g_{L,R}$, was explored. Most recently, a model independent determination of Z' couplings to quarks and leptons and a comparison with the LHC colliders was performed in Ref. [4, 5] .

Probes at the LHC

In the main production channels, $pp \to Z' \to \ell^+\ell^-$ ($\ell = e, \mu$) one would be able to measure the mass $M_{Z'}$, the total width Γ_{tot} and the total cross section $\sigma_{\ell\ell}$. The quantity $\sigma_{\ell\ell}\Gamma_{tot}$ would in turn yield information on an overall strength of the Z' gauge coupling. On the other hand there is a need for probes which are sensitive to the *relative strength* of Z' couplings. The nature of such probes can be classified according to the type of channel in which they can be measured.

- (i) The forward-backward asymmetry [6] ,
- (ii) the ratio of cross-sections in different rapidity bins [7]

constitute distributions, *i.e.*, "refinements" in the main production channel.[2] The forward-backward asymmetry was long recognized as useful to probe a particular combination of quark and lepton couplings. The rapidity ratio is a useful complementary probe separating the Z' couplings to the u and d quarks due to the harder valence u-quark distribution in the proton relative to the d-quark.

- (iii) Rare decays $Z' \to W\ell\nu_\ell$ [13, 14] ,
- (iv) associated productions $pp \to Z'V$ with $V = (Z, W)$[10] and $V = \gamma$ [15]

[2]If the proton polarization were available the corresponding asymmetries would also to be useful [8] . In other two-fermion channels measurements of the τ polarization in the $pp \to Z' \to \tau^+\tau^-$ channel [9] , is also a useful probe. Measurements of the cross section in $pp \to Z' \to$ *jet jet* channel is the only probe available for the left-handed quark coupling [10] . Recent studies [11, 12] indicate that the measurement of the cross-section in this channel may be possible with appropriate kinematic cuts, excellent dijet mass resolution, and detailed knowledge of the QCD backgrounds.

provide another set of useful probes in the the four-fermion final state channels. These probes have suppressed rates compared to the two-fermion channels. Rare decays turn out to have sizable statistics, however only the modes $Z' \to W\ell\nu_\ell$ [14, 16, 17] with appropriate cuts are useful[3] without large standard model and QCD backgrounds. These modes probe a left-handed leptonic coupling. On the other hand the associated productions turn out to be relatively clean signals [10] with slightly smaller statistics than rare decays. They probe a particular combination of the gauge couplings to quarks and are thus complementary to rare decays.

Probes at the NLC

There the cross sections and corresponding asymmetries in the two-fermion final state channels, $e^+e^- \to f\bar{f}$, will be measured. The analysis is based on the following probes:

$$\sigma^\ell, \quad R^{had} = \frac{\sigma^{had}}{\sigma^\ell}, \quad A^\ell_{FB}. \tag{1}$$

In the case that longitudinal polarization of the electron beam is available there are additional probes:

$$A^{\ell,had}_{LR}, \quad A^\ell_{LR,FB} \tag{2}$$

Here σ, A_{FB}, A_{LR} and $A_{LR,FB}$ refer to the corresponding cross-sections, forward-backward asymmetries, left-right asymmetries and left–right–forward–backward asymmetries, respectively. The index ℓ refers to all three leptonic channels (considering only s-channel exchange for electrons) and *had* to all hadronic final states. The above quantities distinguish among different models[2] , however, they do not yield information on all the Z' couplings.

If one assumes[4] an efficient heavy flavor (c, b, t) tagging there are the following additional observables:

$$R^f = \frac{\sigma^f}{\sigma^\ell}, \quad A^f_{FB} \; ; \; f = c, b, t \quad , \tag{3}$$

and with available polarization:

$$A^f_{LR} \, , \quad A^f_{LR,FB} \; ; \; f = c, b, t \; . \tag{4}$$

These additional probes in turn allow for complete determination of the Z' gauge couplings to ordinary fermions.[5]

Determination of gauge couplings at the LHC

We assume the c.m. energy $\sqrt{s} = 16$ TeV[6] and integrated luminosity $\mathcal{L}_{int} = 100$ fb^{-1}. We consider only statistical uncertainties associated with the probes (i-iv), which yield

[3] $Z' \to Z\ell^+\ell^-$ does not significantly discriminate between models.

[4] Note that at the LEP an efficient tagging of the charm and bottom flavors was achieved [18] .

[5] See Ref. [4] for detailed discussions.

[6] For the new projected c.m. energy $\sqrt{s} = 14$ TeV, the cross section in the main production channel decreases by ~30% and thus the statistical error bars on the probes increase by 14% .

sufficient qualitative information. Realistic fits, which include updated structure functions, kinematic cuts, and detector acceptances are expected to give larger uncertainties for the couplings.

We consider the following typical models: Z_χ in $SO_{10} \to SU_5 \times U_{1\chi}$, Z_ψ in $E_6 \to SO_{10} \times U_{1\psi}$, $Z_\eta = \sqrt{3/8}Z_\chi - \sqrt{5/8}Z_\psi$ in superstring inspired models in which E_6 breaks directly to a rank 5 group,and Z_{LR} in LR symmetric models. For conventions in the neutral current interactions see Ref. [19] . In the following we assume family universality, neglect $Z-Z'$ mixing and assume $[Q', T_i] = 0$, which holds for $SU_2 \times U_1 \times U_1'$ and LR models. Here, Q' is the Z' charge and T_i are the SU_{2L} generators which incidentally is satisfied by the above models.

The relevant quantities [10, 7] to distinguish between different models are the charges, $\hat{g}^u_{L2} = \hat{g}^d_{L2} \equiv \hat{g}^q_{L2}$, $\hat{g}^u_{R2}$, $\hat{g}^d_{R2}$, $\hat{g}^\nu_{L2} = \hat{g}^e_{L2} \equiv \hat{g}^\ell_{L2}$, and $\hat{g}^\ell_{R2}$, and the gauge coupling strength

Table 1. [7] Values of the couplings (5) for the typical models. The statistical error bars indicate how well the coupling could be measured at the LHC (c.m. energy $\sqrt{s} = 16$ TeV and integrated luminosity $\mathcal{L}_{int} = 100\,\mathrm{fb}^{-1}$) for $M_{Z'} = 1$ TeV.

	χ	ψ	η	LR
γ^ℓ_L	0.9 ± 0.018	0.5 ± 0.03	0.2 ± 0.015	0.36 ± 0.007
γ^q_L	0.1	0.5	0.8	0.04
$\tilde{U}$	1 ± 0.18	1 ± 0.27	1 ± 0.14	37 ± 8.3
$\tilde{D}$	9 ± 0.61	1 ± 0.41	0.25 ± 0.29	65 ± 14

g_2. The signs of the charges will be hard to determine at hadron colliders. Thus the following four "normalized" observables can be probed [10] :

$$\gamma^\ell_L = \frac{(\hat{g}^\ell_{L2})^2}{(\hat{g}^\ell_{L2})^2 + (\hat{g}^\ell_{R2})^2}, \quad \gamma^q_L = \frac{(\hat{g}^q_{L2})^2}{(\hat{g}^\ell_{L2})^2 + (\hat{g}^\ell_{R2})^2},$$
$$\tilde{U} = \frac{(\hat{g}^u_{R2})^2}{(\hat{g}^q_{L2})^2}, \qquad \tilde{D} = \frac{(\hat{g}^d_{R2})^2}{(\hat{g}^q_{L2})^2}. \tag{5}$$

The values of these couplings for the typical models and the corresponding statistical uncertainties for the γ^ℓ_L, $\tilde{U}$, and $\tilde{D}$ couplings are given in Table I.[7] In particular, γ^ℓ_L can be determined very well, primarily due to the small statistical errors for the rare decay modes. On the other hand the quark couplings have larger uncertainties. In Figure 1 90% confidence level ($\Delta\chi^2 = 6.3$) contours[8] are given in a three-dimensional plot for $\tilde{U}$

[7] γ^q_L could be determined [10] by measuring the branching ratio $B(Z' \to q\bar{q})$. See the footnote on the previous page.

[8] The 90% confidence level contours for projections on the more familiar two-dimensional parameter subspaces correspond to $\Delta\chi^2 = 4.6$.

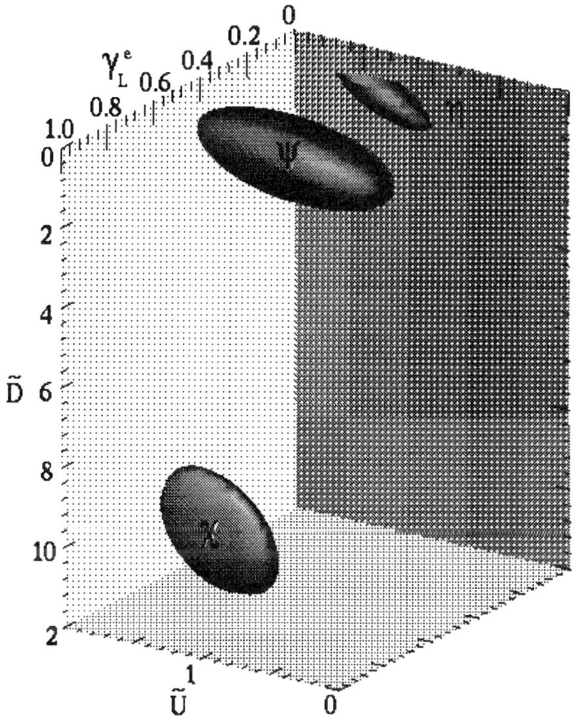

Figure 1. [4] 90% confidence level ($\Delta\chi^2 = 6.3$) contours for the χ, ψ and η models are plotted for $\tilde{U}$, versus $\tilde{D}$, versus γ_L^ℓ. The input data are for $M_{Z'} = 1$ TeV at the LHC ($\sqrt{s} = 16$ TeV and $\mathcal{L}_{int} = 100\,\mathrm{fb}^{-1}$) and include statistical errors only.

versus $\tilde{D}$ versus γ_L^ℓ for the η, ψ and χ models (the LR model is in a different region of the parameter space). Note a clear separation between the models.

For $M_{Z'} \simeq 2$ TeV a reasonable discrimination between models and determination of the couplings may still be possible, primarily from the forward-backward asymmetry and the rapidity ratio. However, for $M_{Z'} \simeq 3$ TeV there is little ability to discriminate between models.

Determination of gauge couplings at the NLC

We assume the c.m. energy $\sqrt{s} = 500$ GeV and the integrated luminosity $\mathcal{L}_{int} = 10$ fb^{-1}. For probes (1-4) we use the exact tree level expressions and assume 100% efficiency for the heavy flavor tagging (probes (3-4)) and 100% longitudinal polarization of the initial electron beam for probes (2) and (4). Only statistical uncertainties are taken into account. If a new Z' is known to exist, a realistic fit should include full radiative

Table 2. [4] The value of the couplings for typical models and statistical error-bars as determined from the probes (1-4) at the NLC (c.m. energy $\sqrt{s} = 500$ GeV and integrated luminosity $\mathcal{L}_{int} = 20\,\text{fb}^{-1}$). $M_{Z'} = 1$ TeV. 100% heavy flavor tagging efficiency and 100% longitudinal polarization of the electron beam is assumed for the first set of error bars, while the error bars in parentheses are for the probes without polarization.

	χ	ψ	η	LR
P_V^ℓ	$2.0 \pm 0.08\,(0.26)$	$0.0 \pm 0.04\,(1.5)$	$-3.0 \pm 0.5\,(1.1)$	$-0.15 \pm 0.018\,(0.072)$
P_L^q	$-0.5 \pm 0.04\,(0.10)$	$0.5 \pm 0.10\,(0.2)$	$2.0 \pm 0.3\,(1.1)$	$-0.14 \pm 0.037\,(0.07)$
P_R^u	$-1.0 \pm 0.15\,(0.19)$	$-1.0 \pm 0.11\,(1.2)$	$-1.0 \pm 0.15\,(0.24)$	$-6.0 \pm 1.4\,(3.3)$
P_R^d	$3.0 \pm 0.24\,(0.51)$	$-1.0 \pm 0.21\,(2.8)$	$0.5 \pm 0.09\,(0.48)$	$8.0 \pm 1.9\,(4.1)$
ϵ_A	$0.071 \pm 0.005\,(0.018)$	$0.121 \pm 0.017\,(0.02)$	$0.012 \pm 0.003\,(0.009)$	$0.255 \pm 0.016\,(0.018)$

corrections, true experimental cuts and detector acceptances, which are expected to increase the uncertainties.

Because the photon couplings are vector-like and the ℓ couplings to Z have the property $\hat{g}_{L1}^\ell \simeq -\hat{g}_{R1}^\ell$ it turns out that probes (1-4) single out the Z' leptonic couplings primarily in the combinations $\hat{g}_{L2}^\ell \pm \hat{g}_{R2}^\ell$. We therefore chose the $\hat{g}_{L2}^\ell - \hat{g}_{R2}^\ell$ combination, which turns out to be a convenient choice for the typical models used in the analysis, to "normalize" the four charges in the following way:

$$P_V^\ell = \frac{\hat{g}_{L2}^\ell + \hat{g}_{R2}^\ell}{\hat{g}_{L2}^\ell - \hat{g}_{R2}^\ell}, \quad P_L^q = \frac{\hat{g}_{L2}^q}{\hat{g}_{L2}^\ell - \hat{g}_{R2}^\ell}, \quad P_R^{u,d} = \frac{\hat{g}_{R2}^{u,d}}{\hat{g}_{L2}^q} \,. \tag{6}$$

In addition, the probes (1-4) are sensitive to the following ratio of an overall gauge coupling strength divided by the "reduced" Z' propagator:

$$\epsilon_A = (\hat{g}_{L2}^\ell - \hat{g}_{R2}^\ell)^2 \frac{g_2^2}{4\pi\alpha} \frac{s}{M_{Z'}^2 - s} \,. \tag{7}$$

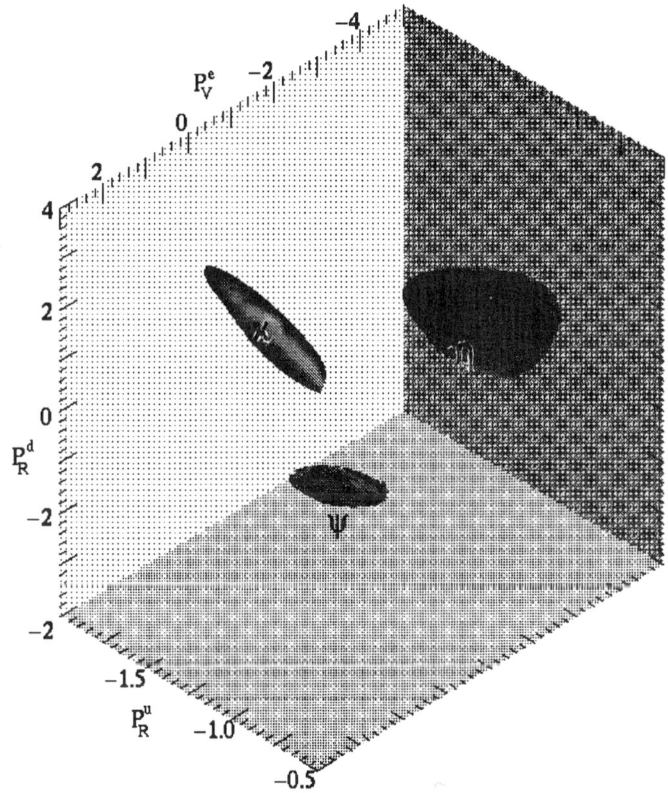

Figure 2. [4] 90% confidence level ($\Delta\chi^2 = 6.3$) regions for the χ, ψ, and η models with $M_{Z'} = 1$ TeV are plotted for P^u_R versus P^d_R versus P^ℓ_V at the NLC ($\sqrt{s} = 500$ GeV, $\mathcal{L}_{int} = 20$ fb^{-1}). Only statistical uncertainties are included.

Here α is the fine structure constant. Note that couplings 5 probed by the LHC, do not determine couplings 6 unambiguously. In particular, determination of γ_L^ℓ, $\tilde{U}$ and $\tilde{D}$ at the LHC would yield an eight-fold ambiguity for the corresponding three couplings in 6.

Statistical uncertainties for couplings (6-7) are given in Table II. The Z' charges can be determined to around $10-20\%$. Poor determination of couplings for the η model is related to the small value of ϵ_A in this case. Note that relative error bars are typically by a factor of ~ 2 bigger than the corresponding ones at the LHC. Without polarization the error bars increase by a factor $2-10$, and thus yield only marginal information about the couplings. The ψ model has particularly poorly determined couplings without polarization. In Fig. 2, 90% confidence level ($\Delta\chi^2 = 6.3$) contours are given in a three dimensional plot of P_R^u versus P_R^d versus P_V^ℓ for the χ, ψ and η models (the LR model is in a different region of the parameter space).

In the case of smaller, say, 25%, heavy flavor tagging efficiency and in the case that the electron beam polarization is reduced to, say, 50%, the determination of the couplings is poorer, however still useful. For $M_{Z'} \sim 2$ TeV, the uncertainties for the couplings in the typical models are too large to discriminate between models.

Conclusions

The analysis demonstrates complementarity between the NLC and the LHC colliders, which in conjunction allow for determination of the $M_{Z'}$, an overall Z' gauge coupling strength as well as a unique determination of *all* the quark and lepton charges with sufficiently small error bars, provided $M_{Z'} \lesssim 1$ TeV.

Acknowledgment

I would like to thank F. del Aguila and P. Langacker for enjoyable collaboration. The work was supported by the Department of Energy Grant # DE-AC02-76-ERO-3071, and the Texas National Research Commission Laboratory.

References

[1] M. Cvetič, F. del Aguila, and P. Langacker, Proceedings of the Workshop on *Physics and Experiments at Linear e^+e^- Colliders*, Waikoloa, Hawaii, April 26-30, 1993, F. Harris et al. eds. (World Scientific 1993), p. 490.

[2] A. Djouadi, A. Leike, T. Riemann, D. Schaile and C. Verzegnassi, in the Proceedings of the *Workshop on Physics and Experiments with Linear e^+e^- Colliders*, September 1991, Saariselkeä, Finland, R. Orava ed., Vol. II, p. 515; A. Djouadi, A. Leike, T. Riemann, D. Schaile and C. Verzegnassi, *Z. Phys.* **C56** (1992) 289.

[3] J. Hewett and T. Rizzo, in the Proceedings of the *Workshop on Physics and Experiments with Linear e^+e^- Colliders*, September 1991, Saariselkeä, Finland, R. Orava ed., Vol. II, p. 489; *ibidem* p. 501.

[4] F. del Aguila and M. Cvetič, U. Pennsylvania preprint, UPR-590-T (December 1993) Phys. Rev.**D** in press.

[5] A. Leike, DESY preprint DESY 91-154 (November 1993), to be published in Z. Phys. C.

[6] P. Langacker, R. Robinett, and J. Rosner, Phys. Rev. **D30**, 1470 (1984).

[7] F. del Aguila, M. Cvetič, and P. Langacker, Phys. Rev. **D48**, R969 (1993).

[8] A. Fiandrino and P. Taxil, Phys. Rev. **D44**, 3409 (1991) and Phys. Lett. **B292**, 242 (1992).

[9] J. Anderson, M. Austern, and B. Cahn, Phys. Rev. Lett. **69**, 25 (1992) and Phys. Rev. **D46**, 290 (1992).

[10] M. Cvetič and P. Langacker, Phys. Rev. **D46**, 4943 (1992).

[11] A. Henriques and L. Poggioli, ATLAS Collaboration, Note PHYS-NO-010 (October 1992); T. Rizzo, ANL-HEP-PR-93-18 (March 1993).

[12] P. Mohapatra, Mod. Phys. Lett. **A8**, 771 (1993).

[13] T. Rizzo, Phys. Lett. **B192**, 125 (1987).

[14] M. Cvetič and P. Langacker, Phys. Rev. **D46**, R14 (1992).

[15] T. Rizzo, Phys. Rev. **D47**, 965 (1993).

[16] F. del Aguila, B. Alles, L. Ametller and A. Grau, Phys. Rev. **D48**, 425 (1993).

[17] J. Hewett and T. Rizzo, Phys. Rev. **D47**, 4891 (1993).

[18] LEP Collaborations, CERN preprint, CERN/PPE/93-157 (August 1993).

[19] P. Langacker and M. Luo, Phys. Rev. **D44**, 817 (1991).

SECTION IV

ASPECTS OF PARTICLE PHYSICS

SIBERIAN SNAKES AND POLARIZED BEAMS

Lazarus G. Ratner

Department of Physics
University of Michigan
Ann Arbor, MI 48109-1120

Polarized proton beams have been accelerated up to 22 GeV in circular accelerators[1], but the techniques used to overcome depolarizing resonances and preserve polarization during acceleration have been difficult and time consuming.[2] A spin 1/2 proton precesses around the magnetic field in the accelerator by an angle $\theta_p = G\gamma\theta_D$ where θ_D is the deflection angle produced by the accelerator field and G is the anomalous magnetic moment of the proton. In a purely vertical magnetic field, the vertical polarization would remain vertical; but all practical accelerators have radial field components, which rotate the spin about a radial axis. When these radial fields occur at a frequency which is in resonance with the spin precession frequency, they add in phase and depolarize the beam. Note that the number of spin precessions in one turn around the accelerator is $G\gamma$. There are two main sources of radial fields: (1) the focussing fields needed to keep the particles in the accelerator and (2) the imperfection fields which cause vertical closed-orbit errors due to the impracticability of having a mid-plane accurate to better than 0.01 mm.

The first resonance type is called an intrinsic resonance; these occur when $G\gamma = nP \pm \nu_y$ where n is an integer, P is the periodicity of the accelerator, and ν_y is the vertical betatron tune. The second type is called an imperfection resonance; these occur when $G\gamma = k$ where k is an integer. At the Alternating Gradient Synchrotron (AGS), pulsed quadrupoles and dipoles were installed to correct these resonances, as shown in Figure 1. One corrected these resonances by making a fast shift in ν_y to avoid each intrinsic resonance and by using a set of dipoles to cancel the horizontal magnetic fields introduced by the magnet misalignments. Unfortunately with all these corrections, it took about two weeks to reach 20 GeV and even the two hardworking tuners, shown in Figure 2, would have a tough time to tune for a 100 weeks to reach 1 TeV.

---

* Work performed under the auspices of the U.S. Department of Energy.

Obviously, new techniques must be used to go much above 20 GeV and fortunately, a new scheme was invented by Ya. S. Derbenev and A. M. Kondratenko[3] of the Institute of Nuclear Physics in Novosibirsk, Siberia. It was named a "Siberian Snake"; as shown in Figure 3 it consists of a configuration of horizontal and vertical transverse dipole magnets, which maintain the spin orientation of a polarized beam, by a series of vertical and horizontal bends. The up, down, and sideways bends, which also preserve the orbit trajectory, give the impression of sinuous motion which led to the "snake" appellation. The net effect of a snake is to rotate the spin by 180° around a horizontal direction while preserving the trajectory. If we had two Siberian snakes installed 180° apart in a circular accelerator with one rotating the

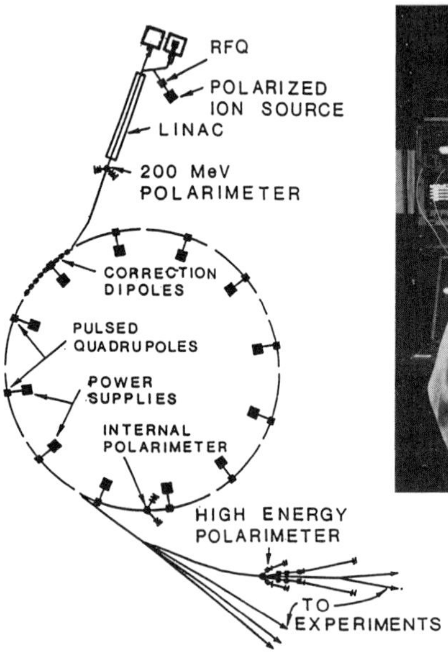

Figure 1. Alternating Gradient Synchrotron showing special hardware needed for accelerating polarized protons.

Figure 2. A.D. Krisch and L.G. Ratner tuning the AGS imperfection and intrinsic resonance corrections.

spin about the longitudinal axis (Type I snake) and the other around the radial axis (Type II snake), the vertical polarization would be stable with the spin up in half the ring and down in the other half. Any depolarization from horizontal fields would be reversed after one revolution due to the snake's spin rotation; the same horizontal field would then cancel the effect from the previous turn. Thus, with the Siberian snake, the spin tune, ν_{sp}, which is the number of spin rotations per revolution would be 1/2, since the tranverse polarization reverses direction after one turn. Therefore, the condition $G\gamma$ = integer or $G\gamma$ = integer $\pm\ \nu_y$ is never satisfied except at $\nu_y = 1/2$ which cannot exist in an accelerator since this causes a betatron resonance which blows up the beam.

Figure 4 shows the spin vector components' motion due to a Type I snake.[4] The vertical, z, and horizontal, x, components repeat every other turn, but the stable spin direction is along the y component, which repeats every turn.

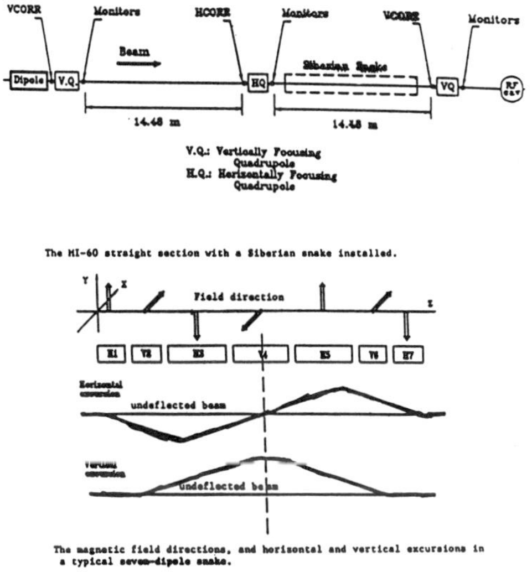

Figure 3. Siberian snake schematic from the FNAL Main Injector Proposal.

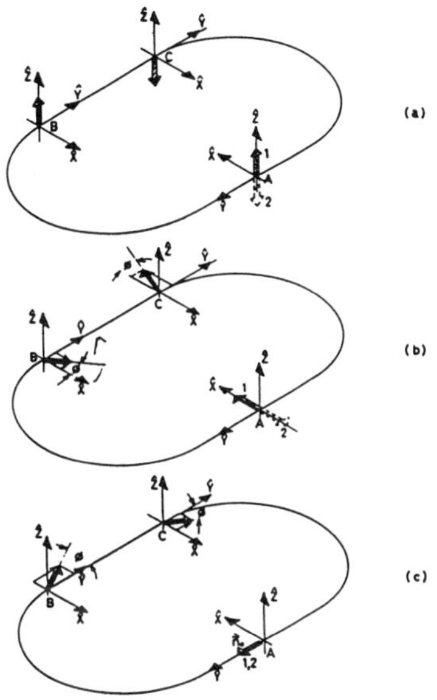

Siberian snake type 1; evolution of spin-vector components over one revolution starting at A. The $\hat{y}$-component at A in (c) is the closed solution n_0.

Figure 4. Action of a type 1 Siberian snake.

The Siberian snake concept seemed very attractive and initial tests were devised to experimentally test the concept at the Indiana University Cyclotron Facility (IUCF). We devised a simpler scheme than two snakes, which are each a sequence of dipoles; we instead used a single solenoid snake to rotate the spin; the stable spin direction (or precession direction) is then in the horizontal plane and depends on energy and location. A solenoid can rotate the spin by about 180° around the longitudinal direction; thus a solenoid is a Type I snake. Such a solenoid was put into the IUCF Cooler Ring[5] and tests[6–11] were made to determine if indeed the polarization of the circulating beam was preserved. Figure 5 shows the calculated stable polarization direction at 108 MeV in the IUCF Cooler Ring.

In Figure 6, we varied the Cooler's net longitudinal magnetic field, by producing a longitudinal imperfection field which varied the resonance strength while keeping the Cooler energy constant. Without a snake, near the $G\gamma = 2$ imperfection resonance, full polarization can only be maintained when the imperfection fields are exactly compensated. There are also clear dips on both sides of the central peak due to synchrotron depolarizing resonances; these are another type of depolarizing resonance; they occur when $\nu_{sp} = 2 \pm \nu_s$ where ν_s is the synchrotron tune. To check this, we varied the rf voltage which changes ν_s; we indeed found the proper correlation between ν_s and ν_{sp}. With the snake turned on, notice that there is no depolarization from any of the resonances.

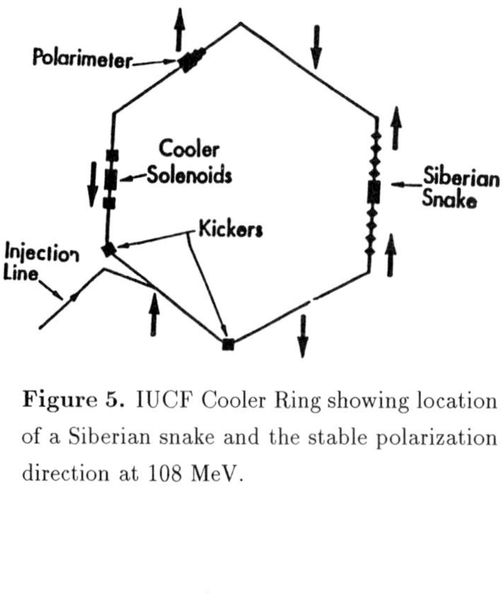

Figure 5. IUCF Cooler Ring showing location of a Siberian snake and the stable polarization direction at 108 MeV.

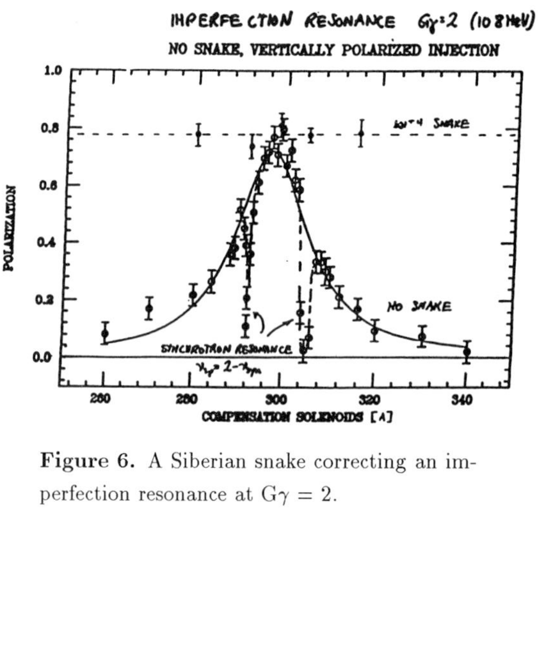

Figure 6. A Siberian snake correcting an imperfection resonance at $G\gamma = 2$.

Similarly, we find that the Siberian snake could correct the intrinsic depolarizing resonance, $G\gamma = \nu_y - .3$, as shown in Figure 7. With the snake off, the beam was totally depolarized at the resonant ν_y of 5.13. We also studied other snake properties; we found that with an rf solenoid, one can flip the spin by passing through a resonant frequency. This spin-flipping would strongly discriminate against systematic errors in a stored polarized beam scattering experiment. We also used the rf solenoid to overlap an rf induced depolarizing resonance with the $G\gamma = 2$ resonance; we found significant interactions which disappeared when the snake was turned on.

The Siberian snake eliminated all observable depolarization in all these cases. Therefore it is of interest to ask whether a less than 180° spin rotation might eliminate weaker depolarizing resonances. A partial Siberian snake, which would rotate the spin by less than 180°, would be very useful in lower energy accelerators ($\leq$ 50 GeV) because a partial snake would reduce orbit excursions and x-y coupling. For a Siberian snake of spin rotation angle δ, the spin tune, ν_s, can be written

$$cos(\pi\nu_s) = cos(\pi G\gamma)cos\left(\frac{\delta}{2}\right)$$

We see that for a full snake, where $\delta = 180°$, the quantity $\cos(\pi\nu_s) = 0$; this implies that $\nu_s = 1/2$. While for no snake, $\delta = 0°$; thus $\cos(\pi\nu_s) = \cos(\pi G\gamma)$. One can show that when δ is less than 180°, a partial Siberian snake generates spin flip at every integer spin resonance and thus produces a polarized beam, either up or down, depending on the number of resonances transversed. In the presence of a resonance of strength ϵ, the condition for spin flip is $\delta \gg 2\pi|\epsilon| + \sqrt{\delta\pi\alpha}$, where α

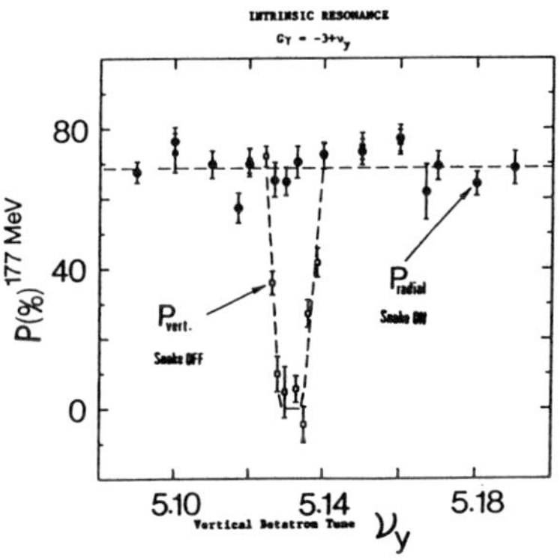

Figure 7. Correcting an intrinsic resonance at $G\gamma = -3 + \nu_y$.

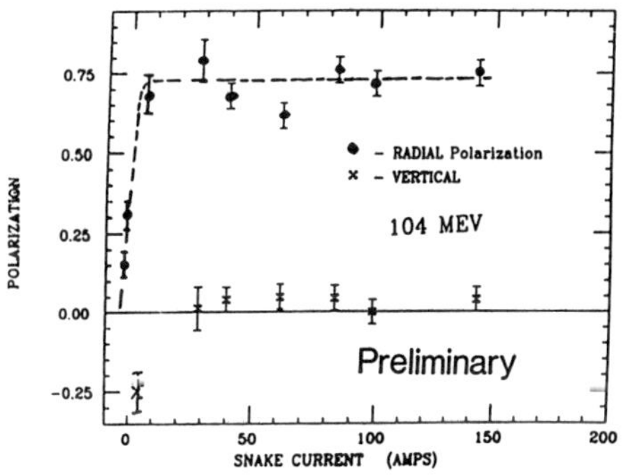

Figure 8. A preliminary look at the action of a partial Siberian snake.

is the crossing speed of the accelerating beam. For example at the AGS, where $\epsilon \approx 10^{-2}$ and $\alpha \approx 4 \times 10^{-5}$, $\delta \gg 4°$ would ensure complete spin flip. The data taken at the IUCF Cooler Ring give some evidence for a partial snake as shown in Figure 8. We have constructed a 9°, 4.7 T·m solenoid (Figure 9) for the AGS as part of a long-range program to provide a polarized beam for RHIC (Relativistic Heavy Ion Collider).[12] If this snake works, we should be able to tune up the AGS for polarized beam in a few days rather than two weeks and use it as an injector for RHIC.

To make RHIC and the FNAL Tevatron-collider operate with polarized protons requires Siberian snakes and partial snakes in the preinjectors and in the colliders. Now we will consider the requirements for these colliders. Since we have already mentioned RHIC, let us continue with it before we take up the FNAL collider. Hopefully, both of these facilities will be built in the 20th century, but I do not expect to see polarization physics until the 21st century. Figure 10 shows the Brookhaven complex with the elements needed for polarized beams in RHIC. The polarized H^- source, Linac, Booster, 200 MeV polarimeter, partial Siberian snake, AGS, pulsed quadrupoles, and the AGS polarimeter exist. Over the next few years, tests to produce an acceptable beam for RHIC will go on. Figure 11 is the latest layout of RHIC showing the STAR and PHENIX detectors, polarimeters, Siberian snakes, and spin rotators to provide longitudinally polarized beam for the experiments. Reference 5 gives details of the RHIC complex, as well as a discussion of the physics that can be done with polarized beams. The RHIC spin collaboration, a group with some 24 international institutions, is actively pursuing the various aspects of making RHIC a polarized beam collider.

Figure 9. 4.7 T-M solenoid snake at the AGS.

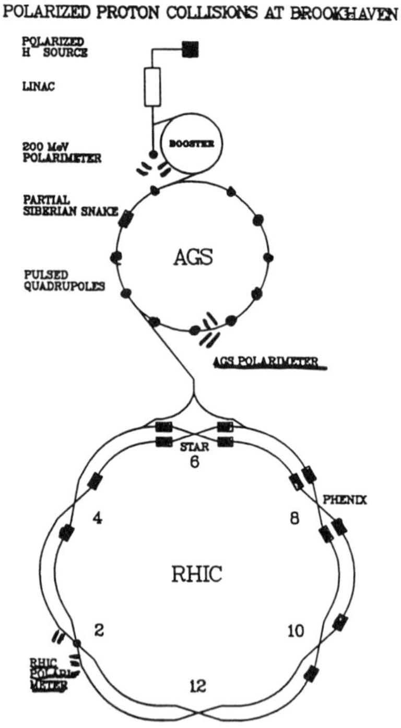

Figure 10. Brookhaven complex with the elements needed for polarized beams in RHIC.

For FNAL, the University of Michigan is heading an international collaboration of some eight institutions funded by FNAL for producing the necessary studies. They have already produced a report on accelerating polarized protons to 120 and 150 GeV in the FNAL Main Injector,[13] which was submitted to FNAL in March of 1992. They are now preparing a report which will outline the necessary items to make the Tevatron a polarized beam facility. There is a difference between FNAL and RHIC, in that the latter will collide polarized beams, the former being a $\bar{p}$ p collider will only have polarized protons. However, the use of a polarized jet target would allow many interesting experiments of $p\uparrow + p\uparrow$ up to 1 TeV and certainly $\bar{p} + p\uparrow$ at 2 TeV collider energies could lead to new physics. Figure 12 shows the FNAL layout showing the necessary additions for polarized beam at 150 GeV.

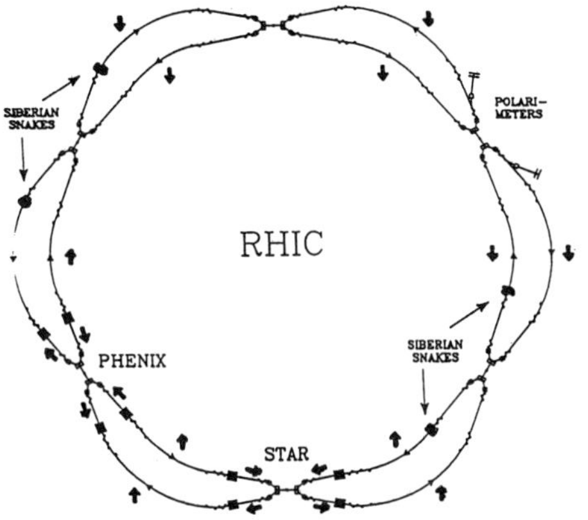

Figure 11. RHIC showing detectors and snakes.

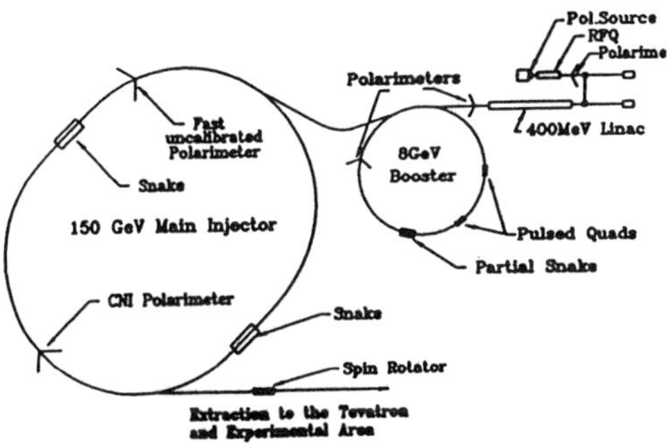

Figure 12. Hardware for the polarized beam in the FNAL Main Injector.

Meanwhile, polarized beams in the GeV range are available[2] at Saturne in Saclay, at 3 GeV; KEK in Japan, at 12 GeV; and have been available[1] in the past at the Brookhaven AGS up to 22 GeV. The new proposed facilities will take us from the present energies of less than 10 GeV c.m. to 400 GeV or 2 TeV c.m. Exploring spin at these high energies may lead us to a new picture of fundamental interactions, just as in the past spin led to new knowledge and new concepts in physics. With the demise of the SSC, this may be the only window where new physics will appear.

References

1. F. Z. Khiari, et al., *Phys. Rev.* D39:45 (1989).
2. T. Khoe, et al., *Part. Accel.* 6:213 (1975); J.L. Laclare, et al., *J. Phys. (Paris), Colloq.* 46:C2-499 (1985); H. Sato, et al., *Nucl. Instrum. Methods Phys. Res.* Sect. A272:617 (1988).

3. Ya.S. Derbenev and A.M. Kondratenko, *Part. Accel.* 8:115 (1978).
4. B.W. Montague, *Physics Reports.* 113:71 (1984).
5. R.E. Pollock, *IEEE Trans. Nucl. Sci.* NS-30:2056 (1983).
6. A.D. Krisch, et al., *Phys Rev. Lett.* 63:1137 (1989).
7. J.E. Goodwin, et al., *Phys. Rev. Lett.* 64:2779 (1990).
8. M.G. Minty, et al., *Phys. Rev.* D44:R1361 (1991).
9. V.A. Anferov, et al., *Phys. Rev.* A46:R7383 (1992).
10. R. Baiod, et al., *Phys. Rev. Lett.* 70:2557 (1993).
11. R.A. Phelps, et al., *Phys. Rev. Lett.* 72:1479 (1994).
12. Proposal on Spin Physics Using the RHIC Polarized Collider. Unpublished Proposal by the RHIC Spin Collaboration (RSC) to Brookhaven National Laboratory (Sept. 1993).
13. Acceleration of Polarized Protons to 120 and 150 GeV in the Fermilab Main Injector, Spin Collaboration Unpub., Univ. of Michigan Report (March 1992).

PROGRESS IN NEUTRINO PHYSICS A SURVEY OF EXPERIMENTS AND THEORY

S.L. Mintz
Physics Department
Florida International University
Miami, Florida 33199

M. Pourkaviani
Massachusetts Institute of Technology
Cambridge, MA 02139

INTRODUCTION

Recently a host of new experiments involving low energy neutrino reactions in nuclei are in progress or being planned. These are being undertaken at LAMPF and by the KARMEN collaboration at the Rutherford-Appleton Laboratory. Part of this resurgence of interest lies in the fact LAMPF faces a possible shutdown so that it may not be possible to undertake many of these experiments in for the foreseeable future if they are not done now. However the greatest impetus to these experiments stems from a desire to clear up the discrepancies which exist in both theory and experiment in this area. In particular, without a full understanding of these processes it is not possible to obtain and interpret solar neutrino data with any degree of reliability.

Before pursuing these topics, we distinguish between two types of neutrino reactions,exclusive and inclusive. Exclusive reactions are reactions to a specified final state. A typical reaction of this type which has been in fact measured is $\nu_e + {}^{12}C \longrightarrow e^- + {}^{12}N_{(g.s.)}$. Inclusive reactions leave the final state unspecified. An example of one which has also been measured is $\nu_e + {}^{12}C \longrightarrow e^- + X$.At present experiment and theory seem to be in reasonable agreement for inclusive reactions for the one case which has been measured1, 2 namely the former reaction given above. However the situation is very different for the inclusive case.

The first inclusive measurement was done for the reaction, $\nu_e + {}^{12}C \longrightarrow e^- + X$, by the Los Alamos[1] group and began in 1987. This experiment made use of electron neutrinos obtained from stopped muons which were originally produced from π^+ decay. The resulting electron neutrino spectrum is the Michel spectrum which peaks at about 39 MeV and for which the maximum neutrino energy is 52.8 MeV. Their detector was a tracking calorimeter made of modules of plastic scintillation counters with a pre-trigger system to eliminate cosmic ray muon events and a post-trigger system to identify the ${}^{12}N_{(gs)}$ which undergoes positron beta decay to ${}^{12}C$. They obtained a spectrum averaged result of:

$$< \sigma_c >= (1.41 \pm 0.16(stat) \pm 0.16(sys)) \times 10^{-41}\ cm^2. \qquad (1)$$

Unified Symmetry: In the Small and in the Large,
Edited by B.N. Kursunoglu *et al.*, Plenum Press, New York, 1995

From this they obtained the the averaged cross section for the reaction, $\nu_e + {}^{12}C \longrightarrow e^- + {}^{12}N^*$, to the excited states of nitrogen given by:

$$< \sigma_c >= (3.6 \pm 20\%) \times 10^{-42}\ cm^2. \tag{2}$$

where the subscript c on σ_c denotes a charged current reaction. At the time there was only one theoretical calculation due to Donnelly which yielded a result of:

$$< \sigma_c >= 3.7 \times 10^{-42}\ cm^2 \tag{3}$$

which seemed in good agreement with experiment.

The KARMEN group performed essentially the same experiment as the Los Alamos group did. They also used neutrinos produced from muon decay at rest,i.e. Michel spectrum electron neutrinos. However they were able to measure the exclusive neutral current reaction,$\nu_e + {}^{12}C \longrightarrow \nu_e + {}^{12}C^*$, as well as the reactions measured at LAMPF. They used a high resolution 56 ton liquid scintillator detector which had a 10^{-4} duty factor which allowed suppression of cosmic ray background. The KARMEN collaboration is a long term collaboration. They began taking data in 1989 and intend to continue for several more years. The most recent value for the reaction, $\nu_e + {}^{12}C \longrightarrow e^- + {}^{12}N^*$, that they have announced[4] is:

$$< \sigma_c >= (8.6 \pm 1.3(stat) \pm 1.2(sys)) \times 10^{-42}\ cm^2. \tag{4}$$

This is almost three times the result obtained at Los Alamos. Thus a serious discrepancy exists which must be explained.

Meanwhile theoretical calculations have kept pace with experimental results. A shell model based calculation[5] making use of a random phase approximation yielded a result for the averaged cross section for $\nu_e + {}^{12}C \longrightarrow e^- + {}^{12}N^*$ of:

$$< \sigma_c >= 6.4 \times 10^{-42}\ cm^2. \tag{5}$$

In addition, a tensor model calculation which we shall describe in detail[6] yields a value of:

$$< \sigma_c >= 9.77 \times 10^{-42}\ cm^2. \tag{6}$$

Thus the theoretical situation is about as murky as the experimental one.

These discrepancies have as mentioned, given rise to a number of new experimental efforts. Several of these efforts are concerned with the nucleus,${}^{127}I$. This nucleus has several features which might make it a suitable detector of solar neutrinos. We discuss these features and the new experiments in the next section of this paper.

NEW EXPERIMENTS

The nucleus, ${}^{127}I$, has a number of features of interest. First it has its giant dipole resonance at a relatively low energy with a center at approximately 15 MeV. The resonance is broad enough that higher energy neutrinos from the reaction:

$${}^8B \longrightarrow {}^8Be + e^+ + \nu_e \tag{7}$$

which is responsible for the largest flux of solar neutrinos in the range of energy from 2 MeV to 10 MeV, might have an enhanced reaction rate with an ${}^{127}I$ detector. Furthermore ${}^{127}I$ begins to emit nucleons for reactions with neutrinos having energies above 7.2 MeV. Finally ^{127}I has a large neutron excess,21, which should also drive up reaction rates. In addition it can be combined with sodium and lithium to make suitable targets. The desirability of constructing better solar neutrino detectors is very clear because in most of the presently used detectors, the target nuclei can be excited to many states some of which decay to the final measured states and some which do not. This makes theoretical calculations very difficult. In fact, no neutrino detector has ever been calibrated with a terrestrial neutrino source. Instead theoretical calculations have been used to predict the expected result and it is increasingly clear that these may not be reliable.

The same features which make ^{127}I a potential candidate for a solar neutrino experiment make it an interesting target for a Michel spectrum experiment. The giant dipole resonance can be excited by a very large fraction of the neutrino flux. This simplifies the theoretical analysis since closure may be invoked with more justification than for the ^{12}C over this same energy range. Also for the same reasons as previously discussed cross sections such be much higher than for the ^{12}C case of the earlier experiments.

One experiment[6] is already in progress at Los Alamos. This experiment is for the reaction $\nu_e +^{127} I \longrightarrow e^- + X$ and averages the cross section over the Michel spectrum for electron neutrinos. Sodium iodide is disolved in water and exposed to the neutrino source. The interaction of ^{127}I with the neutrinos produce ^{127}Xe gas which is collected from the water having been swept out by a flow of helium gas. The gas is then purified and put into a gas proportional counter. The xenon then decays via orbital electron capture back into ^{127}I. The group is currently beginning the construction of a 100 ton iodine detector at Homestake. This target will have approximately 5×10^{29} target atoms of iodine (about one quarter of the number or chlorine atoms in the present detector). Because this detector will be capable of separating daytime events from night events, it will be able to look for neutrino oscillation caused by the earth. When the current calibration at Los Alamos is complete, the group hopes to use a ^{37}Ar source producing 814 KeV neutrinos to calibrate an ^{127}I detector at an energy close to that from ^{7}Be. This would increase the usefulness of an iodine detector greatly as it is very difficult to undertake reliable calculations in this range. This group expects to announce preliminary results very soon. The KARMEN collaboration[7] is also considering a Michel spectrum measurement for ^{127}I. At present there are discussions between these two groups for a differential cross section experiment for the reaction, $\nu_e +^{127} I \longrightarrow e^- + X$, at low energies. This would completely calibrate a solar neutrino detector but will take years to finish.

Another low energy detector is being readied by a group at the university of Maryland[8], this detector is meant for solar neutrinos and makes use of $^{7}Li^{127}I$. Actually the detector makes use of unseparated or ^{6}Li depleted lithium. Natural lithium is about 7.5 percent ^{6}Li and about 92.5 percent ^{7}Li. Iodine is of course virtually 100 percent ^{127}I. This particular target is Europium activated and both the nuclei and electrons are targets. Charged current reactions can occur as:

$$\nu_e +^7 Li \longrightarrow^7 Be + e^-$$

or as:

$$\nu_e +^{127} I \longrightarrow^{127} Xe + e^-.$$

In addition a corresponding set of neutral current reactions:

$$\nu +^7 Li \longrightarrow^7 Li + \nu$$

and

$$\nu +^{127} I \longrightarrow^{127} I + \nu$$

occur where ν is any neutrino and the final state nucleus may be a ground state or excited state. Finally there are reactions of the kind:

$$\nu + e^- \longrightarrow \nu' + e'^{\,-}.$$

These reactions contribute to the cross section which is reasonably large. The target is expected to be around 33 tons in mass which is one to two orders of magnitude lighter than current detectors. This target would give rise to approximately 2 charged current events per day,about 1 neutral current reaction per day, and approximately 40 electron scatterings per day. It is to be hoped that this detector will soon be ready for service.

Finally, another ^{12}C experiment[9] is underway at Los Alamos. This experiment makes use of the LSND at LAMPF but unlike the experiments underway on ^{127}I and unlike the earlier ^{12}C experiments, this is not over the Michel spectrum. This particular experiment makes use of a spectrum of muon neutrinos from π^+ decays in flight. This gives rise to a spectrum with energies from 10 MeV to 250 MeV with a peak at about 70 MeV. However is still large at 140 MeV, well above the muon production threshold, so that the reaction:

$$\nu_\mu + {}^{12}C \longrightarrow \mu^- + X \tag{8}$$

may be observed. This group has already collected some data and is at present still running.

THE THEORETICAL SITUATION

The theoretical situation is presently as noted unclear. In the ^{12}C cases cited earlier, two shell model calculations, differed by almost a factor of two, the more recent which made use of a random phase approximation being the larger of the two. In addition,a tensor based model gave an even larger result. For a range of energy suitable for reaction,Eq.(8), there exists a Fermi gas model calculation which also makes use of a random phase approximation. This model was developed by Vogel, Langanke, and Reeder. Finally there are two other models, one by Mintz and Pourkaviani[10], the other by Kim and Mintz[11]. The former makes use of tensor to describe the hadronic part of the weak current, the terms of which are phenomenologically determined. The latter is a closure approximation treatment. Both models make of total muon capture results as input into the models. We discuss these models in what follows.

We consider as an example the reaction, $\nu_e +^{127} I \longrightarrow e^- + X$. The starting point for both methods is the matrix element:

$$M_{ki} = \frac{G}{\sqrt{2}} \, cos\theta \bar{u}_e \gamma^\lambda (1-\gamma_5) u_\nu < k | J^\dagger_\lambda(0) |^{127} I > \tag{9}$$

where k is a particular final state and:

$$J^\dagger_\mu(0) = V^\dagger_\mu(0) - A^\dagger_\mu(0) \tag{10}$$

We shall only outline the process by which we obtain the cross section as the details have already appeared in the literature[10,11]. We simply note that the cross section may be written as:

$$\sigma_c = \sum_k \frac{m_\nu}{2ME_\nu} \int d^3P_e \, |M_{ki}|^2 \, \frac{m_e}{E_e(2\pi)^3} \frac{d^3P_k}{2E_k(2\pi)^3} (2\pi)^4 \delta^4(P_k + P_e - P_\nu - P_i). \tag{11}$$

The quantity $|M_{ki}|^2$ is written as:

$$|M_{ki}|^2 = \frac{G^2 \cos^2 \theta_C}{2m_\nu m_\nu} L^{\sigma\lambda} < k|J_\sigma^\dagger(0)|^{12}C >< k|J_\lambda^\dagger(0)|^{127}I >^* . \tag{12}$$

The quantity, $L^{\sigma\lambda}$, is the lepton tensor appropriate to this process and is given by:

$$L^{\sigma\lambda} = p_e^\sigma p_\nu^\lambda - p_e \cdot p_\nu g^{\sigma\lambda} + p_\nu^\sigma p_e^\lambda - \epsilon^{\alpha\sigma\beta\lambda} p_{e\alpha} p_{\nu\beta}. \tag{13}$$

In order to work with average quantities, we assume an average nuclear excitation of δ given by:

$$M_x - M_i = \delta \tag{14}$$

where δ is clearly a function of the incoming neutrino energy and must be determined. We also assume on the basis of our knowledge of individual states that the interaction is largely in the forward direction and that:

$$< E_e > \simeq E_\nu - \delta. \tag{15}$$

and:

$$< \vec{p}_e > \simeq \sqrt{(E_\nu - \delta)^2 - m_e^2}. \tag{16}$$

Although the value of δ, will vary with incoming neutrino energy, above the giant dipole resonance at 15 MeV it should increase slowly in our region of interest. This enables us to easily obtain $< E_e >$ and $< \vec{p}_e >$ over a large part of the range of neutrino energy of interest. We can, using these averages obtain the quantity:

$$\sigma_c = \frac{G^2 cos^2 \theta_C}{2ME_\nu} \int d\Omega_e \sum_k < k|J_\sigma^\dagger(0)|^{12}C >< k|J_\lambda^\dagger(0)|^{127}I >^* L^{\sigma\lambda} \times \frac{< |\vec{p_e}| >}{2M - 2E_e + 2E_e \cos\theta_e \frac{<E_e>}{<|\vec{p_e}|>}}. \tag{17}$$

The hadronic part of eq.(17) may be replaced as follows:

$$< k|J_\sigma^\dagger(0)|^{12}C >< k|J_\lambda^\dagger(0)|^{127}I >^* \equiv Q_{\lambda\sigma}(P_i, < q >) \tag{18}$$

which is a tensor. We have previously shown[11,13] that this tensor may be reduced to the form which contains only two unknown functions.

$$Q^{\mu\nu(3)} = \alpha g^{\mu\nu} + \frac{\beta}{M^2} P_i^\mu P_i^\nu. \tag{19}$$

The cross section is then found to be:

$$\sigma_c = \frac{G^2 \cos^2(\theta_C)}{4\pi} \frac{< |\vec{p}_e| >< E_e > D}{M(M + E_\nu)} \tag{20}$$

where:

$$D = \beta - 2\alpha \tag{21}$$

and an impulse approximation based calculation[11] gives $D(q^2)$ as:

$$D = a_o - b_o q^2. \tag{22}$$

We assume this simple q^2 dependence for D, which we may also write as $D = a_o + b_o|q^2|$.

Thus our result depends upon two parameters,a_o and b_o. Low error total muon capture results are available for the process,$\mu^- +^{127} I \longrightarrow \nu_\mu + X$. However a little thought shows that this is not the muon capture process which is appropriate to determine the the hadronic part of the matrix element $\nu_e +^{127} I \longrightarrow e^- + X$. In the neutrino reaction there is an excess of 21 neutrons all of which serve as targets for the neutrino helping to boost the expected rates. We need a total muon capture rate for a nucleus with 21 excess protons but the same total number of nucleons, as ^{127}I. Such a nucleus would not be stable but we shall see that we can construct a muon capture rate for such a nucleus. We therefore note that by proceeding by a calculation very similar to the neutrino reaction case[10] , we find for the total muon capture rate an expression similar to Eq.(20):

$$\Gamma_{TOT} = \frac{C|\Phi(0)|^2 G^2 \cos^2\theta_C < E_\nu >^2 D}{8\pi M_i(M_i + m_\mu)}. \tag{23}$$

As we remarked earlier, the total muon capture rate is needed for a nucleuswhich has 74 protons,53 neutrons and a total $A = 127$. We may call this nucleus ^{127}W and note that it does not exist. However we may make use of the fact that there is an extremely accurate semi-empirical formula for the total muon capture rate in nuclei[12]. This formula is given by:

$$\Gamma = Z_{eff}^4 G_1[1 + G_2\frac{A}{2Z} - G_3\frac{A-2Z}{2Z} - G_4(\frac{A-Z}{2A} + \frac{A-2Z}{8AZ})] \tag{24}$$

where $G_1 = 261, G_2 = -0.040, G_3 = -0.26$, and $G_4 = 3.24$. This formula, Eq.(24) fits all exisitng to within 20 percent and in most cases is much better. We therefore use Eq.(24) to produce a muon capture rate for ^{127}W. We take the ratio of the rates obtained by Eq.(24) for ^{127}W to that for ^{127}I and set it equal to the expression in terms of D given by Eq.(23). When we do this we find that:

$$4.1 \times \frac{^W Z}{^I Z} = \frac{^W D}{^I D} \tag{25a}$$

or

$$5.7\ ^I D =^W D. \tag{25b}$$

Thus from the total muon capture rate for ^{127}I we are able to proceed. This total muon capture rate is known[9] and is given by $\Gamma_{TOT} = 11.2 \pm .11 \times 10^6\ sec^{-1}$. We also need the correction factor,C, in eq.(15). This as is well known,is given by $C = (\frac{Z_{eff}}{Z})^4$ which for the iodine case case yields:

$$C = .093. \tag{25}$$

We note that even a small error in Z_{eff} will have large consequences but the effective values for Z are believed to be well known. Using $E_\nu \simeq 90.62\ MeV$, we obtain:

$$^I D = 1.244\ \times 10^{12} \tag{26a}$$

or

$$^W D = 7.1\ \times 10^{12}. \tag{26b}$$

We have at this point a value for ${}^{W}D$ at q^2 appropriate for muon capture but we need an additional information to determine ${}^{W}D$ completely. In the case of ${}^{12}C$ we relied on inclusive electron scattering data and some impulse approximation results to fully obtain D. Here this is not possible. However an impulse approximation result[12] yields a value for $\tilde{D} = \frac{D(q^2)}{D(0)}$ given by:

$$\tilde{D} = \frac{[1 - (\frac{(A-Z)}{2A})\delta(\vec{q}^2)]}{[1 - \frac{(A-Z)}{2A}\delta(0)} \tag{27a}$$

where:

$$\delta(\vec{q}^{\,2}) = (\frac{d}{r_o})^3(1 - \frac{\vec{q}^{\,2}d^2}{10}). \tag{27b}$$

and where from Eq.(14) we may write:

$$\tilde{D} = (1 - \frac{b_o}{a_o}q^2) \equiv (1 - b'q^2). \tag{27c}$$

This yields:

$$\tilde{D} = 1 + .1898q^2. \tag{27d}$$

and thus with Eq.(26b) determines ${}^{W}D$. We note that $\delta(\vec{q}^{\,2})$ is unrelated to the δ of Eq.(14). Still $\delta(E_\nu)$ is needed as a function of neutrino energy. Above the giant dipole resonance we expect closure to be applicable and so for $E_\nu > 30MeV$ we set $\delta(E_\nu) = 15\ MeV$. We choose this value from previous experience[11,10] with total muon capture rates where good results are obtained at 15 to 20 MeV above the giant dipole resonance with closure. From our experience with the ${}^{12}C$ case, below 30 MeV we use a decreasing δ given by:

$$\delta(E_\nu) = 1.667 \times 10^{-2}E_\nu^2 - 6.0023 \times 10^{-3}. \tag{28}$$

We have tried several different forms for $\delta(E_\nu)$ and the precise form of the function does not have more than an effect of a few percent on the value of $< \sigma_c >$. We are now able to evaluate the cross section.

We next consider a model[12] developed by Kim and Mintz for treating neutrino reactions in nuclei. This is strictly a closure approximation and should begin to be accurate at about 15 to 20 MeV about the giant dipole resonance which is assumed to saturate the matrix element. We do not include here a full derivation which is already in the literature but simply mention the salient points. As already mentioned, the starting point is Eq.(9). The quantity $S_\mu^\pm$ is defined by:

$$S_{\mu\nu}^{\pm} \equiv \sum_k \int \frac{d\vec{P}_k}{4\pi} < i|Q_\nu^{(+)}(\vec{q})|k >< k|Q_\mu^{(-)}(\vec{q})|i > \tag{29}$$

with:

$$Q_\mu^{(\pm)}(\vec{q}) \equiv \int d\vec{x} J_\mu^{\pm}(\vec{x}, 0)e^{i\vec{x}\cdot\vec{q}}. \tag{30}$$

We assume that low lying states (particularly those associated with the giant dipole resonance) saturate the sum in Eq.(21) so that we may write, applying closure:

$$S_{\mu\nu} =< i|Q_\mu^{(+)}(\vec{q})Q_\nu^{(-)}(\vec{q})|i > . \tag{31}$$

Making use of translational invariance, we can show that the neutrino cross section and the total muon capture rate depend upon essentially the same matrix element:

$$[< i|\vec{Q}^{(+)} \cdot \vec{Q}^{(-)}|i > + < i|Q_o^{(+)} Q_o^{(-)}|i >$$

The matrix elements themselves can be evaluated by making use of an impulse approximation for the weak current, namely:

$$J_\alpha^{(\pm)}(\vec{x}, 0) \simeq \sum_{a=1}^{A} (\Gamma_\alpha)_a \delta(\vec{x} - \vec{r}_a) \tag{32}$$

where:

$$(\Gamma_\alpha)_a = i g_V \delta_{\alpha,4} - (1 - \delta_{\alpha,4} g_A)(\sigma_\alpha)_a. \tag{33}$$

With these definitions we may write:

$$\int \frac{d\Omega}{4\pi} [< i|\vec{Q}^{(+)} \cdot \vec{Q}^{(-)}|i > + < i|Q_o^{(+)} Q_o^{(-)}|i >= Z[g_V^2 + 3g_A^2](1 - \frac{A - Z}{2A}\delta(\vec{q}^2)). \tag{34}$$

This quantity occurs as mentioned above in the expressions for Γ_{TOT} and for σ_c. From Eq.(34) the only difference is the q^2 at which the matrix element is evaluated. As we have produced an artificial value in Eq.(24) for the quantity, Γ_{TOT}, we may write σ_c as:

$$\sigma_c = \frac{2\pi \Gamma_{TOT} E_e p_e}{C_i (\alpha Z)^3 m_\mu^5 < \eta^2 >} \frac{1 - (\frac{A-Z}{2A})\delta(\vec{q}^2)}{1 - 3.13(\frac{A-Z}{2A})}. \tag{35}$$

In Eq.(35) $\delta(\vec{q}^{\,2})$ is given by Eq.(27). We know that Eq.(35) is valid only well above closure so we expect it to produce an underestimate for $< \sigma_c >$. We therefore take a worst case (lowest estimate) value for $\delta(\vec{q}^{\,2})$, namely:

$$\delta(\vec{q}^{\,2}) \simeq 3.38(1 + .091 \frac{\vec{q}^{\,2}}{m_\mu^2}). \tag{36}$$

We now have everything necessary to calculate cross sections for the reaction, $\nu_e + ^{127}I \longrightarrow e^- + X$. Finally we note that from reference 11 we may write the cross section for inclusive neutral current neutrino scattering as:

$$\sigma_N = \frac{G^2 < E' >^2}{8M^2\pi}(0.832D). \tag{37}$$

We are therefore able to evaluate the charged current cross sections in both models and to find the neutral current cross section. In figure 1 we plot the results of the tensor model for the charged current cross section,σ_c. As already noted, at and above 30 MeV we assume $\delta(E_\nu)$ is given by the constant,15 MeV, but below 30 MeV we write it as given in Eq.(28). We note that $< \sigma_c >$ is not greatly affected by any reasonable choice of functions. From figure 1 and the Michel spectrum we obtain the spectrum averaged value:

$$< \sigma_c >= 636 \times 10^{-42}\ cm^2. \tag{38}$$

where we expect a 35 percent error to be associated with this number.

In figure 2, we plot the charged current cross section using the second model here. This model yields zero below the giant dipole resonance at 15 MeV as it must from the assumptions used in its derivation. From this result we obtain a Michel spectrum averaged cross section:

$$< \sigma_c >= 423 \times 10^{-42}\ cm^2. \tag{39}$$

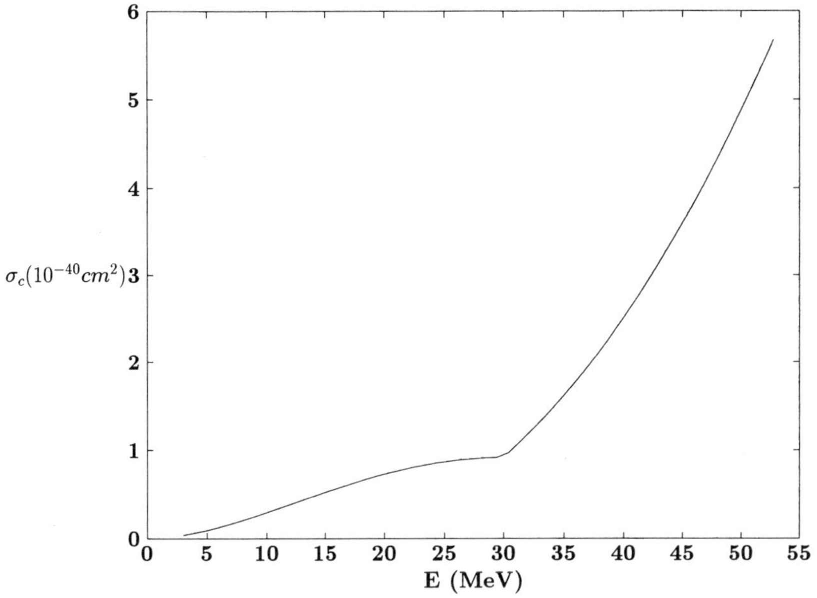

Figure 1 Cross section for the reaction $\nu_e +^{127} I \longrightarrow e^- + X$ as a function of neutrino energy. This calculation is via the tensor model.

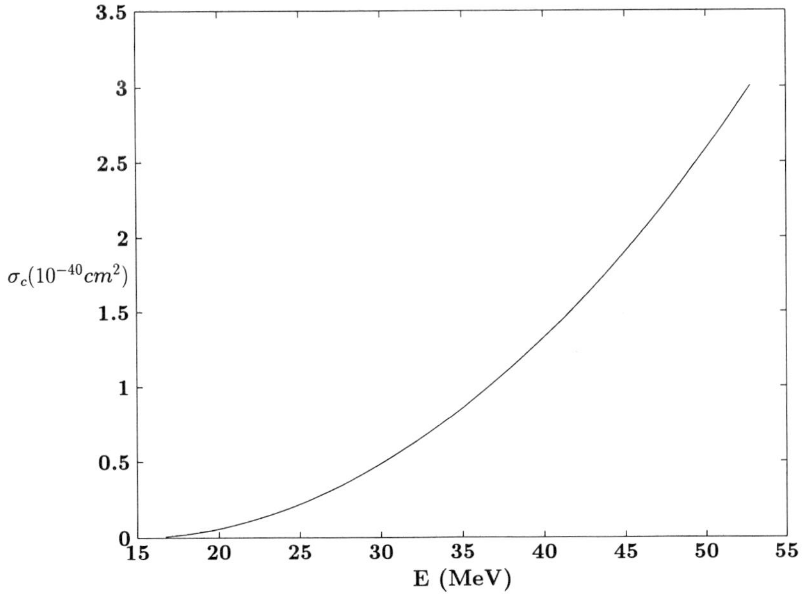

Figure 2 Cross section for the reaction $\nu_e +^{127} I \longrightarrow e^- + X$ as a function of neutrino energy. The solid curve is the actual cross section. This calculation is via the closure approximation model.

DISCUSSION

At present there are no published estimates of the Michel spectrum averaged inclusive neutrino cross sections in ^{127}I. However there are some informal estimates circulating which are based to a large extent on the substantial neutron excess in ^{127}I. These estimates are of the order of 600 $\times$ 10^{-42} cm^2. This is in the range of our result and may indicate that Fermi gas type models may be useful for large nuclei. However our result, and this is true for both models described here,is strongly tied to the total muon capture result. This connection ties the value down but clearly is a weakness in the present calculation where the muon capture result must be inferred.

Thus we believe that a value for $< \sigma_c >$ of the order of 450 *to* 750 $\times$ 10^{-42} might be roughly expected. For the region below the giant dipole resonance,i.e., the region of interest for solar neutrino observation, we expect our value to be less reliable than for higher energies but still to be a reasonable rough estimate.

Finally, we wish to note that some early results for the reaction, $\nu_\mu + {}^{12}C \longrightarrow \mu^- + X$, have been reported[14] at the most recent American Physical Society meeting (April, 1994) from Los Alamos. These preliminary results are in good agreement with the two models described here which yield essentially the same result for the energy range for which the reaction is possible (since the spectrum is peaked well above the giant dipole resonance) but in strong disagreement with Fermi gas model calculation. This gives us some slight added confidence concerning these models.

Thus we see that we that although the theoretical and experimental questions in this area are at present unanswered, many new attempts both experimental and theoretical are underway. Perhaps in the next five years we will have a much better understanding of low energy neutrino reactions in nuclei. However what we already know is sufficient to make us pause before drawing any hasty conclusions concerning the solar neutrino problem.

REFERENCES

[1] R.C. Allen,Phys Rev Lett.**64**,1871(1990)

[2] R.Maschuw et al.,**Neutrino Physics Proceedings of the International Workshop, Heidelberg**,eds. H.V. Klapdor and B. Pohv,Springer,Berlin,147(1988).

[3] T.W. Donnelly,Phys. Lett.**B43**,93(1973).

[4] B.Zeitnitz,Invited talk,Amer. Phys. Soc.,Washington,April,1993.

[5] E.Kolbe,K. Langanke,S.Krewald and F.-K. Thielmann,Nucl. Phys.**A540**,599(1992).

[6] K. Lande, private communication.

[7] J.A. Edgington,private communication.

[8] G.Chang,private communication

[9] C.Athanassopoulos et al,contributed talk,Amer. Phys. Soc.,Washington,April 1994.

[10] M. Pourkaviani and S.L. Mintz, Inclusive Inelastic neutral-current Neutrino Reactions in ^{12}C and the Charged Current Reaction, to appear in Nucl. Phys. A.

[11] C.W. Kim and S.L. Mintz, Phys. Rev. **C31**,274(1985).

[12] T.Suzuki et al.,Phys. Rev. **C35**,2212(1987).

SECTION V

FURTHER INSPIRATIONS FROM THE ELECTROWEAK THEORY, SUPERSYMMETRY, SUPERGRAVITY

IMPLICATIONS OF SUPERSYMMETRIC GRAND UNIFICATION*

V. Barger, M. S. Berger, and P. Ohmann

Physics Department
University of Wisconsin
Madison, WI 53706

ABSTRACT

Unification of the interactions of the Standard Model is possible in its simplest supersymmetric extension. The implications of the λ_t fixed-point solution on the top mass and on Higgs phenomenology is discussed. Expected correlations between the masses of various supersymmetric particles are detailed.

1. INTRODUCTION

The search for symmetries beyond those in the Standard Model is a constant task in modern particle physics. Since there is no compelling disagreement between the Standard Model and experiment, why then should one look for the physics beyond the Standard Model? The compelling reason is that the mechanism behind electroweak symmetry breaking is completely unknown.

The attempts to describe the electroweak symmetry breaking of the Standard Model fall largely into two broad classes: a weakly-interacting symmetry breaking sector and a strongly-interacting symmetry breaking sector. What is a natural value for the mass of the Higgs bosons that characterize the first case? It is easy to describe the requirements on the Higgs sector in the minimal version of the supersymmetric standard model, commonly referred to by its acronym MSSM. In this case the constraints from supersymmetry on the Higgs sector leads to an upper bound on the mass of the lightest physical Higgs boson.

The improvement in the precision data from LEP calls for a reevaluation of the viability of grand unified theories in the context of supersymmetry. Research has concentrated recently on including two-loop contributions in the renormalization group equations, the investigation of the impact of threshold corrections at both the electroweak and grand unified scales, and the estimation of the effects of non-renormalizable operators at the GUT scale.

2. PHENOMENOLOGICAL MOTIVATIONS FOR SUPERSYMMETRY[1]

The major motivations for supersymmetry are the following

*Talk presented by V. Barger

Unified Symmetry: In the Small and in the Large,
Edited by B.N. Kursunoglu *et al.*, Plenum Press, New York, 1995

- Unification of couplings[2] — With SUSY, couplings evolve to an intersection at $M \sim 10^{16}$ GeV. In the standard model, the gauge coupling "triangle" fails to close and unification of gauge couplings cannot be rescued even by large threshold corrections. See Figures 1a, b.
- The problems with technicolor — The problems that flavor changing neutral currents (FCNCs) and the generation of fermion masses pose for technicolor theories are well known. The simplest technicolor theories have problems when confronted with precision measurements of radiative corrections in the electroweak theory.
- Dark Matter — A candidate for cold dark matter of the Universe arises naturally in supersymmetry. The lightest supersymmetric particle (LSP) is absolutely stable if a certain symmetry (known as R-parity) exists.
- Radiative Breaking of the electroweak symmetry[3–14] — The Higgs mechanism can be understood in the context of supersymmetric GUTs as a negative contribution to a Higgs mass-squared by a large logarithm of the ratio of the GUT to electroweak scales.
- Proton Decay[15,16] — In the context of grand unified theories the heavy states mediate transitions between quarks and leptons, thus violating lepton and baryon number conservation. Since the rates for these transitions are governed in part by the mass of the GUT scale states, the sensitive searches for proton decay can impose severe restrictions on GUT models. In fact the minimal SU(5) model predicts proton decay at a rate already excluded by experiment. The supersymmetric models have a higher unification scale and the dimension six operators that plague the non-supersymmetric models are suppressed, but the situation is complicated by the introduction of dimension five operators in supersymmetry.

3. EVOLUTION OF COUPLINGS: RGE

The unification of gauge groups is not a radical idea. In fact this idea has already been partially realized in nature as the Standard Model. The only radical idea introduced by many grand unified theories is that there is a "desert" from the electroweak scale upward to almost the Planck scale.

The gauge couplings evolve according to ordinary differential equations derived from renormalization group ideas. Large logarithms which depend on the scale of a

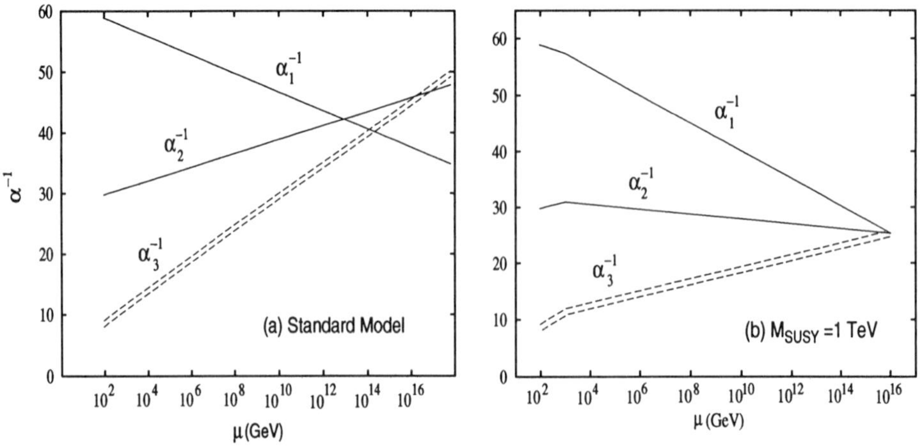

Figure 1. Gauge coupling evolution (a) in the SM and (b) in a SUSY-GUT example.

process can be absorbed into the gauge couplings. This gives rise to a "running" gauge coupling that depends on scale. At the one-loop level these equations are not coupled to each other, and the solutions for the reciprocal of the parameters $\alpha_i \equiv g_i^2/4\pi$ are just linear functions of $t = \ln(Q/M_G)$ where Q is the running mass scale and M_G is the GUT unification mass,

$$\alpha_i^{-1}(Q) = \alpha_i^{-1}(M_G) - \frac{b_i}{2\pi}t\ . \tag{1}$$

At the two-loop level the gauge couplings obey the RGE,[17-19]

$$\frac{dg_i}{dt} = \frac{g_i}{16\pi^2}\left[b_i g_i^2 + \frac{1}{16\pi^2}\left(\sum_{j=1}^{3} b_{ij} g_i^2 g_j^2 - \sum_{j=t,b,\tau} a_{ij} g_i^2 \lambda_j^2\right)\right]\ . \tag{2}$$

The quantities b_i, b_{ij}, and a_{ij} are determined by the particle content in the effective theory.

Although unification can be restored in the non-supersymmetric case by adding extra Higgs doublets that change the evolution of the electroweak gauge couplings α_1 and α_2, this also lowers the scale M_G at which unification occurs and thereby exacerbates the violation of the proton decay bound rather than solving it as in the supersymmetric case. Another possibility is to put an intermediate scale between the GUT scale and the electroweak scale. Then gauge coupling unification occurs at the expense of adding new physics at this intermediate scale, thereby adding to the complexity of unification and making the resulting theory less predictive.

4. YUKAWA COUPLING EVOLUTION AND THE λ_t FIXED-POINT SOLUTION

The relationship between m_b and m_τ in grand unified theories incorporating a desert is the most generic example of Yukawa coupling evolution. It is interesting to note that this relationship can be made to work in supersymmetric grand unified theories, but it implies a strong constraint on the parameter space. In particular it gives a relationship between the top quark mass and the angle β that describes the alignment of the vacuum in two Higgs doublet models (and the MSSM).

The Yukawa couplings are related to the fermions masses in our convention[20] by

$$\lambda_b(m_t) = \frac{\sqrt{2}\,m_b(m_b)}{\eta_b v\cos\beta}\,, \qquad \lambda_\tau(m_t) = \frac{\sqrt{2}m_\tau(m_\tau)}{\eta_\tau v\cos\beta}\,, \qquad \lambda_t(m_t) = \frac{\sqrt{2}m_t(m_t)}{v\sin\beta}\ . \tag{3}$$

The scaling factors η_b and η_τ relate the Yukawa couplings to their values at the scale m_t. The evolution of these Yukawa couplings is described by the RGEs,

$$\frac{d\lambda_t}{dt} = \frac{\lambda_t}{16\pi^2}\left[-\frac{13}{15}g_1^2 - 3g_2^2 - \frac{16}{3}g_3^2 + 6\lambda_t^2 + \lambda_b^2\right]\,, \tag{4}$$

$$\frac{dR_{b/\tau}}{dt} = \frac{R_{b/\tau}}{16\pi^2}\left[\frac{4}{3}g_1^2 - \frac{16}{3}g_3^2 + \lambda_t^2 + 3\lambda_b^2 - 3\lambda_\tau^2\right]\ . \tag{5}$$

where the ratio $R_{b/\tau} \equiv \frac{\lambda_b}{\lambda_\tau}$. A well-known prediction of many GUT theories is that $R_{b/\tau}$ is equal to unity at the GUT scale[21] when the b and τ are in the same representation of the GUT gauge group. Figures 2 and 3 show the solution of these renormalization group equations for values of the bottom quark mass. One sees that the top Yukawa coupling tends to be driven to its infrared fixed point.[20,22-29]

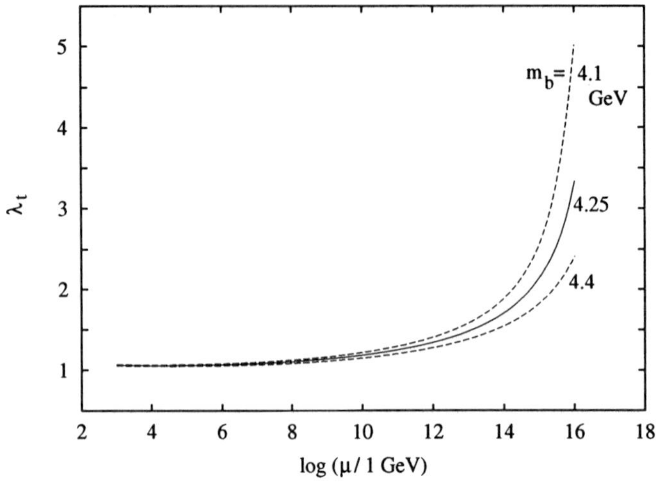

Figure 2. If λ_t is large at M_G, then the renormalization group equation causes $\lambda_t(Q)$ to evolve rapidly towards an infrared fixed point as $Q \to m_t$ (from Ref. 20).

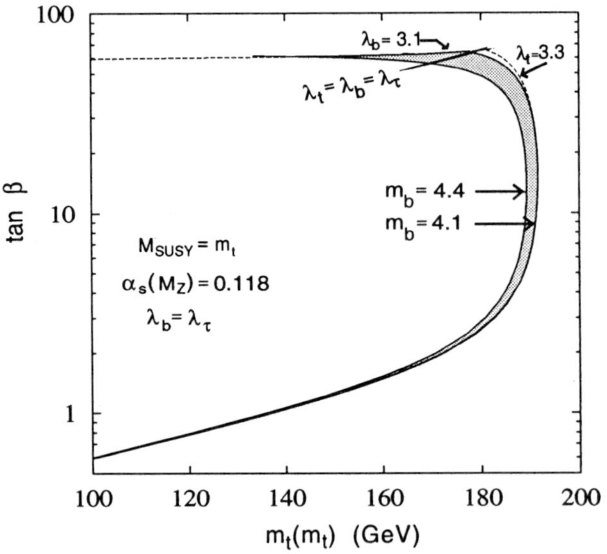

Figure 3. Contours of constant $m_b(m_b)$ in the $m_t(m_t), \tan\beta$ plane (from Ref. 20).

The fixed-point solution leads to the following relation between the top quark mass (in the $\overline{DR}$ dimensional reduction scheme[30] with minimal subtraction) and $\tan\beta$.

$$\lambda_t(m_t) = \frac{\sqrt{2}m_t(m_t)}{v\sin\beta} \quad \Rightarrow \quad m_t(m_t) \approx \frac{v}{\sqrt{2}}\sin\beta = (192\text{GeV})\sin\beta \tag{6}$$

Converting this relation to the top quark pole mass yields[20]

$$m_t^{\text{pole}} \approx (200\text{GeV})\sin\beta\ . \tag{7}$$

If one takes the λ_t fixed-point solution and also assumes that the top quark mass m_t^{pole} is less than about 160 GeV, important consequences result for the Higgs sector of the MSSM. From Fig. 4 it is clear that given these assumptions $\tan\beta$ is very near one. Since $\tan\beta = 1$ is a flat direction in the Higgs potential, for which the associated Higgs boson is massless at tree level, and the true mass of the lightest Higgs is given almost entirely by the one-loop radiative corrections, m_h tends to be at the light end of its range. This case was discussed in detail by Diaz and Haber.[31] In this low $\tan\beta$ region the Higgs mass is particularly sensitive to higher order corrections.[32–34] The upper bound on m_h that results is shown by the boundary of the theoretically disallowed region in Fig. 5.

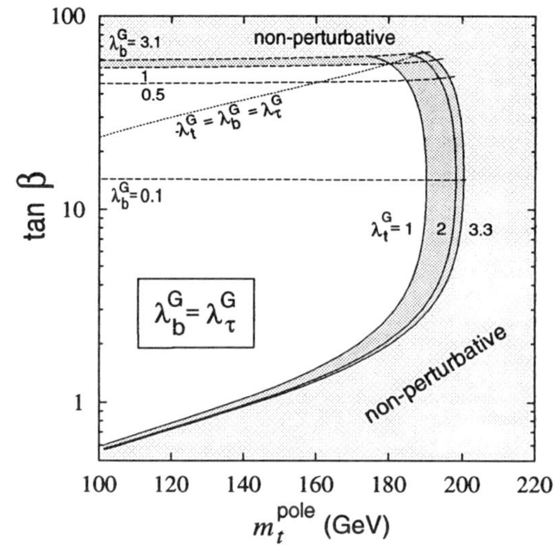

Figure 4. The fixed-point regions are given by Yukawa couplings at the GUT scale being larger than about 1 ($\lambda_i^G \gtrsim 1$). Even larger values of the Yukawa couplings results in a breakdown of perturbation theory.

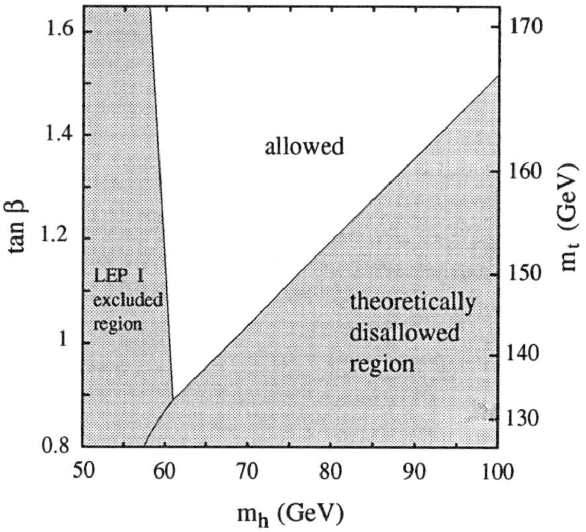

Figure 5. The λ_t-fixed-point solution regions allowed by the LEP I data in the $m_h, \tan\beta$ plane.[27] The top quark masses on the right hand side are m_t(pole), correlated to $\tan\beta$ by the fixed-point solution.

If the top quark is sufficiently light, the fixed-point solution dictates that the Higgs potential is such that the SUSY Higgs (h, H, A, $H^{\pm}$) searches are much more constrained. The Higgs searches at LEP have excluded a Standard Model Higgs boson with mass $m_{H_{SM}}$ less than 62.5 GeV. The MSSM couplings give the relation

$$\Gamma(Z \to Z^* h) = \sin^2(\beta - \alpha)\Gamma(Z \to Z^* H_{SM}) \tag{8}$$

for $m_h = m_{H_{SM}}$. Here the angle α describes the mixing between the CP-even components of the two Higgs doublets. For a top quark mass less than about 170 GeV, the fixed point gives $\tan\beta \approx 1$. Then in this region of parameter space

$$\sin^2(\beta - \alpha) \sim 1$$

so that the bound on the MSSM lightest Higgs is close to the bound on the Standard Model Higgs.

5. THRESHOLD CORRECTIONS

As discussed above, the requirement of gauge and Yukawa coupling unification can be successful in the minimal versions of supersymmetric grand unified theories and can result in strong constraints on the parameter space of the model. With the improved electroweak data recently made available, a number of groups have been attempting to incrementally refine the theoretical predictions. The one-loop renormalization group equations have been extended to two-loops in the Yukawa sector[18,20] and in the soft-supersymmetry breaking parameters.[35] Moreover threshold corrections at the GUT and electroweak (and electroweak-scale SUSY thresholds) have been investigated.[26,28,34,36–39,41] The RGE evolution yields the generic features that result from the large separation of scales typical in GUT theories. Threshold effects are typically model-dependent and sensitive to the detailed spectrum (the supersymmetric spectrum, the top mass, etc. at the electroweak scale and the superheavy spectrum at the GUT scale)[†]. Threshold effects have been incorporated into the predictions from gauge and Yukawa coupling unification and in the Higgs potential, and eventually will be included in the full supersymmetric spectrum as well.[42]

The theory below the grand unified theory is an effective theory with the heavy GUT-scale particles integrated out. Since the heavy particles do not completely fill the representations of the grand unified gauge group, the group is broken and the RGEs of the effective symmetry exhibit this broken symmetry and the gauge couplings diverge below the GUT scale. The process of integrating out the heavy particles gives rise to threshold corrections that depend on the details of the grand unified theory. Since the threshold corrections at the GUT scale depend on the superheavy spectrum, one therefore expects these corrections to be constrained by the proton decay limits. In a similar way there are threshold corrections at the electroweak scale from the effective theories that are introduced there (the scale of supersymmetry can be chosen to be different than the electroweak scale). Typically one makes some assumptions about the supersymmetric spectrum to simplify the problem (such as in a supergravity-based scenario). Ultimately one hopes that these model-dependent features can be used to distinguish between the various realization of the supersymmetric models.

The threshold corrections to Yukawa coupling unification are also relevant[41] to the analyses of GUT scale mass matrix ansätze. Relations between fermion masses and mixing angles that arise in these scenarios will be modified by model-dependent effects.

[†]The Yukawa coupling unification condition $\lambda_b^G = \lambda_\tau^G$ is itself a model-dependent feature satisfied in simple GUT scenarios.

6. WHERE ARE THE SPARTICLES?

Sometimes supersymmetry is criticized because of a proliferation of parameters. This is not necessarily a fair criterion, since in some minimal versions of supergravity theories the complete mass spectrum and couplings can be explained with the addition of as few as three or five parameters. At the present time our ignorance of the mechanism of supersymmetry breaking should make us cautious about sweeping statements about the supersymmetric particles that rely on some GUT-scale assumptions; however, it is not unreasonable to expect there will be correlations between the supersymmetric particle masses and couplings since we hope that the ultimate theory that describes them near the Planck scale is a simple and economical one. Figure 6 shows representative results for RGE evolution of the sparticle masses.

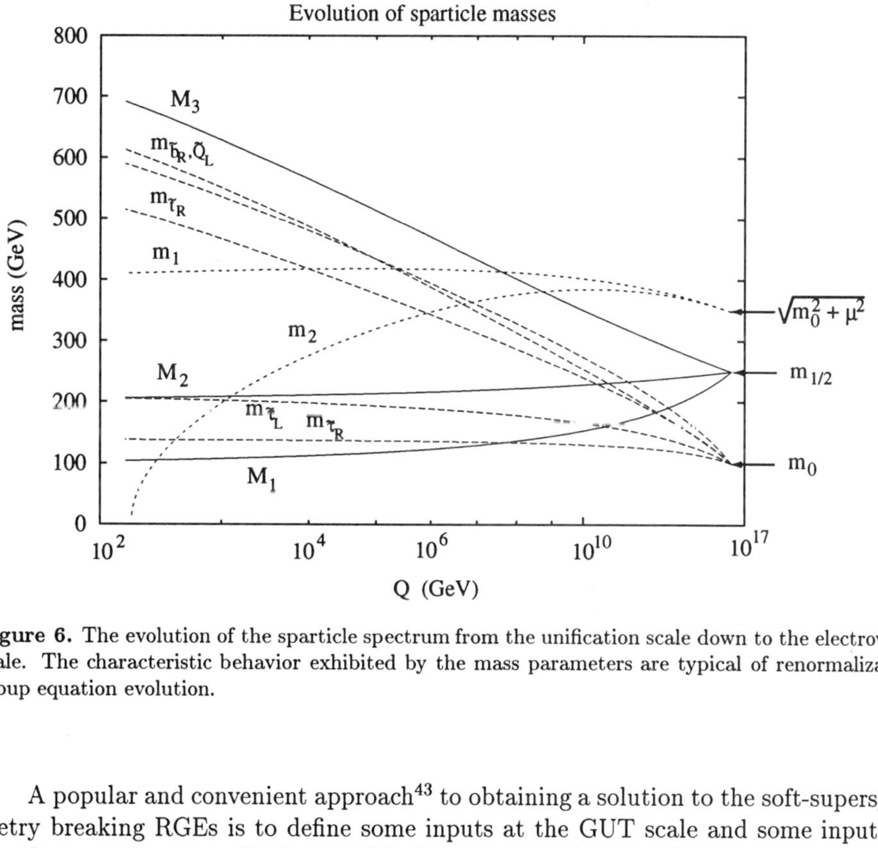

Figure 6. The evolution of the sparticle spectrum from the unification scale down to the electroweak scale. The characteristic behavior exhibited by the mass parameters are typical of renormalization group equation evolution.

A popular and convenient approach[43] to obtaining a solution to the soft-supersymmetry breaking RGEs is to define some inputs at the GUT scale and some inputs at the electroweak scale. We have dubbed this approach the ambidextrous approach[13] to distinguish it from the bottom-up[9] and top-down[44] approaches where all inputs are defined at the same scale. Common to all of these approaches is the requirement that correct electroweak symmetry breaking (EWSB) be achieved. This is accomplished by imposing two minimization conditions obtained by minimizing the Higgs potential.

The tree-level Higgs potential is given by

$$\begin{aligned} V_0 &= (m_{H_1}^2 + \mu^2)|H_1|^2 + (m_{H_2}^2 + \mu^2)|H_2|^2 + m_3^2(\epsilon_{ij} H_1{}^i H_2{}^j + \text{h.c.}) \\ &\quad + \frac{1}{8}(g^2 + g'^2)\left[|H_1|^2 - |H_2|^2\right]^2 + \frac{1}{2}g^2|H_1^{i*}H_2^i|^2 , \end{aligned} \tag{9}$$

where m_{H_1}, m_{H_2}, and m_3 are soft-supersymmetry breaking parameters and ϵ_{ij} is the total antisymmetric tensor. The minimum of the Higgs potential must occur by the acquisition of vacuum expectation values. Minimizing V_0 with respect to the two neutral CP-even Higgs degrees of freedom yields

$$\frac{1}{2}M_Z^2 = \frac{m_{H_1}^2 - m_{H_2}^2 \tan^2\beta}{\tan^2\beta - 1} - \mu^2 . \tag{10}$$

$$-B\mu = \frac{1}{2}(m_{H_1}^2 + m_{H_2}^2 + 2\mu^2)\sin 2\beta . \tag{11}$$

The masses in these equations are running masses that depend on the scale Q in the RGEs that describe their evolution. Hence the solutions obtained are functions of the scale Q. Equations (10) and (11) are a particularly convenient form since the gauge couplings dependence (the D-terms in the language of supersymmetry) is isolated in Eq. (10). This equation also clearly shows the fine-tuning problem that may be present in the radiative breaking of the electroweak symmetry. For large values of $|\mu|$, there must be a cancellation between large terms on the right hand side to obtain the correct experimentally measured M_Z (or equivalently the electroweak scale). For $\tan\beta$ near one, a cancellation of large terms must occur. Finally, these minimization equations illustrate the power of the ambidextrous approach. For EWSB to be satisfied one need only specify m_t, M_Z, $\tan\beta$ at the electroweak scale and the common gaugino mass $m_{1/2}$, scalar mass m_0, and the trilinear scalar coupling A at the GUT scale. Then one solves the minimization equations given above to obtain μ (up to a sign) and B, thereby implicitly satisfying the EWSB requirement. For more details, see Ref. 13.

A heavy top quark produces large corrections to the Higgs potential of the MSSM.[45] Gamberini, Ridolfi, and Zwirner showed[5] that the tree-level Higgs potential is inadequate for the purpose of analyzing radiative breaking of the electroweak symmetry because the tree-level Higgs vacuum expectation values v_1 and v_2 are very sensitive to the scale at which the renormalization group equations are evaluated. The one-loop corrections to the Higgs potential effectively moderates this sensitivity to the scale Q. The one-loop corrections are conveniently calculated using the tadpole method.[13,46,47] The corrections to Eqs. (10) and (11) can be obtained by calculating the two tadpoles with two independent CP-even Higgs as external lines as in Fig. 7. The one-loop corrected minimization conditions can then be used to generate a complete supersymmetric particle spectrum which satisfies EWSB.

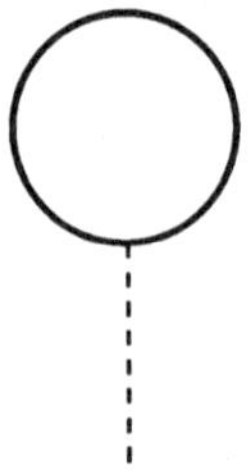

Figure 7. The tadpole diagrams offer a simple method to obtain the one-loop modifications to the minimization conditions. The loop consists of all matter and gauge-Higgs contributions, and the external lines are the two CP-even Higgs fields.

Including only the leading contribution coming from the top quark loop (and neglecting the D-term contributions to the squark masses) one obtains the expressions

$$\frac{1}{2}M_Z^2 = \frac{m_{H_1}^2 - m_{H_2}^2\tan^2\beta}{\tan^2\beta - 1} - \mu^2 - \frac{3g^2m_t^2}{32\pi^2M_W^2\cos 2\beta}\Bigg[2f(m_t^2) - f(m_{\tilde{t}_1}^2) - f(m_{\tilde{t}_2}^2) + \frac{f(m_{\tilde{t}_1}^2) - f(m_{\tilde{t}_2}^2)}{m_{\tilde{t}_1}^2 - m_{\tilde{t}_2}^2}\left((\mu\cot\beta)^2 - A_t^2\right)\Bigg], \tag{12}$$

$$-B\mu = \frac{1}{2}(m_{H_1}^2 + m_{H_2}^2 + 2\mu^2)\sin 2\beta - \frac{3g^2m_t^2\cot\beta}{32\pi^2M_W^2}\Bigg[2f(m_t^2) - f(m_{\tilde{t}_1}^2) - f(m_{\tilde{t}_2}^2) - \frac{f(m_{\tilde{t}_1}^2) - f(m_{\tilde{t}_2}^2)}{m_{\tilde{t}_1}^2 - m_{\tilde{t}_2}^2}(A_t + \mu\cot\beta)(A_t + \mu\tan\beta)\Bigg], \tag{13}$$

where

$$f(m^2) = m^2\left(\ln\frac{m^2}{Q^2} - 1\right). \tag{14}$$

The extra one-loop contribution included above renders the solution less sensitive to the scale Q,[8–10,13,48] as can be shown explicitly by examining the relevant renormalization group equations for the parameters that enter into the minimization conditions. The complete expressions for the one-loop contributions can be found in Ref. 13. The fine-tuning problem is alleviated somewhat, but not entirely, by the inclusion of one-loop corrections to the Higgs potential. As our naturalness criterion we require

$$|\mu(m_t)| < 500\ \text{GeV}. \tag{15}$$

The gaugino masses are related (through one-loop order) by the same ratios that describe the gauge couplings at the electroweak scale. This observation, together with the fact that $|\mu|$ is large, yields simple correlations between the lightest chargino and neutralinos and the gluino,[6,10,49] namely

$$M_{\chi_1^0} \simeq M_1, \tag{16}$$

$$M_{\chi_1^\pm} \simeq M_{\chi_2^0} \simeq M_2 = \frac{\alpha_2}{\alpha_1}M_1 \simeq 2M_1 \simeq 2M_{\chi_1^0}, \tag{17}$$

$$m_{\tilde{g}} = M_3 = \frac{\alpha_3}{\alpha_2}M_2 = \frac{\alpha_3}{\alpha_1}M_1. \tag{18}$$

In our analysis the quantities in these equations are evaluated at scale m_t. The heaviest chargino and the two heaviest neutralino states are primarily Higgsino with

$$M_{\chi_2^\pm} \simeq M_{\chi_3^0} \simeq M_{\chi_4^0} \simeq |\mu|. \tag{19}$$

As previously noted the mass of the lightest Higgs h arises mainly from radiative corrections.[27,31,34,50] The heavy Higgs states are (approximately) degenerate $\approx M_A$ because at tree-level $M_A = -\frac{B\mu}{\sin 2\beta} \approx -B\mu$ is large. The squark and slepton masses also display simple asymptotic behavior at large $|\mu|$. The first and second squark generations are approximately degenerate. The splitting of the stop quark masses grows as $|\mu|$ increases. The splitting of the sbottom states does not change much with μ for small $\tan\beta$.

The approximate experimental bounds that we impose are listed in Table 1. Together with our naturalness criteria $|\mu(m_t)| < 500$ GeV, these bounds give the allowed region in the $m_0, m_{1/2}$ plane shown as the shaded areas in Fig. 8.

Table 1. Approximate experimental bounds.

Particle	Experimental Limit (GeV)
gluino	120
squark, slepton	45
chargino	45
neutralino	20
light higgs	60

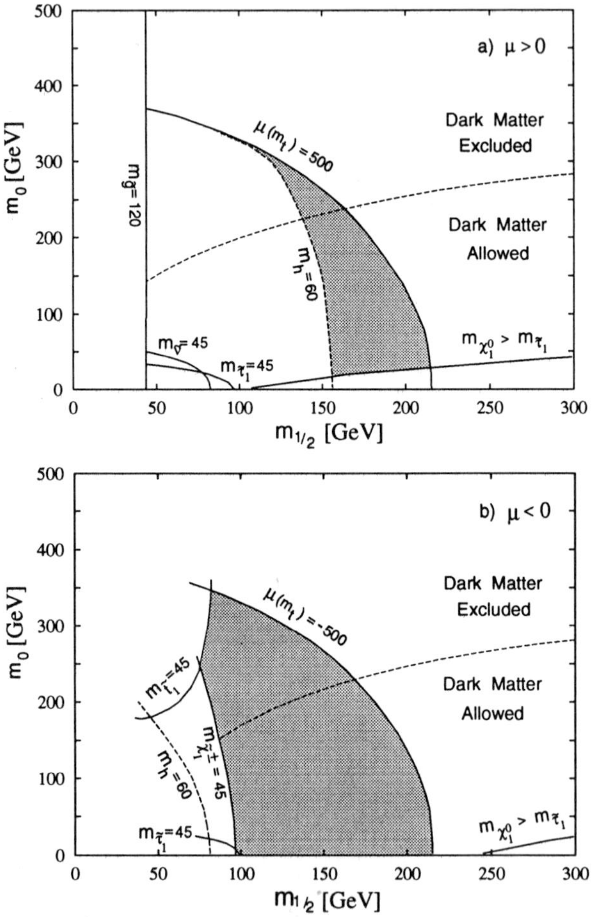

Figure 8. Allowed regions of parameter space for $m_t(m_t) = 160$ GeV, $\tan\beta = 1.47$ (a low-$\tan\beta$ fixed-point solution); from Ref. 13.

The prediction for m_h in the low-$\tan\beta$ region is particularly sensitive to two-loop corrections.[34] Hence the precise location of the $m_h = 60$ GeV contour is somewhat uncertain. The MSSM has conserved R-parity so the lightest supersymmetric particle (LSP) is stable. Usually the LSP is the lightest neutralino, but for small values of m_0 the supersymmetric partner of the tau lepton is sometimes lighter. For the lightest SUSY particle to be neutral there is an upper bound on the value of $m_{1/2}$ for small m_0. In particular such an upper bound exists for no-scale models ($m_0 = 0$), and is more stringent for $\mu > 0$ due to the mixing between the left and right handed $\tilde{\tau}$, giving a stau lighter than the lightest neutralino. The LSP can also account for the dark matter of the Universe.[51,52] The large values of μ obtained from the low $\tan\beta$ solution result in the lightest neutralino being predominantly gaugino (see Figure 9). This leads to a reduced rate of annihilation of neutralinos and can provide too much relic abundance and overclose the Universe. This constraint is shown as the dashed line in Figure 8; this line should be regarded as a semi-quantitative one only since the contributions of s-channel poles that can enhance the annihilation rate have been neglected.

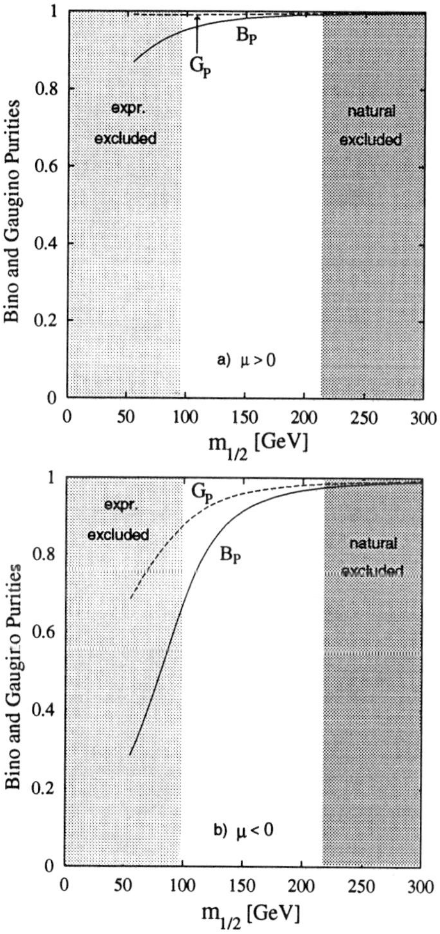

Figure 9. Bino and gaugino purities for low $\tan\beta$ fixed-point solution.

Figure 10 shows the dependence of the sparticle masses on the parameter $m_{1/2}$. The large $|\mu|$ obtained from the low-$\tan\beta$ solutions lead to highly correlated masses. Figure 11 shows the squark masses, which are quite degenerate for the light families.

Figure 12 shows the supersymmetric particle mass dependence on the parameter m_0 for a fixed value of $m_{1/2}$. The lighter top squark eigenstate $\tilde{t}_1$ has an approximately constant mass with increasing m_0. This occurs because $|\mu|$ increases with m_0 giving rise to increased mixing between the left- and right-handed top squarks (lowering the lightest mass eigenstate).

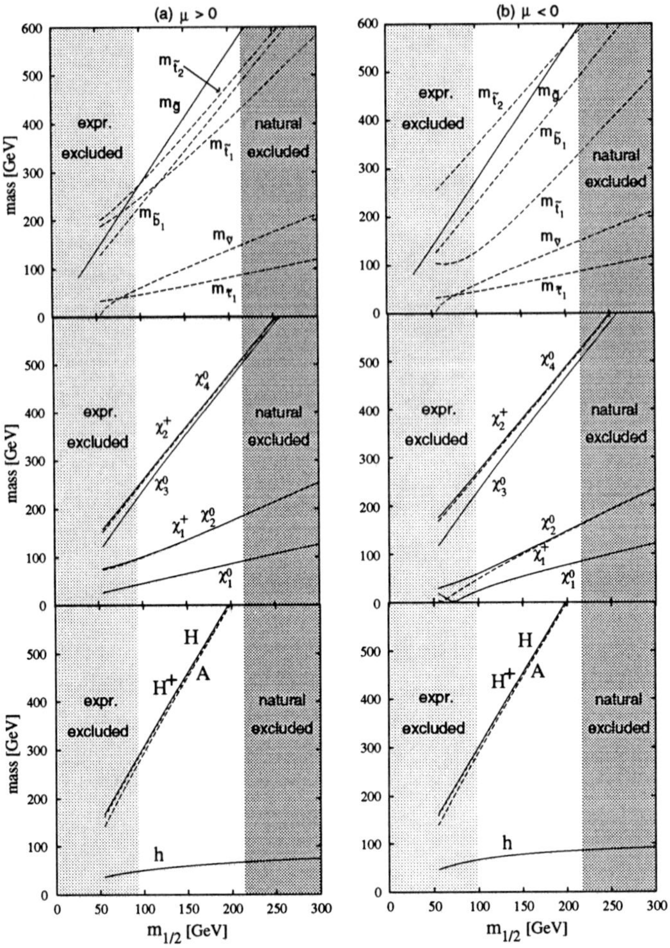

Figure 10. Particle spectra as a function of $m_{1/2}$. The shaded bands at the left are regions excluded by the experimental constraints. The shaded bands at the right labeled "natural excluded" require $|\mu(m_t)| > 500$ GeV (from Ref. 13).

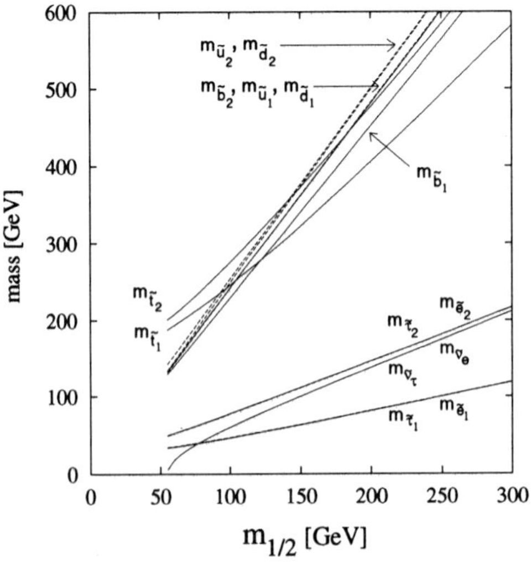

Figure 11. Supersymmetric scalar masses showing asymptotic behavior (from Ref. 13).

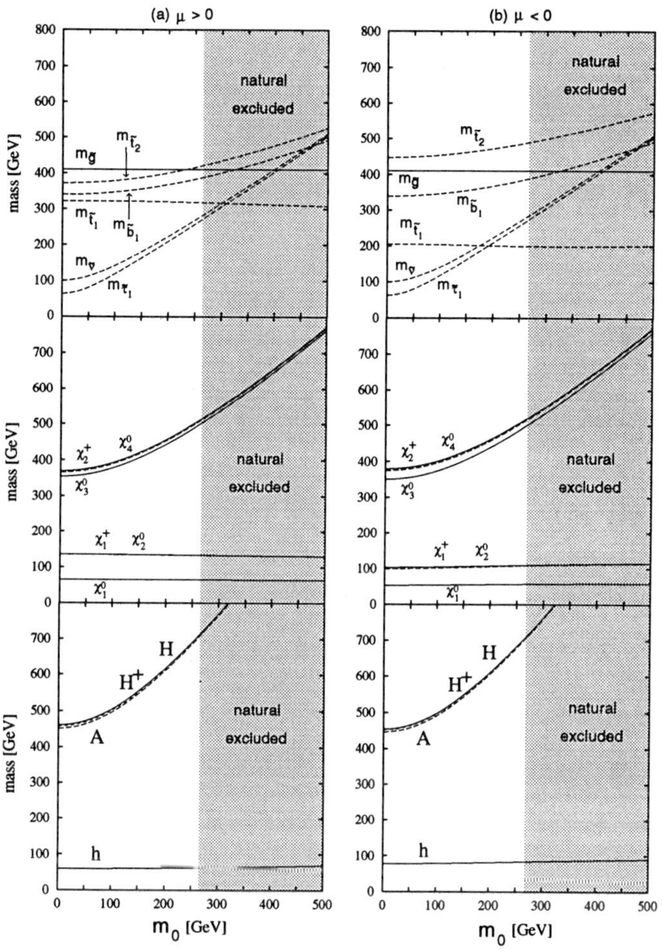

Figure 12. The supersymmetric particle mass dependence on the parameter m_0 for (a) $\mu > 0$ and (b) $\mu < 0$ with $m_{1/2} = 150$ GeV and $A^G = 0$ (from Ref. 13).

7. SUSY SIGNALS

At hadron colliders a plethora of sparticle production processes are possible, as illustrated in Fig. 13. The SUSY particles must be created in pairs (for theories with an R-symmetry). Gluinos should be copiously produced at a future hadron collider. As the mass of the gluino increases, new decay channels open up. Figure 14 shows typical branching fraction for gluinos assuming that $m_{\tilde{g}} < m_{\tilde{q}}$[53] (in this figure it is not assumed that μ is large as required by radiative breaking of the electroweak symmetry). The predicted cascade gluino decays provide multiple signals for experimental searches. Gluino decays in the Yukawa unified supergravity scenario with radiative electroweak symmetry breaking have been investigated in Ref. 54. A phenomenological discussion of the Yukawa unified no-scale model can be found in Ref. 55.

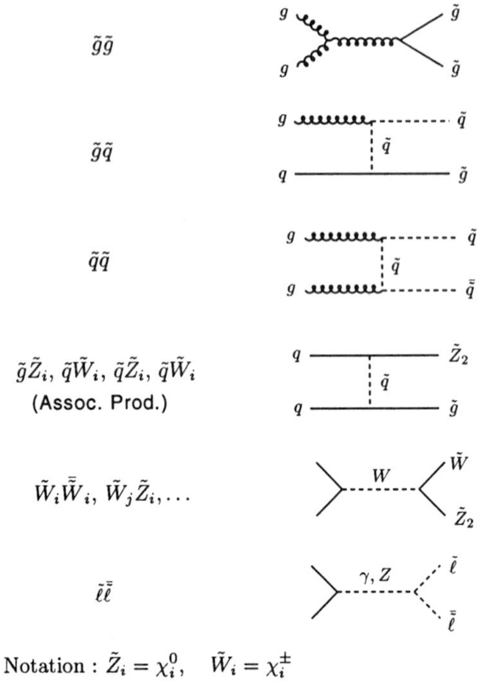

Figure 13. SUSY production processes at hadron colliders.

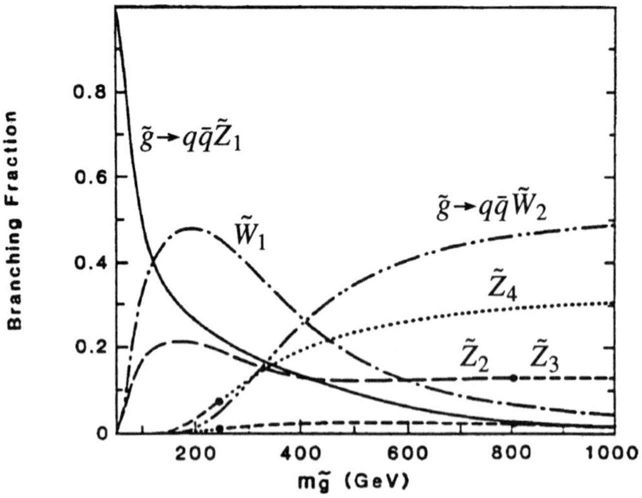

Figure 14. Gluino branching fractions assuming that $m_{\tilde{g}} < m_{\tilde{q}}$ (from Ref. 53).

8. CONCLUSIONS

The continued viability of supersymmetric grand unified theories with respect to the increased precision of the low energy data calls for more refined theoretical analyses. The following observations summarize the principal points of this review.

- A low-energy supersymmetry is consistent with a desert unification scenario in grand unified theories.
- The observed ratio m_b/m_τ is consistent with SUSY GUTs. In fact, this ratio indicates that the top quark Yukawa coupling is near its infrared fixed point; this situation has significant implications for SUSY Higgs searches if the top quark is lighter than about 165 GeV. In that case the upper bound on m_h is of order 100 GeV.
- Solutions with a λ_t fixed point, $m_t \lesssim 175$ GeV and radiative breaking of the electroweak symmetry breaking are allowed by our naturalness criterion $|\mu(M_Z)| \simeq |\mu(m_t)| < 500$ GeV for both signs of the supersymmetric Higgs mass parameter μ. These solutions are characterized by relatively large values of $|\mu|$, which implies that the supersymmetric particle spectrum displays a simple asymptotic behavior in the simplest supergravity models.
- In the early universe the LSP annihilates sufficiently (at least in the approximation that s-channel pole annihilation is neglected) over most of the parameter space $m_0 \lesssim 300$ GeV, so as not to overclose the universe.
- The one-loop corrections to the Higgs potential somewhat ameliorate the fine-tuning problem.
- The tadpole method is a convenient way to calculate the one-loop minimization conditions. We have obtained these conditions in an analytic form including all contributions from the gauge-Higgs sector and matter multiplets. This method is easily extended to non-minimal Higgs sectors or to models with additional low-lying states.

ACKNOWLEDGMENTS

This research was supported in part by the University of Wisconsin Research Committee with funds granted by the Wisconsin Alumni Research Foundation, in part by the U.S. Department of Energy under contract no. DE-AC02-76ER00881, and in part by the Texas National Laboratory Research Commission under grant nos. RGFY93-221 and FCFY9302. MSB was supported in part by an SSC Fellowship. PO was supported in part by an NSF Graduate Fellowship.

REFERENCES

1. For reviews see J. Wess and J. Bagger, "Supersymmetry and Supergravity," Princeton University Press (1983); P. Fayet and S. Ferrara, Phys. Rep. **C32**, 249 (1977); P. van Nieuwenhuizen, Phys. Rep. **C68**, 189 (1981); H. P. Nilles, Phys. Rep. **C110**, 1 (1984); H. Haber and G. Kane, Phys. Rep. **C117**, 75

(1985); M. F. Sohnius, Phys. Rep. **C128**, 39 (1985); X. Tata, UH-511-713-90, Lectures presented at Mt. Sorak Symp. on the Standard and Beyond, Mt. Sorak, Korea, Aug 1990; 20 Years of Supersymmetry Conference (1993); International Workshop on Supersymmetry and Unification of Fundamental Interactions (SUSY-93),Northeastern University, Boston, MA, ed. P. Nath (1993); J. Lopez, Texas A&M preprint CTP-TAMU-42/93 from the Proceedings of the International School of Subnuclear Physics 31st Course, Erice, Italy, hep-ph 9308323; D. R. T. Jones, TASI-93 Lectures, Boulder, Co (1993).

2. U. Amaldi, W. de Boer, and H. Furstenau, Phys. Lett. **B260** (1991) 447; J. Ellis, S. Kelley and D. V. Nanopoulos, Phys. Lett. **B260** 131 (1991); P. Langacker and M. Luo, Phys. Rev. **D44** 817 (1991).
3. For a review see P. Nath, R. Arnowitt, and A. H. Chamseddine, "Applied $N = 1$ Supergravity," World Scientific (1984).
4. L. E. Ibañez and G. G. Ross, Phys. Lett. **B110**, 215 (1982); L. Alvarez-Gaumé, M. Claudson and M. B. Wise, Nucl. Phys. **207**, 96 (1982); J. Ellis, D. V. Nanopoulos, and K. Tamvakis, Phys. Lett. **B121**, 123 (1983); L. E. Ibañez, Nucl. Phys. **218**, 514 (1983); L. Alvarez-Gaumé, J. Polchinski and M. B. Wise, Nucl. Phys. **221**, 495 (1983); J. Ellis, J. S. Hagelin, D. V. Nanopoulos, and K. Tamvakis, Phys. Lett. **B125**, 275 (1983); L. E. Ibañez and C. Lopez, Phys. Lett. **B126**, 54 (1983); Nucl. Phys. **B233**, 511 (1984); C. Kounnas, A. B. Lahanas, D. V. Nanopoulos, and M. Quiros, Nucl. Phys. **B236**, 438 (1984); U. Ellwanger, Nucl. Phys. **B238**, 665 (1984); L. E. Ibañez, C. Lopez, and C. Muñoz, Nucl. Phys. **B256**, 218 (1985); A. Bouquet, J. Kaplan, and C. A. Savoy, Nucl. Phys. **B262**, 299 (1985); H. J. Kappen, Phys. Rev. **D35**, 3923 (1987); M. Drees, Phys. Rev. **D38**, 718 (1988); H. J. Kappen, Phys. Rev. **D38**, 721 (1988); B. Gato, Z. Phys. **C37**, 407 (1988); G. F. Giudice and G. Ridolfi, Z. Phys. **C41**, 447 (1988); S. P. Li and H. Navelet, Nucl. Phys. **B319**, 239 (1989).
5. G. Gamberini, G. Ridolfi, and F. Zwirner, Nucl. Phys. **B331**, 331 (1990).
6. R. Arnowitt and P. Nath, Phys. Rev. **69**, 725 (1992); Phys. Lett. **B289**, 368 (1992).
7. G. G. Ross and R. G. Roberts, Nucl. Phys. **B377**, 571 (1992).
8. S. Kelley, J. L. Lopez, D. V. Nanopoulos, H. Pois, and K. Yuan, Nucl. Phys. **B398**, 3 (1993).
9. M. Olechowski and S. Pokorski, Nucl. Phys. **B404**, 590 (1993).
10. P. Ramond, Institute for Fundamental Theory Preprint UFIFT-HEP-93-13 (1993), hep-ph 9306311; D. J. Castaño, E. J. Piard, and P. Ramond, Institute for Fundamental Theory Preprint UFIFT-HEP-93-18 (1993), hep-ph 9308335.
11. W. de Boer, R. Ehret, and D. I. Kazakov, Inst. für Experimentelle Kernphysik preprint IEKP-KA/93-13, Contribution to the International Symposium on Lepton Photon Interactions, Ithaca, NY (1993), hep-ph 9308238.
12. M. Carena, M. Olechowski, S. Pokorski, and C. E. M. Wagner, CERN preprint CERN-TH.7060/93 (1993), hep-ph 9311222.
13. V. Barger, M. S. Berger, and P. Ohmann, University of Wisconsin preprint MAD/PH/801 (1993), hep-ph 9311269.
14. G. Kane, C. Kolda, L. Roszkowski, and J. D. Wells, University of Michigan preprint UM-TH-93-24 (1993), hep-ph 9312272.
15. R. Arnowitt and P. Nath, Phys. Rev. **70**, 3696 (1993).

16. J. Hisano, H. Murayama and T. Yanagida, Phys. Rev. Lett. **69**, 1014 (1992); Nucl. Phys. **B402**, 46 (1993).
17. M. B. Einhorn and D. R. T. Jones, Nucl. Phys. **196**, 475 (1982).
18. J. E. Björkman and D. R. T. Jones, Nucl. Phys. **B259**, 533 (1985).
19. K. Inoue, A. Kakuto, H. Komatsu and S. Takeshita, Prog. Theor. Phys. **67**, 1889 (1982).
20. V. Barger, M.S. Berger, and P. Ohmann, Phys. Rev. **D47**, 1093 (1993).
21. M. Chanowitz, J. Ellis, and M. K. Gaillard, Nucl. Phys. **B128**, 506 (1977).
22. B. Pendleton and G. G. Ross, Phys. Lett. **98B**, 291 (1981); C. T. Hill, Phys. Rev. **D24**, 691 (1981); W. Zimmermann, Commun. Math. Phys. **97**, 211 (1985); J. Kubo, K. Sibold, and W. Zimmerman, Phys. Lett. **B200**, 191 (1989).
23. C. D. Froggatt, I. G. Knowles, and R. G. Moorhouse, Phys. Lett. **B249**, 273 (1990); **B298**, 356 (1993).
24. S. Dimopoulos, L. Hall and S. Raby, Phys. Rev. Lett. **68**, (1992) 1984; Phys. Rev. **D45**, 4192 (1992).
25. V. Barger, M. S. Berger, T. Han, and M. Zralek, Phys. Rev. Lett. **68**, 3394 (1992).
26. M. Carena, S. Pokorski, and C. E. M. Wagner, Nucl. Phys. **B406**, 59 (1993); W. Bardeen, M. Carena, S. Pokorski, and C. E. M. Wagner, Phys. Lett. **B320**, 110 (1994).
27. V. Barger, M. S. Berger, P. Ohmann, and R. J. N. Phillips, Phys. Lett. **B314**, 351 (1993).
28. P. Langacker and N. Polonsky, Phys. Rev. **D49**, 1454 (1994); N. Polonsky, Talk presented at the XVI Kazimierz Meeting on Elementary Particle Physics, Penn U. preprint UPR-0588-T, Kazimierz, Poland (1993), hep-ph 9310292.
29. B. C. Allanach and S. F. King, University of Southampton preprint SHEP-93/94-15 (1994), hep-ph 9403212.
30. W. Siegel, Phys. Lett. **B84**, 193 (1979); D. M. Capper D. R. T. Jones, and P. van Nieuwenhuizen, Nucl. Phys. **B167**, 479, (1980).
31. M. Diaz and H. Haber, Phys. Rev. **D46**, 3086 (1992).
32. J. Kodaira, Y. Yasui, and K. Sasaki, Hiroshima preprint HUPD-9316 (1993), hep-ph 9311366.
33. R. Hempfling and A. H. Hoang, DESY preprint DESY-93-162 (1993), hep-ph 9401219.
34. P. Langacker and N. Polonsky, Univ. of Pennsylvania preprint UPR-0594T (1994), hep-ph 9403306.
35. Y. Yamada, KEK preprint KEK-TH-383 (1994), hep-ph 9401241; S. P. Martin and M. T. Vaughn, Northeastern University preprint NUB-3081-93-TH (1994), hep-ph 9311340.
36. R. Barbieri and L. J. Hall, Phys. Rev. Lett. **68**, 752 (1992).
37. L. J. Hall and U. Sarid, Phys. Rev. Lett. **70**, 2673 (1993).
38. L. J. Hall, R. Rattazzi, and U. Sarid, Lawrence Berkeley Lab preprint LBL-33997 (1993), hep-ph 9306309.
39. R. Hempfling, DESY preprint DESY-93-092 (1993).
40. M. Carena, M. Olechowski, S. Pokorski, and C. E. M. Wagner, Max Planck Institute preprint MPI-Ph/93-103 (1994), hep-ph 9402253.

41. B. D. Wright, University of Wisconsin preprint MAD/PH/812 (1994), hep-ph 9404217.
42. The GUT scale threshold corrections to gaugino masses have been investigated in J. Hisano, H. Murayama, and T. Goto, Phys. Rev. **D49**, 1446 (1994).
43. S. Kelley, J. L. Lopez, D. V. Nanopoulos, H. Pois, K. Yuan, Phys. Lett. **B273**, 423 (1991).
44. R. Hempfling, DESY preprint DESY 94-057 (1994), hep-ph 9404226.
45. M. S. Berger, Phys. Rev. **D41**, 225 (1990).
46. S. Weinberg, Phys. Rev. **D7**, 2887 (1973); S. Y. Lee and A. M. Sciaccaluga, Nucl. Phys. **B96**, 435 (1975); K. T. Mahanthappa and M. A. Sher, Phys. Rev. **D22**, 1711 (1980).
47. M. Sher, Phys. Rep. **C179**, 273 (1989).
48. B. de Carlos and J. A. Casas, Phys. Lett. **B309**,320 (1993); B. de Carlos, CERN preprint CERN-TH-7024-93, Talk given at the *XVI International Warsaw Meeting on Elementary Particle Physics* (1993), hep-ph 9310232.
49. J. L. Lopez, D. V. Nanopoulos, and H. Pois Phys. Rev. **D47**, 2468 (1993).
50. J. L. Lopez, D. V. Nanopolous, H. Pois, X. Wang, and A. Zichichi, Phys. Lett. **B306**, 73 (1993).
51. For an overview of dark matter in these models, see e. g. L. Roszkowski, Phys. Lett. **B262**, 59 (1991); J. L. Lopez, D. V. Nanopoulos, and K. Yuan, Nucl. Phys. **B370**, 445 (1992); S. Kelley, J. L. Lopez, D. V. Nanopoulos, H. Pois, and K. Yuan, Phys. Rev. **D47**, 2461 (1993); L. Roszkowski, University of Michigan preprint UM-TH-93-06 (1993), hep-ph 9302259.
52. R. G. Roberts and L. Roszkowski, Phys. Lett. **B309**, 329 (1993).
53. H. Baer, V. Barger, D. Karatas, and X. Tata, Phys. Rev. **D36**, 96 (1987); see also R.M. Barnett, J.F. Gunion and H.E. Haber, Phys. Rev. **D37**, 1892 (1988).
54. H. Baer, M. Drees, C. Kao, M. Nojiri, and X. Tata, University of Wisconsin preprint MAD/PH/825 (1994), hep-ph 9403307.
55. J. F. Gunion and H. Pois, University of California at Davis preprint UCD-94-1, hep-ph 9402268.

REALISTIC SUPERSTRING MODELS

Alon E. Faraggi†
School of Natural Science, Institute for Advanced Study
Olden Lane, Princeton, NJ 08540

ABSTRACT

I discuss the construction of realistic superstring standard–like models in the four dimensional free fermionic formulation. I discuss the massless spectrum of the superstring standard–like models and the texture of fermion mass matrices. These models suggest an explanation for the top quark mass hierarchy. At the cubic level of the superpotential only the top quark get a mass term. The lighter quarks and leptons obtain their mass terms from nonrenormalizable terms that are suppressed relative to the cubic order term. A numerical estimate yielded $m_t \sim 175 - 180\ GeV$. The suppression of the lightest generation masses results from the horizontal symmetries in the superstring models. The problems of neutrino masses, gauge coupling unification and hierarchical SUSY breaking are discussed. I argue that the realistic features of these models are due to the underlying $Z_2 \times Z_2$ orbifold, with standard embedding, at the free fermionic point in toroidal compactification space.

1. Introduction

The most fundamental problem in high energy physics is the nature of the mechanism responsible for breaking the electroweak symmetry and for generating the fermion masses. Two main school of thoughts were developed to address this problem. The first assumes that the origin of symmetry breaking is dynamic and that the scalars doublets are composite. The second, assumes the existence of fundamental scalar representations and tries to incorporate the Standard Model into a fundamental theory, which unifies the known interactions at a much higher scale. In view of LEP precision data, the second approach is more successful [1]. Ultimately, future hadron colliders will determine the nature of the electroweak symmetry breaking mechanism.

---

† SSC fellow. e–mail address: faraggi@sns.ias.edu

Unified Symmetry: In the Small and in the Large,
Edited by B.N. Kursunoglu *et al.*, Plenum Press, New York, 1995

If we accept the notion of fundamental scalar representations and unification, we must question what is the fundamental scale of unification. Slow evolution of the Standard Model gauge couplings and proton lifetime support a big desert scenario. It is very plausible that the fundamental scale of unification is the Planck scale, at which none of the known interactions can be neglected. Many of the observations at low energies will arise from a fundamental theory at the Planck scale. The most developed Planck scale theories, to date, are superstring theories. In this talk I discuss the construction of realistic superstring models and their phenomenological implications.

Initially it was hoped that the uniqueness of the heterotic string [2] in ten dimensions would lead to a unique heterotic string theory in four dimensions. However, soon thereafter it was realized that in four dimensions there is a large number of consistent theories. Viable models can be constructed by compactifying the extra dimensions on a Calabi–Yau manifold [3] or on an orbifold [4]. Alternatively, one can formulate consistent string theories by identifying the extra degrees of freedom as an internal conformal field theory in the form of free bosons [5], free fermions [6], or as a product of minimal models [7]. Thus the initial hope that consistency alone will determine the string vacuum did not materialize.

Further progress can be made by pursuing a dual approach. On the one hand we must study the theoretical aspects of superstring theory and understand its fundamental principles at the perturbative and nonperturbative levels. We then hope to learn how the true string vacuum is selected. Alternatively, we may try to construct realistic string models by imposing phenomenological constraints. The realistic string models may then be used as a testing ground to test our ideas on string theory and to study how Planck scale physics may determine the parameters of the Standard Model.

Two approaches can be pursued to connect superstring theory with the Standard Model. One is through a GUT model at an intermediate energy scale [8,9]. The second possibility is to derive the Standard Model directly from superstring theory [10–14]. Due to proton decay considerations the second possibility is preferred. Consider the dimension four operators, $\eta_1 u_L^C d_L^C d_L^C + \eta_2 d_L^C QL$, that exist in the most general supersymmetric Standard Model. Unless η_1 and η_2 are highly suppressed, these dimension four operators mediate rapid proton decay. In the minimal supersymmetric Standard Model one imposes a discrete symmetry, R parity, that forbids these terms. In the context of superstring theories these discrete symmetries are usually not present. If $B-L$ is gauged as in $SO(10)$, these dimension four operators are forbidden by gauge invariance. However they may still be induced from the nonrenormalizable terms,

$$\eta_1(u_L^C d_L^C d_L^C N_L^C)\Phi + \eta_2(d_L^C Q L N_L^C)\Phi, \tag{1}$$

where Φ is a string of $SO(10)$ singlets that fixes the string selection rules and gets a VEV of $O(M_{Pl})$. N_L^C is the Standard Model singlet in the 16 of $SO(10)$. It is seen that the ratio $\langle N_L^c\rangle/M_{Pl}$ controls the rate of proton decay. Consequently, the VEV $\langle N_L^c\rangle$ has to be suppressed. In superstring GUT models, that have been constructed to date, $\langle N_L^c\rangle$ is used to break the GUT symmetry because there are no adjoint representations in the massless spectrum. Next, consider proton decay from dimension five operators. Dimension five operators are induced in SUSY GUT models by exchange of Higgsino color triplets [15]. Proton lifetime constraint requires that Higgsino color multiplets are sufficiently heavy, of the order of 10^{16} GeV. Supersymmetric GUT models must admit some doublet–triplet splitting mechanism, which satisfies these requirements. Although, such a mechanism has been constructed in different supersymmetric GUT models, in general, further assumptions have to be made on the matter content and interactions of the supersymmetric GUT models. If the Standard Model gauge group is obtained directly at the string level then we can construct models in which the Higgsino color triplets are projected out from the massless spectrum by the GSO projections. Thus, the proton lifetime considerations motivate us to conjecture that in a realistic string model the Standard Model nonabelian gauge group must be obtained directly at the string level.

In view of the large number of, a priori, possible string models, trying to construct one realistic model may not seem very meaning full. It is very plausible that models with some realistic features may be constructed in different regions of the compactification space. What would then tell us why one is preferred over the other. However, not all the points in the compactification space are alike. String theory exhibits a new kind of symmetry, usually referred to as target space duality [25], which is a generalization of the $R \to 1/R$ duality in the case of S^1. At the self–dual point, $R_j = 1/R_j$, space–time symmetries are enhanced. For appropriate choices of the background fields the space–time symmetries are maximally enhanced. At the maximally symmetric point the internal degrees of freedom that are needed to cancel the conformal anomaly may be represented in terms of internal free fermions propagating on the string world–sheet. It is not outrageous to assume that if string theory has anything to do with nature the true string model will be located near this highly symmetric point. Thus, we are led to consider superstring standard–like models in the free fermionic formulation.

2. Superstring standard–like models

The superstring standard–like models are constructed in the free fermionic formulation. In the free fermionic formulation [6] of the heterotic string in four dimensions all the world–sheet degrees of freedom required to cancel the conformal anomaly are represented in terms of free fermions propagating on the string world–sheet. For the left–movers (world–sheet supersymmetric) one has the usual space–time fields X^μ, ψ^μ, $(\mu = 0, 1, 2, 3)$, and in addition the following eighteen real free fermion fields: χ^I, y^I, ω^I $(I = 1, \cdots, 6)$, transforming as the adjoint representation of $SU(2)^6$. The supercurrent is given in terms of these fields as follows

$$T_F(z) = \psi^\mu \partial_z X_\mu + \sum_{i=1}^{6} \chi^i y^i \omega^i.$$

For the right movers we have $\bar{X}^\mu$ and 44 real free fermion fields: $\bar{\phi}^a$, $a = 1, \cdots, 44$. Under parallel transport around a noncontractible loop the fermionic states pick up a phase. A model in this construction is defined by a set of basis vectors of boundary conditions for all world–sheet fermions. These basis vectors are constrained by the string consistency requirements (e.g. modular invariance) and completely determine the vacuum structure of the model. The physical spectrum is obtained by applying the generalized GSO projections. The low energy effective field theory is obtained by S–matrix elements between external states. The Yukawa couplings and higher order nonrenormalizable terms in the superpotential are obtained by calculating correlators between vertex operators. For a correlator to be nonvanishing all the symmetries of the model must be conserved. Thus, the boundary condition vectors determine the phenomenology of the models.

The first five vectors (including the vector **1**) in the basis consist of the NAHE* set

$$S = (\underbrace{1, \cdots, 1}_{\psi^\mu, \chi^{1..6}}, 0, \cdots, 0 | 0, \cdots, 0). \tag{2a}$$

$$b_1 = (\underbrace{1, \cdots\cdots\cdots, 1}_{\psi^\mu, \chi^{12}, y^{3,\dots,6}, \bar{y}^{3,\dots,6}}, 0, \cdots, 0 | \underbrace{1, \cdots, 1}_{\bar{\psi}^{1,\dots,5}, \bar{\eta}^1}, 0, \cdots, 0). \tag{2b}$$

$$b_2 = (\underbrace{1, \cdots\cdots\cdots\cdots, 1}_{\psi^\mu, \chi^{34}, y^{1,2}, \omega^{5,6}, \bar{y}^{1,2} \bar{\omega}^{5,6}}, 0, \cdots, 0 | \underbrace{1, \cdots, 1}_{\bar{\psi}^{1,\dots,5}, \bar{\eta}^2}, 0, \cdots, 0). \tag{2c}$$

$$b_3 = (\underbrace{1, \cdots\cdots\cdots\cdots, 1}_{\psi^\mu, \chi^{56}, \omega^{1,\cdots,4}, \bar{\omega}^{1,\cdots,4}}, 0, \cdots, 0 | \underbrace{1, \cdots, 1}_{\bar{\psi}^{1,\dots,5}, \bar{\eta}^3}, 0, \cdots, 0). \tag{2d}$$

with the choice of generalized GSO projections $c\begin{pmatrix} b_i \\ b_j \end{pmatrix} = c\begin{pmatrix} b_i \\ S \end{pmatrix} = c\begin{pmatrix} 1 \\ 1 \end{pmatrix} = -1$, and the others given by modular invariance.

The gauge group after the NAHE set is $SO(10) \times E_8 \times SO(6)^3$ with $N = 1$ space–time supersymmetry, and 48 spinorial 16 of $SO(10)$, sixteen from each sector b_1, b_2 and b_3. The NAHE set divides the internal world–sheet fermions in the following way: $\bar{\phi}^{1,\cdots,8}$ generate the hidden E_8 gauge group, $\bar{\psi}^{1,\cdots,5}$ generate the $SO(10)$ gauge group, and $\{\bar{y}^{3,\cdots,6}, \bar{\eta}^1\}$, $\{\bar{y}^1, \bar{y}^2, \bar{\omega}^5, \bar{\omega}^6, \bar{\eta}^2\}$, $\{\bar{\omega}^{1,\cdots,4}, \bar{\eta}^3\}$ generate the three horizontal $SO(6)^3$ symmetries. The left–moving $\{y, \omega\}$ states are divided to $\{y^{3,\cdots,6}\}$, $\{y^1, y^2, \omega^5, \omega^6\}$, $\{\omega^{1,\cdots,4}\}$ and χ^{12}, χ^{34}, χ^{56} generate the left–moving $N = 2$ world–sheet supersymmetry.

The internal fermionic states $\{y, \omega | \bar{y}, \bar{\omega}\}$ correspond to the six left–moving and six right–moving compactified dimensions in a geometric formulation. This correspondence is illustrated by adding the vector X to the NAHE set, with periodic boundary conditions for the set $(\bar{\psi}^{1,\cdots,5}, \bar{\eta}^{1,2,3})$ and antiperiodic boundary conditions for all other world–sheet fermions. This boundary condition vector extends the gauge symmetry to $E_6 \times U(1)^2 \times E_8 \times SO(4)^3$ with $N = 1$ supersymmetry and twenty-four chiral 27 of E_6. The same model is generated in the orbifold language [4] by moding out an $SO(12)$ lattice by a $Z_2 \times Z_2$ discrete symmetry with standard embedding. In the construction of the standard–like models beyond the NAHE set, the assignment of boundary conditions to the set of internal fermions $\{y, \omega | \bar{y}, \bar{\omega}\}$ determines many of the properties of the low energy spectrum, such as the number of generations, the presence of Higgs doublets, Yukawa couplings, etc.

In the realistic free fermionic models the boundary condition vector X is replaced by the vector 2γ in which $\{\bar{\psi}^{1,\cdots,5}, \bar{\eta}^1, \bar{\eta}^2, \bar{\eta}^3, \bar{\phi}^{1,\cdots,4}\}$ are periodic and the remaining left– and right–moving fermionic states are antiperiodic. The set $\{1, S, 2\gamma, \xi_2\}$ generates a model with $N = 4$ space–time supersymmetry and $SO(12) \times SO(16) \times SO(16)$ gauge group. The b_1 and b_2 twist are applied to reduce the number of supersymmetries from $N = 4$ to $N = 1$ space–time supersymmetry. The gauge group is broken to $SO(4)^3 \times U(1)^3 \times SO(10) \times E_8$. The $U(1)$ combination $U(1) = U(1)_1 + U(1)_2 + U(1)_3$ has a non–vanishing trace and the trace of the two orthogonal combinations vanishes. The number of generations is still 24, eight from each sector b_1, b_2 and b_3 The chiral generations are

* This set was first constructed by Nanopoulos, Antoniadis, Hagelin and Ellis (NAHE) in the construction of the flipped $SU(5)$. *nahe*=pretty, in Hebrew.

now 16 of $SO(10)$ from the sectors b_j ($j = 1, 2, 3$). The $10+1$ and the E_6 singlets from the sectors $b_j + X$ are replaced by vectorial 16 of the hidden $SO(16)$ gauge group from the sectors $b_j + 2\gamma$. As I will show below the structure of the sector $b_j + 2\gamma$ with respect to the sectors b_j plays an important role in the texture of fermion mass matrices.

The standard–like models are constructed by adding three additional vectors to the NAHE set [11,12,13,14]. The the $SO(10)$ symmetry is broken in two stages, first to $SO(6) \times SO(4)$ and next to $SU(3) \times SU(2) \times U(1)^2$. One example is presented in the table, where only the boundary conditions of the "compactified space" are shown. In the gauge sector $\alpha, \beta\{\bar{\psi}^{1,\cdots,5}, \bar{\eta}^{1,2,3}, \bar{\phi}^{1,\cdots,8}\} = \{1^3, 0^5, 1^4, 0^4\}$ and $\gamma\{\bar{\psi}^{1,\cdots,5}, \bar{\eta}^{1,2,3}, \bar{\phi}^{1,\cdots,8}\} = \{\frac{1}{2}^9, 0, 1^2, \frac{1}{2}^3, 0\}$ break the symmetry to $SU(3) \times SU(2) \times U(1)_{B-L} \times U(1)_{T_{3_R}} \times SU(5)_h \times SU(3)_h \times U(1)^2$. The choice of generalized GSO coefficients is: $c\begin{pmatrix} b_j \\ \alpha, \beta, \gamma \end{pmatrix} = -c\begin{pmatrix} \alpha \\ 1 \end{pmatrix} = c\begin{pmatrix} \alpha \\ \beta \end{pmatrix} = -c\begin{pmatrix} \beta \\ 1 \end{pmatrix} = c\begin{pmatrix} \gamma \\ 1, \alpha \end{pmatrix} = -c\begin{pmatrix} \gamma \\ \beta \end{pmatrix} = -1$ (j=1,2,3), with the others specified by modular invariance and space–time supersymmetry.

3. The number of generations

Free fermionic models with the NAHE set correspond to $Z_2 \times Z_2$ orbifold at a special point in toroidal compactification space. At this point the internal compactified dimensions can be represented in terms of free world–sheet fermions. At this specific point the symmetries due to the compactified dimensions are enhanced from $U(1)^6$ to $SO(12)$. The enhancement is due both to compactification at the self–dual point $R_j = 1/R_j$ and due to specific values of the background fields. The structure of the $Z_2 \times Z_2$ orbifold, with standard embedding, at the specific point in compactification space is the root of the realistic properties of the free fermionic models. The first requirement from any superstring model is that the low energy spectrum contains a net chirality of three generations. A general $Z_2 \times Z_2$ orbifold would not produce three generation models. However, miraculously, at the most symmetric point in compactification space, three generations are obtained very naturally. The reason is that at this point the number of fixed points, in each twisted sector, can be simultaneously reduced to one fixed point, from each twisted sector. The three generations are then aligned along the three orthogonal complex planes of the $Z_2 \times Z_2$ orbifold. At the level of the NAHE set each sector b_1, b_2 and b_3 has eight generations. Three additional vectors are needed to reduce the number of generations to one generation from each sector b_1, b_2 and b_3. Each generation has horizontal symmetries that constrain the allowed interactions. Each generation has two gauged $U(1)$ symmetries $U(1)_{R_j}$ and $U(1)_{R_{j+3}}$. For every right–moving $U(1)$ symmetry there is a corresponding left–moving global $U(1)$ symmetry $U(1)_{L_j}$ and $U(1)_{L_{j+3}}$. Finally, each generation has two Ising model operators that are obtained by pairing a left–moving real fermion with a right–moving real fermion.

Table 1. A three generations $SU(3) \times SU(2) \times U(1)^2$ model [13].

	y^3y^6, $y^4\bar{y}^4$, $y^5\bar{y}^5$, $\bar{y}^3\bar{y}^6$	$y^1\omega^6$, $y^2\bar{y}^2$, $\omega^5\bar{\omega}^5$, $\bar{y}^1\bar{\omega}^6$	$\omega^1\omega^3$, $\omega^2\bar{\omega}^2$, $\omega^4\bar{\omega}^4$, $\bar{\omega}^1\bar{\omega}^3$
α	1, 1, 1, 0	1, 1, 1, 0	1, 1, 1, 0
β	0, 1, 0, 1	0, 1, 0, 1	1, 0, 0, 0
γ	0, 0, 1, 1	1, 0, 0, 0	0, 1, 0, 1

4. Higgs doublets

Higgs doublets in the standard–like models are obtained from two distinct sectors. The first type are obtained from the Neveu–Schwarz sector, which produces three pairs of electroweak doublets $\{h_1, h_2, h_3, \bar{h}_1, \bar{h}_2, \bar{h}_3\}$. The Neveu–Schwarz sector corresponds to the untwisted sector of the orbifold models. Each pair of Higgs doublets can couple at tree level only to the states from the sector b_j. This results from the horizontal symmetries, $U(1)_j$, $(1,2,3)$ and is a reflection of the structure of the $Z_2 \times Z_2$ twisting. There is a stringy doublet–triplet splitting mechanism that projects out the color triplets and leaves the electroweak doublets in the spectrum. Thus, the superstring standard–like models resolve the GUT hierarchy problem. The second type of Higgs doublets are obtained from the vector combination $b_1 + b_2 + \alpha + \beta$. The states in this sector are obtained by acting on the vacuum with a single fermionic oscillator and transform only under the observable sector.

In addition to electroweak doublets and color triplets the Neveu–Schwarz sector and the sector $b_1 + b_2 + \alpha + \beta$ produce singlets of $SO(10) \times E_8$. These singlets play an important role in the phenomenology of the superstring standard–like models. The VEVs of the $SO(10)$ singlet fields in the massless spectrum of the superstring models determine the light Higgs representations and generate the fermion mass hierarchy.

The Neveu–Schwarz sector and the sector $b_1 + b_2 + \alpha + \beta$ produce four [12] or five [13] pairs of electroweak doublets. Several pairs receive heavy mass from the VEVs of Standard Model singlets in the massless spectrum. At the cubic level there are two pairs of electroweak doublets. The light Higgs doublets are combinations of (h_1, h_2, h_{45}) and $(\bar{h}_1, \bar{h}_2, \bar{h}_{45})$. The Higgs doublets h_3 and $\bar{h}_3$

obtain a large mass from $SO(10)$ singlet VEVs. This results from requiring F–flatness of the cubic level superpotential [20]. The absence of h_3 and $\bar{h}_3$ from the light eigenstates results in b_3 being identified with the lightest generation. At the nonrenormalizable level one additional pair receives a superheavy mass and one pair remains light to give masses to the fermions at the electroweak scale. Requiring F–flatness imposes that the light Higgs representations are $\bar{h}_1$ or $\bar{h}_2$ and h_{45}.

5. The sectors $b_j + 2\gamma$

As mentioned above the realistic free fermionic models contain massless states from the sectors $b_j+2\gamma$ $(j = 1, 2, 3)$. These states arise due to the $Z_2 \times Z_2$ twisting on a gauge lattice with $SO(16) \times SO(16)$ rather than $E_8 \times E_8$. Thus, the realistic free fermionic models correspond to $(2,0)$ rather than $(2,2)$ compactification. The sectors $b_j + 2\gamma$ produce the vectorial 16 representation of the hidden $SO(16)$ gauge group, decomposed under the final hidden gauge group. The number of 16 of the hidden $SO(16)$ is equal to the number of 16 of the observable $SO(10)$ gauge group. The horizontal charges in the sectors $b_j + 2\gamma$ are similar to the ones in the sectors b_j. The VEVs of the states from the sectors $b_j + 2\gamma$ are responsible for generating texture zeroes in the fermion mass matrices.

The massless spectrum described until now results from the $Z_2 \times Z_2$ twist with standard embedding. Therefore, it generally holds for all the free fermionic models that are based on $Z_2 \times Z_2$ orbifold with standard embedding. In addition to the states from the sectors mentioned above there are massless sectors that arise due to the sectors that correspond to Wilson line breaking. The states in these sectors usually do not have the standard $SO(10)$ embedding. In particular the weak hypercharge and $U(1)_{Z'}$ charge usually differs from the standard $SO(10)$ assignment.

6. Top quark mass hierarchy

Trilinear and nonrenormalizable contributions to the superpotential are obtained by calculating correlators between vertex operators [16]

$$A_N \sim \langle V_1^f V_2^f V_3^b \cdots V_N^b \rangle, \tag{3}$$

where V_i^f (V_i^b) are the fermionic (scalar) components of the vertex operators. The non–vanishing terms are obtained by applying the rules of Ref. [16]. The cubic level Yukawa couplings for the quarks and leptons are determined by the boundary conditions in the vector γ according to the following rule [14,17]

$$\Delta_j = |\gamma(U(1)_{L_{j+3}}) - \gamma(U(1)_{R_{j+3}})| = 0, 1 \qquad (j = 1, 2, 3) \tag{4a}$$

$$\Delta_j = 0 \rightarrow d_j Q_j h_j + e_j L_j h_j; \tag{4b}$$

$$\Delta_j = 1 \rightarrow u_j Q_j \bar{h}_j + N_j L_j \bar{h}_j, \tag{4c}$$

where $\gamma(U(1)_{R_{j+3}})$, $\gamma(U(1)_{L_{j+3}})$ are the boundary conditions of the world–sheet fermionic currents that generate the $U(1)_{R_{j+3}}$, $U(1)_{L_{j+3}}$ symmetries.

The superstring standard–like models contain an anomalous $U(1)$ gauge symmetry. The anomalous $U(1)$ generates a Fayet–Iliopoulos term by the VEV of the dilaton field that breaks supersymmetry and destabilizes the vacuum [19]. Supersymmetry is restored by giving VEVs to Standard Model singlets in the massless spectrum of the superstring models. However, as the charge of these singlets must have $Q_A < 0$ to cancel the anomalous $U(1)$ D–term equation, in many models a phenomenologically realistic solution does not exist. In fact a very restricted class of standard–like models with $\Delta_j = 1$ for $j = 1, 2, 3$, were found to admit a solution to the F and D flatness constraints. Consequently, the only models that were found to admit a solution are models which have tree level Yukawa couplings only for $+\frac{2}{3}$ charged quarks.

This result suggests an explanation for the top quark mass hierarchy relative to the lighter quarks and leptons. At the cubic level only the top quark gets a mass term and the mass terms for the lighter quarks and leptons are obtained from nonrenormalizable terms. To study this scenario we have to examine the nonrenormalizable contributions to the doublet Higgs mass matrix and to the fermion mass matrices [20,21].

At the cubic level there are two pairs of electroweak doublets. At the nonrenormalizable level one additional pair receives a superheavy mass and one pair remains light to give masses to the fermions at the electroweak scale. Requiring F–flatness imposes that the light Higgs representations are $\bar{h}_1$ or $\bar{h}_2$ and h_{45}.

The nonrenormalizable fermion mass terms of order N are of the form $cgf_if_jh\phi^{N-3}$ or $cgf_if_j\bar{h}\phi^{N-3}$, where c is a calculable coefficient, g is the gauge coupling at the unification scale, f_i, f_j are the fermions from the sectors b_1, b_2 and b_3, h and $\bar{h}$ are the light Higgs doublets, and ϕ^{N-3} is a string of Standard Model singlets that get a VEV and produce a suppression factor $(\langle\phi\rangle/M)^{N-3}$ relative to the cubic level terms. Several scales contribute to the generalized VEVs. The leading one is the scale of VEVs that are used to cancel the "anomalous" $U(1)$ D–term equation. The next scale is generated by Hidden sector condensates. Finally, there is a scale which is related to the breaking of $U(1)_{Z'}$, $\Lambda_{Z'}$. Examination of the higher order nonrenormalizable terms reveals that $\Lambda_{Z'}$ has to be suppressed relative to the other two scales.

At the cubic level only the top quark gets a nonvanishing mass term. Therefore only the top quark mass is characterized by the electroweak scale. The remaining quarks and leptons obtain their mass terms from nonrenormalizable terms. The cubic and nonrenormalizable terms in the superpotential are obtained by calculating correlators between the vertex operators. The top quark Yukawa coupling is generically given by

$$g\sqrt{2} \tag{5}$$

where g is the gauge coupling at the unification scale. In the model of Ref. [13], bottom quark and tau lepton mass terms are obtained at the quartic order,

$$W_4 = \{d^c_{L_1} Q_1 h'_{45} \Phi_1 + e^c_{L_1} L_1 h'_{45} \Phi_1 + d^c_{L_2} Q_2 h'_{45} \bar{\Phi}_2 + e^c_{L_2} L_2 h'_{45} \bar{\Phi}_2\}. \tag{6}$$

The VEVs of Φ are obtained from the cancelation of the anomalous D–term equation. The coefficient of the quartic order mass terms were calculated by calculating the quartic order correlators and the one dimensional integral was evaluated numerically. Thus after inserting the VEV of $\bar{\Phi}_2$ the effective bottom quark and tau lepton Yukawa couplings are given by [13],

$$\lambda_b = \lambda_\tau = 0.35 g^3. \tag{7}$$

They are suppressed relative to the top Yukawa by

$$\frac{\lambda_b}{\lambda_t} = \frac{0.35 g^3}{g\sqrt{2}} \sim \frac{1}{8}. \tag{8}$$

To evaluate the top quark mass, the three Yukawa couplings are run to the low energy scale by using the MSSM RGEs. The bottom mass is then used to calculate $\tan\beta$ and the top quark mass is found to be [13],

$$m_t \sim 175 - 180 GeV. \tag{9}$$

The fact that the top Yukawa is found near a fixed point suggests that this is in fact a good prediction of the superstring standard–like models. By varying $\lambda_t \sim 0.5 - 1.5$ at the unification scale, it is found that λ_t is always $O(1)$ at the electroweak scale.

7. Fermion mass matrices

An analysis of fermion mass terms up to order $N = 8$ revealed the general texture of fermion mass matrices in these models. The sectors b_1 and b_2 produce the two heavy generations. Their mass terms are suppressed by singlet VEVs that are used in the cancellation of the anomalous $U(1)$ D–term equation. The sector b_3 produces the lightest generation. The diagonal mass terms for the states from b_3 can only be generated by VEVs that break $U(1)_{Z'}$. This is due to the horizontal $U(1)$ charges and because the Higgs pair h_3 and $\bar{h}_3$ necessarily gets a Planck scale mass [20]. The suppression of the lightest generation mass terms is seen to be a result of the structure of the vectors α and β with respect to the sectors b_1, b_2 and b_3. The mixing between the generations is obtained from exchange of states from the sectors $b_j + 2\gamma$. The general texture of the fermion mass matrices in the superstring standard–like models is of the following form,

$$M_U \sim \begin{pmatrix} \epsilon, a, b \\ \tilde{a}, A, c \\ \tilde{b}, \tilde{c}, \lambda_t \end{pmatrix} ; \quad M_D \sim \begin{pmatrix} \epsilon, d, e \\ \tilde{d}, B, f \\ \tilde{e}, \tilde{f}, C \end{pmatrix} ; \quad M_E \sim \begin{pmatrix} \epsilon, g, h \\ \tilde{g}, D, i \\ \tilde{h}, \tilde{i}, E \end{pmatrix},$$

where $\epsilon \sim (\Lambda_{Z'}/M)^2$. The diagonal terms in capital letters represent leading terms that are suppressed by singlet VEVs, and $\lambda_t = O(1)$. The mixing terms are generated by hidden sector states from the sectors $b_j + 2\gamma$ and are represented by small letters. They are proportional to $(\langle TT \rangle / M^2)$.

8. Quark flavor mixing

In Ref. [21] it was shown that if the states from the sectors $b_j + 2\gamma$ obtain VEVs in the application of the DSW mechanism, then a Cabibbo angle of the correct order of magnitude can be obtained in the superstring standard–like models. For one specific choice of singlet VEVs that solve the cubic level F and D constraints the down mass matrix M_D is given by

$$M_d \sim \begin{pmatrix} \epsilon & \frac{V_2 \bar{V}_3 \Phi_{45}}{M^3} & 0 \\ \frac{V_2 \bar{V}_3 \Phi_{45} \xi_1}{M^4} & \frac{\bar{\Phi}_2^- \xi_1}{M^2} & 0 \\ 0 & 0 & \frac{\Phi_1^+ \xi_2}{M^2} \end{pmatrix} v_2, \tag{10}$$

where $v_2 = \langle h_{45} \rangle$ and we have used $\frac{1}{2} g \sqrt{2\alpha'} = \sqrt{8\pi}/M_{Pl}$, to define $M \equiv M_{Pl}/2\sqrt{8\pi} \approx 1.2 \times 10^{18} GeV$ [16]. The undetermined VEVs of $\bar{\Phi}_{13}$ and ξ_2 are used to fix m_b and m_s such that $\langle \xi_1 \rangle \sim M$. We also take $tan\beta = v_1/v_2 \sim 1.5$. Substituting the values of the VEVs above and diagonalizing M_D by a biunitary transformation we obtain the Cabibbo mixing matrix

$$|V| \sim \begin{pmatrix} 0.98 & 0.2 & 0 \\ 0.2 & 0.98 & 0 \\ 0 & 0 & 1 \end{pmatrix}. \tag{11}$$

Since the running from the scale M down to the weak scale does not affect the Cabibbo angle by much [22], we conclude that realistic mixing of the correct order of magnitude can be obtained in this scenario. The analysis was extended to show that reasonable values for the entire CKM matrix parameters can be obtained for appropriate flat F and D solutions. For one specific solution the up and down quark mass matrices take the form

$$M_u \sim \begin{pmatrix} \epsilon & \frac{V_3 \bar{V}_2 \Phi_{45} \bar{\Phi}_3^+}{M^4} & 0 \\ \frac{V_3 \bar{V}_2 \Phi_{45} \bar{\Phi}_2^+}{M^4} & \frac{\bar{\Phi}_i^- \bar{\Phi}_i^+}{M^2} & \frac{V_1 \bar{V}_2 \Phi_{45} \bar{\Phi}_2^+}{M^4} \\ 0 & \frac{V_1 \bar{V}_2 \Phi_{45} \bar{\Phi}_1^+}{M^4} & 1 \end{pmatrix} v_1, \tag{12}$$

and

$$M_d \sim \begin{pmatrix} \epsilon & \frac{V_3 \bar{V}_2 \Phi_{45}}{M^3} & 0 \\ \frac{V_3 \bar{V}_2 \Phi_{45} \xi_1}{M^4} & \frac{\bar{\Phi}_2^- \xi_1}{M^2} & \frac{V_1 \bar{V}_2 \Phi_{45} \xi_i}{M^4} \\ 0 & \frac{V_1 \bar{V}_2 \Phi_{45} \xi_i}{M^4} & \frac{\Phi_1^+ \xi_2}{M^2} \end{pmatrix} v_2, \tag{13}$$

with v_1, v_2 and M as before. The up and down quark mass matrices are diagonalized by bi–unitary transformations

$$U_L M_u U_R^\dagger = D_u \equiv \mathrm{diag}(m_u, m_c, m_t), \tag{14a}$$

$$D_L M_d D_R^\dagger = D_d \equiv \mathrm{diag}(m_d, m_s, m_b), \tag{14b}$$

with the CKM mixing matrix given by

$$V = U_L D_L^\dagger. \tag{15}$$

The VEVs of ξ_1 and ξ_2 are fixed to be $\langle \xi_1 \rangle \sim M/12$ and $\langle \xi_2 \rangle \sim M/4$ by the masses m_s and m_b respectively. Substituting the VEVs and diagonalizing M_u and M_d by a bi–unitary transformation, we obtain the mixing matrix

$$|V| \sim \begin{pmatrix} 0.98 & 0.205 & 0.002 \\ 0.205 & 0.98 & 0.012 \\ 0.0004 & 0.012 & 0.99 \end{pmatrix}. \tag{16}$$

The texture and hierarchy of the mass terms in Eqs. (11–12) arise due to the set of singlet VEVs in Eqs. (29). The zeroes in the 13 and 31 entries of the mass matrices are protected to all orders of nonrenormalizable terms. To obtain a non–vanishing contribution to these entries either V_1 and $\bar{V}_3$ or V_3 and $\bar{V}_1$ must obtain a VEV simultaneously. Thus, there is a residual horizontal symmetry that protects these vanishing terms. The 11 entry in the mass matrices, e.g. the diagonal mass terms for the lightest generation states, can only be obtained from VEVs that break $U(1)_{Z'}$ [18]. We assume that $U(1)_{Z'}$ is broken at an intermediate energy scale that is suppressed relative to the scale of scalar VEVs [20]. In Ref. [26] we showed that $U(1)_{Z'}$ is broken by hidden sector matter condensates at $\Lambda_{Z'} \leq 10^{14} GeV$. Consequently, we have taken $\epsilon \leq (\Lambda_{Z'}/M)^2 \sim 10^{-8}$.

Texture zeroes in the fermion mass matrices are obtained if the VEVs of some states from the sectors $b_j + 2\gamma$ vanish. These texture zeroes are protected by the symmetries of the string models to all order of nonrenormalizable terms [21]. For example in the above mass matrices the 13 and 31 vanish because $\{V_1, V_3\}$ get a VEV but $\bar{V}_1$ and $\bar{V}_3$ do not. Therefore these mass matrix terms cannot be formed because they would not be invariant under all the string symmetries. Other textures are possible for other choices of VEVs for the states from the sectors $b_j + 2\gamma$.

9. Neutrino masses

A seesaw type neutrino mass matrix can be constructed from analysis of non-renormalizable terms and for specific choices of singlet VEVs [26]. The neutrino seesaw mass matrix takes the general form The neutrino mass matrix therefore takes the following form for each generation in the basis (ν_L, N^C, Φ)

$$\begin{pmatrix} 0 & km_u & 0 \\ km_u & 0 & m_\chi \\ 0 & m_\chi & m_\phi \end{pmatrix}, \tag{17}$$

with $m_\chi \sim \left(\frac{\Lambda_{Z'}}{M}\right)^3 \left(\frac{\langle\phi\rangle}{M}\right)^n M$ and $m_\phi \sim \left(\frac{\Lambda_{Z'}}{M}\right)^4 \left(\frac{\langle\phi\rangle}{M}\right)^m M$. n and m are the orders at which the terms are obtained. The mass eigenstates are mainly ν, N and ϕ with a small mixing and with the eigenvalues

$$m_\nu \sim m_\phi \left(\frac{km_u}{m_\chi}\right)^2 \qquad m_N, M_\phi \sim m_\chi \tag{18}$$

The constant k gives the effects of Yukawa coupling renormalization. The seesaw scale m_χ is determined by the $U(1)_{Z'}$ breaking scale and by the order at which the nonrenormalizable seesaw terms are obtained. In Ref. [26] the $U(1)_{Z'} \sim 10^{14}\ GeV$ breaking scale was obtained from condensates of the hidden $SU(5)$ gauge group with nontrivial $U(1)_{Z'}$ charges. The order of nonrenormalizable terms that contribute to the seesaw terms in the neutrino mass matrix depends highly on the choice of flat flat directions. Neutrino masses that are in agreement with experimental constraints can be obtained. A novel feature of the superstring seesaw mechanism is that although the $U(1)_{Z'}$ breaking scale may be large (e.g. $\Lambda_{Z'} \approx 10^{14} GeV$) the effective see-saw scale can be much smaller.

9. Gauge coupling unification

While LEP results indicate that the gauge coupling in the minimal supersymmetric Standard Model unify at $10^{16} GeV$, superstring theory predicts that the unification scale is at $10^{18} GeV$. The superstring standard–like models may resolve this problem due to the existence of color triplets and electroweak doublets from exotic sectors that arise from the additional vectors α, β and γ. These exotic states carry fractional charges and do not fit into standard $SO(10)$ representations. Therefore, they contribute less to the evolution of the $U(1)_Y$ beta function than standard $SO(10)$ multiplets. For example in Ref. [23] representations with the following beta function coefficients, in a $SU(3) \times SU(2) \times U(1)_Y$ basis, were found

$$b_{D_1,\bar{D}_1,D_2,\bar{D}_2} = \begin{pmatrix} \frac{1}{2} \\ 0 \\ \frac{1}{5} \end{pmatrix} ; b_{D_3,\bar{D}_3} = \begin{pmatrix} \frac{1}{2} \\ 0 \\ \frac{1}{20} \end{pmatrix} ; b_{\ell,\bar{\ell}} = \begin{pmatrix} 0 \\ \frac{1}{2} \\ 0 \end{pmatrix} .$$

The standard–like models predict $\sin^2\theta_W = 3/8$ at the unification scale due to the embedding of the weak hypercharge in $SO(10)$. In Ref. [23], I showed that provided that the additional exotic color triplets and electroweak doublets exist at the appropriate scales, the scale of gauge coupling unification is pushed to $10^{18} GeV$, with the correct value of $\sin^2\theta_W$ at low energies.

11. Hierarchical SUSY breaking

In Ref. [24] we address the following question: Given a supersymmetric string vacuum at the Planck scale, is it possible to obtain hierarchical supersymmetry breaking in the observable sector? A supersymmetric string vacuum is obtained by finding solutions to the cubic level F and D constraints. We take a gauge coupling in agreement with gauge coupling unification, thus taking a fixed value for the dilaton VEV. We then investigate the role of nonrenormalizable terms and strong hidden sector dynamics. The hidden sector contains two non–Abelian hidden gauge groups, $SU(5) \times SU(3)$, with matter in vector–like representations. The hidden $SU(3)$ group is broken near the Planck scale. We analyze the dynamics of the hidden $SU(5)$ group. The $SU(5)$ hidden matter mass matrix is given by

$$\mathcal{M} = \begin{pmatrix} 0 & C_1 & 0 \\ B_1 & A_2 & C_2 \\ 0 & C_3 & A_1 \end{pmatrix} , \qquad (19)$$

where A, B, C arise from nonrenormalizable terms of orders $N = 5, 8, 7$ respectively and are given by

$$A_1 = \frac{\langle \Phi_{45} \bar{\Phi}_1^- \xi_2 \rangle}{M^2}, \qquad A_2 = \frac{\langle \Phi_{45} \Phi_2^+ \xi_1 \rangle}{M^2}, \tag{20a, b}$$

$$B_1 = \frac{\langle V_3 \bar{V}_2 \Phi_{45} \Phi_{45} \bar{\Phi}_{13} \xi_1 \rangle}{M^5}, \tag{20c}$$

$$C_1 = \frac{\langle V_3 \bar{V}_2 \Phi_{45} \Phi_{45} \bar{\Phi}_{13} \rangle}{M^4}, \quad C_2 = \frac{\langle V_1 \bar{V}_2 \Phi_{45} \Phi_{45} \xi_1 \rangle}{M^4}, \tag{20d, e}$$

$$C_3 = \frac{\langle V_1 \bar{V}_2 \Phi_{45} \Phi_{45} \xi_2 \rangle}{M^4} \quad . \tag{20f}$$

Taking generically $\langle \phi \rangle \sim gM/4\pi \sim M/10$ we obtain $A_i \sim 10^{15}\ GeV$, $B_i \sim 10^{12}\ GeV$, and $C_i \sim 10^{13}\ GeV$. From Eqs. (19–20) we observe that to insure a nonsingular hidden matter mass matrix, we must require $C_1 \neq 0$ and $B_1 \neq 0$. This imposes $\bar{V}_3 \neq 0$ and $V_2 \neq 0$. Thus, the nonvanishing VEVs that generates the Cabibbo mixing also guarantee the stability of the supersymmetric vacuum. The gaugino and matter condensates are given by the well known expressions for supersymmetric $SU(N)$ with matter in $N + \bar{N}$ representations [27],

$$\frac{1}{32\pi^2} \langle \lambda\lambda \rangle = \Lambda^3 \left(det \frac{\mathcal{M}}{\Lambda} \right)^{1/N}, \tag{21a}$$

$$\Pi_{ij} = \langle \bar{T}_i T_j \rangle = \frac{1}{32\pi^2} \langle \lambda\lambda \rangle \, \mathcal{M}_{ij}{}^{-1}, \tag{21b}$$

where $\langle \lambda\lambda \rangle$, $\mathcal{M}$ and Λ are the hidden gaugino condensate, the hidden matter mass matrix and the $SU(5)$ condensation scale, respectively. Modular invariant generalization of Eqs. (20a,b) for the string case were derived in Ref. [28]. The nonrenormalizable terms can be put in modular invariant form by following the procedure outlined in Ref. [29]. Approximating the Dedekind η function by $\eta(\hat{T}) \approx e^{-\pi \hat{T}/12}(1 - e^{-2\pi \hat{T}})$ we verified that the calculation using the modular invariant expression from Ref. [28] (with $\langle \hat{T} \rangle \approx M$) differ from the results using Eq. (20), by at most an order of magnitude. The hidden $SU(5)$ matter mass matrix is nonsingular for specific F and D flat solutions. In Ref. [24] a specific cubic level F and D flat solution was found. The gravitino mass due to the gaugino and matter condensates was estimated to be of the order $1 - 10\ TeV$. The new aspect of our scenario for supersymmetry breaking is the following. As long as only states from the Neveu–Schwarz sector or the sector $b_1 + b_2 + \alpha + \beta$ receive VEVs in the application of the DSW mechanism then one can find exact flat directions at the cubic level of the superpotential. These flat directions will be exact and will not be spoiled by nonrenormalizable terms. The states from the Neveu–Schwarz sector and the sector $b_1 + b_2 + \alpha + \beta$ correspond to untwisted and twisted moduli. However, once some hidden sector matter states obtain a nonvanishing VEV, the cubic level flat directions are no longer exact. Supersymmetry is broken by the inclusion of nonrenormalizable terms. Hidden sector strong dynamics at an intermediate scale may then be responsible for generating the hierarchy in the usual fashion.

12. Conclusion

The Standard Model is in agreement with all current experiments. Furthermore, present day experiments seem to support the big desert scenario and the notion of unification. The Planck scale is the ultimate scale of unification at which none of the known interactions can be ignored. Many properties of the Standard Model will arise from the fundamental Planck scale theory. Superstring theory stands out as the only known theory that can consistently unify gravity with the gauge interactions. The heterotic string is the only string theory that can produce realistic phenomenology. Its consistency requires twenty–six critical dimensions in the bosonic sector and ten critical dimensions in the supersymmetric sector. In the bosonic sector sixteen degrees of freedom are compactified on a flat torus and produce the observable and hidden gauge degrees of freedom. Six degrees of freedom from the bosonic sector, combined with six degrees of freedom from the supersymmetric sector, are compactified on a Calabi–Yau manifold or on an orbifold. String theory exhibits a new kind of symmetry: "target space duality". At the self–dual point, the compactified degrees of freedom can be represented in terms of free world–sheet fermions. At this point space–time symmetries are maximally enhanced. The most realistic superstring models constructed to date were constructed at this point in the compactification space. The underlying structure of the $Z_2 \times Z_2$ orbifold at the free fermionic point in toroidal compactification space is the origin of the realistic nature of free fermionic models. We believe that if string unification is relevant in nature, then the underlying structure of the $Z_2 \times Z_2$ orbifold at the free fermionic point in toroidal compactification space will be intrinsic to the eventual "true" heterotic string model. Thus, it makes sense, in our opinion, to try to build realistic models specifically at this point in the huge compactification space.

The superstring standard–like models contain in their massless spectrum all the necessary states to obtain realistic low energy phenomenology. They resolve the problems of proton decay through dimension four and five operators that are endemic to other superstring and GUT models. The existence of only three generations with standard $SO(10)$ embedding is understood to arise naturally from $Z_2 \times Z_2$ twisting at the free fermionic point in compactification space. Better understanding of the correspondence with other superstring formulations will provide further insight into the realistic properties of these models. In this context it is especially interesting to try to understand the significance of the self-dual point in the compactification space. Finally, the free fermionic standard-like models provide a highly constrained and phenomenologically realistic laboratory to study how the Planck scale may determine the parameters of the Standard Model.

12. Acknowledgments

This work is supported by an SSC fellowship. Part of the work described in this talk was done in collaboration with Edi Halyo.

REFERENCES

1. G. Altarelli, unpublished talk given at the conference *Around the Dyson Sphere*, Princeton, NJ, April 8-9, 1994.
2. D.J.Gross, J.A.Harvey, J.A.Martinec and R.Rohm, Phys.Rev.Lett. **54** (1985) 502; Nucl.Phys.B **256** (1986) 253.
3. P. Candelas, G.T. Horowitz, A. Strominger and E. Witten, Nucl.Phys.B **258** (1985) 46.
4. L.Dixon, J.A.Harvey, C.Vafa and E.Witten, Nucl.Phys.B **274** (1986) 285.
5. K.S. Narain, Phys.Lett.B **169** (1986) 41; W. Lerche, D. Lűst and A.N. Schellekens, Nucl.Phys.B **287** (1987) 477.
6. I.Antoniadis, C.Bachas, and C.Kounnas, Nucl.Phys.B **289** (1987) 87; H.Kawai, D.C.Lewellen, and S.H.-H.Tye, Nucl.Phys.B **288** (1987) 1.
7. D. Gepner, Phys.Lett.B **199** (1987) 380;Nucl.Phys.B **296** (1988) 57.
8. I. Antoniadis, J. Ellis, J. Hagelin, and D.V.Nanopoulos, Phys.Lett.B **231** (1989) 65; I. Antoniadis, G. K. Leontaris and J. Rizos, Phys.Lett.B **245** (1990) 161; J. Lopez, D.V. Nanopoulos and K. Yuan, Nucl.Phys.B **399** (1993) 654, hep-th/9203025.
9. B. Greene *el al.*, Phys.Lett.**B180** (1986) 69; Nucl.Phys.**B278** (1986) 667; **B292** (1987) 606; R. Arnowitt and P. Nath, Phys.Rev.**D39** (1989) 2006; **D42** (1990) 2498; Phys.Rev.Lett. **62** (1989) 222.
10. L.E. Ibañez *et al.*, Phys.Lett. **B191**(1987) 282; A. Font *et al.*, Phys.Lett. **B210** (1988) 101; A. Font *et al.*, Nucl.Phys. **B331** (1990) 421; D. Bailin, A. Love and S. Thomas, Phys.Lett.**B194** (1987) 385; Nucl.Phys.**B298** (1988) 75; J.A. Casas, E.K. Katehou and C. Muñoz, Nucl.Phys.**B317** (1989) 171.
11. A.E.Faraggi, D.V.Nanopoulos and K.Yuan, Nucl.Phys.B **335** (1990) 347.
12. A.E.Faraggi, Phys.Lett.B **278** (1992) 131.
13. A.E.Faraggi, Phys.Lett.B **274** (1992) 47.
14. A.E.Faraggi, Nucl.Phys.B **387** (1992) 239, hep-th/9208024.
15. S. Weinberg, Phys.Rev.D **26** (1982) 475; S. Sakai and T. Yanagida, Nucl.Phys.B **197** (1982) 533; R. Arnowitt, A.H. Chamsdine and P. Nath, Phys.Rev.D **32** (1985) 2348.
16. S.Kalara, J.Lopez and D.V.Nanopoulos, Nucl.Phys.B **353** (1991) 650.
17. A.E.Faraggi, Phys.Rev.D **47** (1993) 5021.

18. A.E.Faraggi, Nucl.Phys.B **407** (1993) 57, hep-ph/9210256; Phys.Lett.B **326** (1994) 62, hep-ph/9311312.
19. M.Dine, N.Seiberg and E.Witten, Nucl.Phys.**B289** (1987) 589.
20. A.E.Faraggi, Nucl.Phys.B **403** (1993) 101, hep-th/9208023.
21. A.E.Faraggi and E.Halyo, Phys.Lett.B **307** (1993) 305, hep-ph/9301261; Nucl.Phys.B **416** (1994) 63, hep-ph/9306235.
22. B. Grzadkowski, M. Lindner and S. Theisen, Phys.Lett.B **198** (1987) 64.
23. A.E.Faraggi, Phys.Lett.B **302** (1993) 202, hep-ph/9301268.
24. A.E.Faraggi and E.Halyo, IASSNS–HEP–94/17, hep-ph/9405223.
25. For a recent review see, A. Giveon, M. Porrati and E. Rabinovici, RI–1–94, NYU–TH.94/01/01, hep-th/9401139.
26. A.E.Faraggi and E. Halyo, Phys.Lett.B **307** (1993) 311.
27. D. Amati *et. al.*, Phys.Rep. **162** (1988) 169.
28. D. Lüst and T. Taylor, Phys.Lett.B **253** (1991) 335.
29. S. Kalara, J.L. Lopez, D.V.Nanopoulos, Phys.Lett.B **275** (1991) 304, hep-ph/9110023.

FLAVOUR MIXING AND THE GENERATION OF MASS

Harald Fritzsch*+

Theory Division, CERN, Geneva
and
Max-Planck-Institut für Physik und Astrophysik
Werner-Heisenberg-Institut für Physik, Munich

ABSTRACT

It is shown that a simple breaking of the subnuclear democracy leads to a successful description of the mixing between the second and third family. In the lepton channel the $\nu_\mu - \nu_\tau$ oscillations are expected to be described by a mixing angle of 2.65° which might be observed soon in neutrino experiments

* On leave from Sektion Physik, Universität München

+ Supported in part by DFG-contract 412/22-1

In the standard electroweak model, both the masses of the quarks as well as the weak mixing angles enter as free parameters, and any further insight into the yet unknown dynamics of mass generation would imply a step beyond the physics of the electroweak standard model. At present it seems far too early to attempt an actual solution of the dynamics of mass generation, and one is invited to follow a strategy similar to the one which led eventually to the solution of the strong interaction dynamics by QCD, by looking for specific patterns and symmetries as well as specific symmetry violations.

The mass spectra of the quarks are dominated strongly by the masses of the members of the third family, i. e. by t and b. Furthermore, the masses of the first family are small compared to those of the second one. Thus a clear hierarchical pattern exists. Also, the CKM–mixing matrix exhibits a hierarchical pattern – the transitions between the second and third family as well as between the first and the third family are small compared to those between the first and the second family.

About 15 years ago it was emphasized[1] that the observed hierarchies signify that nature seems to be close to the so–called "rank–one" limit in which all mixing angles vanish and both the u– and d–type mass matrices are proportional to the rank-one matrix

$$M_0 = \text{const.} \cdot \begin{pmatrix} 0 & 0 & 0 \\ 0 & 0 & 0 \\ 0 & 0 & 1 \end{pmatrix} . \tag{1}$$

Whether the dynamics of the mass generation allows this limit to be achieved in a consistent way remains an unsolved issue. Encouraged by the observed hierarchical pattern of the masses and the mixing parameters, we shall assume that this is the case. In itself it is a non-trivial constraint and can be derived from imposing a chiral symmetry, as emphasized in ref. (2). This symmetry ensures that an electroweak doublet which is massless remains unmixed and is coupled to the W–boson with full strength. As soon as mass is introduced, at least for one member of the doublet, the symmetry is violated and mixing phenomena are expected to show up. That way a chiral evolution of the CKM matrix can be considered.[2] At the first stage only the t and b quark masses are introduced, due to their non-vanishing coupling to the scalar "Higgs" field. The CKM–matrix is unity in this limit. At the next stage the second generation acquires a mass also. Since the (u, d)–doublet is still massless, only the second and the third generations mix, and the CKM–matrix is given by a real 2×2 rotation matrix in the $(c, s) - (t, b)$ subsystem, describing e. g. the mixing between s and b. Only at the next step, at which the u and d masses are introduced, does the full CKM–matrix appear, described in general by three angles and one phase.

It has been emphasized some time ago[3] that the rank-one mass matrix (see eq. (1)) can be expressed in terms of a "democratic mass matrix":

$$M_0 = c \begin{pmatrix} 1 & 1 & .1 \\ 1 & 1 & 1 \\ 1 & 1 & 1 \end{pmatrix} , \tag{2}$$

which exhibits an $S(3)_L \times S(3)_R$ symmetry. Writing down the mass eigenstates in terms of the eigenstates of the "democratic" symmetry, one finds e.g. for the lepton channel:

$$\begin{aligned} e^0 &= \frac{1}{\sqrt{2}}(l_1 - l_2) \\ \mu^0 &= \frac{1}{\sqrt{6}}(l_1 + l_2 - 2l_3) \\ \tau^0 &= \frac{1}{\sqrt{3}}(l_1 + l_2 + l_3) \end{aligned} \tag{3}$$

(l_i: symmetry eigenstates). Note that e^0 and μ^0 are massless in the limit considered here, and any linear combination of the first two state vectors given in eq. (3) would fulfil the same purpose, i. e. the decomposition is not unique, only the wave function of the coherent state τ^0 is uniquely defined. This ambiguity will disappear as soon as the symmetry is violated, and as a result, the masses for the members of the second family are introduced.

Introducing the symmetry eigenstates does not, on its own, imply introducing new physics concepts. However, a new basis might be very welcome if one is trying to find new patterns and symmetries in the quark and lepton spectrum. I like to quote from Feynman's Nobel talk: "Theories of the known, which are described by different physical ideas, may be equivalent in all their predictions and are hence scientifically indistinguishable. However, they are not psychologically identical when trying to move from that base into the unknown. For different views suggest different kinds of modifications which might be made. I, therefore, think that a good theoretical physicist today might find it useful to have a wide range of physical viewpoints and mathematical expressions of the same theory available to him[4]."

The wave functions given in eq. (3) are reminiscent of the wave functions of the neutral pseudoscalar mesons in QCD in the $SU(3)_L \times SU(3)_R$ limit:

$$\begin{aligned} \pi_0^0 &= \frac{1}{\sqrt{2}}(\bar{u}u - \bar{d}d) \\ \eta_0 &= \frac{1}{\sqrt{6}}(\bar{u}u + \bar{d}d - 2\bar{s}s) \\ \eta_0' &= \frac{1}{\sqrt{3}}(\bar{u}u + \bar{d}d + \bar{s}s). \end{aligned} \tag{4}$$

(Here the lower index denotes that we are considering the chiral limit.) Also the mass spectrum of these mesons is identical to the mass spectrum of the leptons and quarks in the "democratic" limit: two mesons (π_0^0, η_0) are massless and act as Nambu–Goldstone bosons, while the third coherent state η_0' is <u>not</u> massless due to the QCD anomaly.

In the chiral limit, the (mass)^2–matrix of the neutral pseudoscalar mesons is also a "democratic" mass matrix when written in terms of the $(\bar{q}q)$– eigenstates $(\bar{u}u)$, $(\bar{d}d)$ and $(\bar{s}s)$[5]:

$$M^2(ps) = \lambda \begin{pmatrix} 1 & 1 & 1 \\ 1 & 1 & 1 \\ 1 & 1 & 1 \end{pmatrix} \tag{5}$$

where the strength parameter λ is given by $\lambda = M^2(\eta_0')\,/\,3$. The mass matrix (5) describes the result of the QCD–anomaly which causes strong transitions between the quark eigenstates (due to gluonic annihilation effects enhanced by topological effects). Likewise, one may argue that analogous transitions are the reason for the lepton–quark mass hierarchy. Here we shall not speculate about a detailed mechanism of this type, but merely study the effect of symmetry breaking.

In the case of the pseudoscalar mesons, the breaking of the symmetry down to $SU(2)_L \times SU(2)_R$ is provided by a direct mass term $m_s\bar{s}s$ for the s–quark. This implies a modification of the (3,3) matrix element in eq. (5), where λ is replaced by $\lambda + M^2(\bar{s}s)$ where $M^2(\bar{s}s)$ is given by $2M_k^2$, which is proportional to $< \bar{s}s >_0$, the expectation value of $\bar{s}s$ in the QCD vacuum. This direct mass term causes the violation of the symmetry and generates at the same time a mixing between η_0 and η_0', a mass for the η_0, and a mass shift for the η_0'.

It would be interesting to see whether an analogue of the simplest violation of the "democratic" symmetry which describes successfully the mass and mixing pattern of the $\eta - \eta'$–system is also able to describe the observed mixing and mass pattern of the second and third family of leptons and quarks. Let us replace the (3,3) matrix element in eq. (2) by $1 + \varepsilon_i$; (i = l (lepton), u (u–quark), d (d–quark)) respectively. The small real parameter ε describes the departure from democratic symmetry and leads

a) to a generation of mass for the second family and

b) to flavour mixing between the third and the second family. Since ε is directly related (see below) to a fermion mass and the latter is not restricted to be positive, ε can be positive or negative. (Note that a negative Fermi–Dirac mass can always be turned into a positive one by a suitable γ_5–transformation of the spin $\frac{1}{2}$ field.) Since the original mass term is represented by a symmetric matrix, we take ε to be real.

First we study the mass and mixing pattern of the charged leptons. The mass operator (trace Θ_μ^μ of the energy–momentum tensor $\Theta_{\mu\nu}$) can be written as

$$\Theta_\mu^\mu = \Theta_\mu^{0\mu} + c_l \varepsilon_l \bar{l}_3 l_3 \tag{6}$$

where $\Theta_\mu^{0\mu}$ describes the mass term in the symmetry limit. The modification of the spectrum and the induced mixing can be obtained by considering the matrix elements:

$$\begin{aligned} < \mu^0 | c_l \varepsilon_l \bar{l}_3 l_3 | \mu^0 > &= +\frac{2}{3} c_l \varepsilon_l \\ < \tau^0 | c_l \varepsilon_l \bar{l}_3 l_3 | \tau^0 > &= +\frac{1}{3} c_l \varepsilon_l \\ < \mu^0 | c_l \varepsilon_l \bar{l}_3 l_3 | \tau^0 > &= -\frac{\sqrt{2}}{3} c_l \varepsilon_l \ . \end{aligned} \tag{7}$$

One observes that

a) the muon acquires a mass given by $c_l \cdot \varepsilon_l$ i. e. $m(\mu)/m(\tau) \cong \frac{2}{9}\varepsilon_l$;

b) the τ–lepton mass is changed slightly $(m(\tau)/m(\tau^0) \cong 1 + \frac{1}{9}\varepsilon_l)$;

c) the flavour mixing is induced by the fact that the perturbation proportional to $\bar{l}_3 l_3$ leads to a non–vanishing transition matrix element between μ^0 und τ^0.

S This phenomenon is analogous to the chiral symmetry violation of QCD, where the s–quark mass term $m_s \bar{s} s$ leads to a mass for the η–meson, a mass shift for the η'–meson and a mixing between η and η'.

It is instructive to rewrite the mass matrix in the hierarchical basis, where one obtains, using the relations (7):

$$M = c_l \begin{pmatrix} 0 & 0 & 0 \\ 0 & +\frac{2}{3}\varepsilon_l & -\frac{\sqrt{2}}{3}\varepsilon_l \\ 0 & -\frac{\sqrt{2}}{3}\varepsilon_l & 3 + \frac{1}{3}\varepsilon_l \end{pmatrix} . \tag{8}$$

In lowest order of ε one finds the mass eigenvalues $m_\mu = \frac{2}{9}\varepsilon_l \cdot m_\tau$, $m_\tau = m_{\tau^0}$, $\Theta_{\mu\tau} = |\sqrt{2} \cdot \varepsilon_l / 9|$.

The exact mass eigenvalues and the mixing angle are given by:

$$\begin{aligned} m_1/c_l &= \frac{3+\varepsilon_l}{2} - \frac{3}{2}\sqrt{1 - \frac{2}{9}\varepsilon_l + \frac{1}{9}\varepsilon_l^2} \\ m_2/c_l &= \frac{3+\varepsilon_l}{2} + \frac{3}{2}\sqrt{1 - \frac{2}{9}\varepsilon_l + \frac{1}{9}\varepsilon_l^2} \\ \sin\Theta_l &= \frac{1}{\sqrt{2}} \left(1 - \frac{1 - \frac{1}{9}\varepsilon_l}{(1 - \frac{2}{9}\varepsilon_l + \frac{1}{9}\varepsilon_l^2)^{1/2}} \right)^{1/2} . \end{aligned} \tag{9}$$

The ratio m_μ/m_τ, observed to be 0.0595, gives $\varepsilon_l = 0.286$ and a $\mu - \tau$ mixing angle of 2.65°. Whether this mixing angle is directly relevant for neutrino oscillations or not depends on the neutrino sector. For massless neutrinos the mixing angle generated in the charged lepton channel by the introduction of the muon mass cannot be observed, i. e. it can be rotated away. If neutrinos have a mass, the neutrino mass matrix will in general induce further mixing angles. A general discussion will not be attempted here.

However, we should like to consider an interesting scenario which is being discussed in connection with cosmological aspects. Let us suppose that the τ–neutrino mass is of the order of 10 eV in order to be relevant for the "missing matter problem" in cosmology, the muon neutrino is in the milli–eV range, i. e. $m(\nu_\mu) < 10^{-2} eV$, and the electron neutrino mass is neglected. The mass generation for the ν_μ–mass proceeds in an analogous way as discussed above for the muon mass. However, the ε–parameter for the neutrino sector turns out to be tiny ($< 5 \cdot 10^{-2}$), and the mixing angle induced via the ν_μ–mass generation can safely be neglected. Thus the angle relevant for the $\nu_\mu - \nu_\tau$ oscillations remains 2.65°, i. e. $\sin^2 2\Theta = 0.0085$. This value is essentially the lowest limit given by the Charm II experiment[7], i. e. it is not ruled out for any value of $\Delta m^2 = m(\nu_\tau)^2 - m(\nu_\mu)^2$.

However, the E531 experiment[8] gives a limit of about $16 eV^2$ for Δm^2, i. e. $m(\nu_\tau) < 4eV$. This limit seems to rule out a cosmological role with respect to the "missing matter" for the τ–neutrino. However, one might caution this conclusion since our mixing angle of 2.65° is not far from the limit of ($\sin^2 2\Theta = 0.004$), at which, according to the E531 experiment, all values of $m(\nu_\tau)$ are allowed. New experiments, e. g. the CHORUS and NOMAD experiments now or soon under way at CERN, will clarify this issue. If the mixing angle is 2.65° as argued above and the ν_τ–mass above 10 eV, one should observe the $\nu_\mu - \nu_\tau$– oscillations within one year[9].

Replacing ε_l by ε_u, ε_d respectively, we can determine the symmetry breaking parameters for the quark sector. The ratio m_s/m_b is allowed to vary in the range 0.022...0.044 (see ref. (9)). According to eq. (9) one finds ε_u to vary from $\varepsilon_d = 0.11$ to 0.21. The associated $s-b$ mixing angle varies from $\Theta(s,b) = 1.0°$ $(\sin\Theta = 0.018)$ and $\Theta(s,b) = 1.95°$ $(\sin\Theta = 0.034)$. As an illustrative example we use the values $m_b(1GeV) = 5200MeV$, $m_s(1GeV) = 220MeV$. One obtains $\varepsilon_d = 0.20$ and $\sin\Theta(s,b) = 0.032$.

To determine the amount of mixing in the (c,t)–channel, a knowledge of the ratio m_c/m_t is required. As an illustrative example we take $m_c(1GeV) = 1.35GeV$, $m_t(1GeV) = 260GeV$ (i. e. $m_t(m_t) = 160GeV$), which gives $m_c/m_t \cong 0.005$. In this case one finds $\varepsilon_u = 0.023$ and $\Theta(c,t) = 0.21°$ $(\sin\Theta(c,t) = 0.004)$.

The actual weak mixing between the third and the second quark family is combined effect of the two family mixings described above. The symmetry breaking given by the ε–parameter can be interpreted, as done in eq. (7), as a direct mass term for the $l_3(u_3, d_3)$ fermion system. However, a direct fermion mass term need not be positive, since its sign can always be changed by a suitable γ_5–transformation. What counts for our analysis is the relative sign of the m_s–mass term in comparison to the m_c–term, discussed previously. Thus two possibilities must be considered:

a) Both the m_s– and the m_c–term have the same relative sign with respect to each other, i. e. both ε_d and ε_u are positive, and the mixing angle between the second and third family is given by the difference $\Theta(sb) - \Theta(ct)$. This possibility seems to be ruled out by experiment, since it would lead to $V_{cb} < 0.03$.

b) The relative signs of the breaking terms ε_d and ε_u are different, and the mixing angle between the (s, b) and (c, t) systems is given by the sum $\Theta(sb) + \Theta(ct)$. Thus we obtain $V_{cb} \cong \sin(\Theta(sb) + \Theta(ct))$.

According to the range of values for m_s discussed above, one finds $V_{cb} \cong$ 0.022...0.038. For example, for $m_s(1GeV) = 220MeV$, $m_c(1GeV) = 1.35GeV$, $m_t(1GeV) = 260GeV$ one finds $V_{cb} \cong 0.036$.

Before discussing the experimental situation, we add a comment about the mass generation for the first family, which at the same time will also generate the other mixing elements, e.g. V_{us} and V_{ub}, of the CKM matrix. These masses can be generated by a further breakdown of the symmetry, e. g. in the matrix of eq. (5) by a small departure of a second diagonal matrix element from unity. (This would correspond to a direct mass term for that state.) Due to the small values of the masses of the first family in comparison to the λ-scale, given by the mass of the third generation fermion (e.g. $m_e/\lambda = 0.0009$), the strength of this symmetry breaking is much smaller than the primary symmetry breaking, which leads to the masses for the second family. (The situation is analogous to the one in hadronic physics, where the breaking of the chiral symmetry is given primarily by the mass of the s–quark, and the m_u/m_d mass terms can be neglected to a good approximation.) In general it is expected, both from the arguments considered here and more generally from the analysis on chiral symmetry given in ref. (2), that the matrix elements V_{cb} and V_{ts} will be affected only by small corrections of order 10^{-3} or less in absolute magnitude (of order $\frac{m_d}{m_b}, \frac{m_u}{m_t}$ respectively). Thus the primary breaking of the democratic symmetry leads solely to a mixing between the second and the third family, and the secondary breaking, responsible for the Cabibbo angle etc., will not affect the 2×2 submatrix of the CKM–matrix describing the $s - b$ mixing in a significant way.

The experiments give $V_{cb} = 0.032 \ldots 0.054$[10]. We conclude from the analysis given above that our ansatz for the symmetry breaking reproduces the lower part of the experimental range. According to a recent analysis the experimental data are reproduced best for $V_{cb} = 0.038 \pm 0.003$[11], i. e. it seems that V_{cb} is lower than previously thought, consistent with our expectation. Nevertheless we obtain consistency with experiment only if the ration m_s/m_b is relatively large, implying $m_s(1GeV) \geq 180MeV$.

It is remarkable that the simplest ansatz for the breaking of the "democratic symmetry", one which nature follows in the case of the pseudoscalar mesons, is able to reproduce the experimental data on the mixing between the second and third family. We interpret this as a hint that the eigenstates of the symmetry l_i, q_i respectively, and not the mass eigenstates, play a special rôle in the physics of flavour dynamics, a rôle which needs to be investigated further.

REFERENCES

1. H. Fritzsch, Nucl. Phys. B155 (1979), 189;

See also: H. Fritzsch, in: Proc. Europhysics Conf. on Flavor Mixing, Erice, Italy (1984).

2. H. Fritzsch, Phys. Lett. B184 (1987) 391.

3. H. Harari, H. Haut and J. Weyers, Phys. Lett. 78B (1978) 459;
Y. Chikashige, G. Gelmini, R.P. Peccei and M. Roncadelli, Phys. Lett. 94B (1980) 499;
C. Jarlskog, in: Proc. of the Int. Symp. on Production and Decay of Heavy Flavors, Heidelberg, Germany (1986);
P. Kaus and S. Meshkov, Mod. Phys. Lett. A3 (1988) 1251; A4 (1989) 603 (E);
H. Fritzsch and J. Plankl, Phys. Lett. B237 (1990) 451.
G.C. Branco, J. I. Silva–Marcos and M.N. Rebelo, Phys. Lett. B237 (1990) 446.

4. R. P. Feynman, "The Development of the Space Time View of Quantum Electrodynamics", Nobel Lecture, reprinted in: Physics Today, Aug. 1966, 31.

5. H. Fritzsch and P. Minkowski, Nuovo Cimento 30A (1975) 393;
H. Fritzsch and D. Jackson, Phys. Lett. 66B (1977) 365.

6. H. Fritzsch and D. Holtmannspötter, CERN preprint CERN–TH 7236/94 (1994).

7. M. Gruwé et al., Phys. Lett. B309 (1993) 463.

8. N. Ushida et al., Phys. Rev. Lett. 57 (1986) 2897.

9. K. Winter, private communication.

10. J. Gasser and H. Leutwyler, Physics Reports 87 (1982) 77.

11. S. Stone, Syracuse preprint HERSY 93–11 (1993)

UNIFICATION OF FUNDAMENTAL INTERACTIONS IN SUPERSYMMETRY [1]

Pran Nath

Theoretical Physics Division, CERN
CH-1211 Geneva 23, Switzerland
and
Department of Physics, Northeastern University
Boston, MA 02115, USA[2]

and

R. Arnowitt

Center for Theoretical Physics, Department of Physics
Texas A&M University, College Station, TX 77843, USA

Abstract

A review is given of recent developments on the implications of supergravity grand unification with SU(5)-type proton decay under the condition $M_{H_3}/m_G < 10$ (where M_{H_3} is the Higgs triplet mass) and the naturalness condition that the universal scalar mass m_0 and the gluino mass are < 1 TeV. It is shown that the maximum achievable lifetime limits on proton lifetime at Super Kamiokande and ICARUS will exhaust the full parameter space of the model under the constraint $m_{\tilde{W}_1} > 100$ GeV. Thus the model predicts the observation of either a light chargino with mass $\lesssim 100$ GeV, or the observation of a $\bar{\nu}K^+$ mode at Super Kamiokande and ICARUS within the above naturalness constraints. Analysis of the $b \to s\gamma$ branching ratio within this model is also discussed. It is shown that there is a significant region of the parameter space where the branching ratio predicted by the model lies within the current experimental bounds. It is pointed out that improved measurements of $B(b \to s\gamma)$ will significantly delineate the parameter space of the model and allow for a more stringent determination of their allowed ranges.

---

[1]Invited Talk at the Conference on "Unified Symmetry in the Small and in the Large", at Coral Gables, Florida, Jan 27–30, 1994.

[2]Permanent address.

I. INTRODUCTION

One of the important developments over the past three years has been the demonstration that the high precision LEP data [1], when extrapolated to high energy using renormalization group equations, leads to unification of the $SU(3)_C \times SU(2)_L \times U(1)$ coupling constants α_3, α_2 and α_1 (where $\alpha_1 = \frac{5}{3}\alpha_Y$ and α_Y is the hypercharge coupling constant) within the standard supersymmetric SU(5) theory [2]. Supergravity grand unification provides an attractive mechanism for the unification of the electroweak and the strong interactions as well as a framework where supersymmetry can be broken in a consistent fashion [3],[4]. Thus the spontaneous breaking of supersymmetry in unified supergravity models leads to an enormous reduction [3], [5]–[7] in the number of arbitrary SUSY-breaking parameters that one encounters in globally supersymmetric theories [8].

We begin by reviewing briefly the ideas of supergravity grand unification. Our starting assumption is the existence of an $N = 1$ supergravity unified theory below the Planck scale $M_{Pl} = 2.4 \times 10^{18}$ GeV. This unified theory can be completely specified in terms of three functions. These are: (1) a superpotential $W(z_a)$, which is a holomorphic function of the chiral fields, (2) a Kähler potential $K(z_a, z_a^*)$, which depends both on the chiral fields and on their complex conjugates, and (3) a gauge kinetic energy function $f_{\alpha\beta}$, which also is a function of the fields and their complex conjugates. One obtains a gaugino mass from supergravity couplings, which is of the form

$$\frac{1}{4}\bar{e}^{\frac{G}{2}}(G^{-1})^a_b\, G,^a\ f^*_{\alpha\beta,b}\bar{\lambda}^\alpha\lambda^\beta \tag{1.1}$$

where $G = -\ln[\kappa^6 WW^*] - \kappa^2 K$, and $\kappa = (M_{Pl})^{-1}$. We shall make the standard assumptions that appear in the formulation of superunified models. First we shall assume that supersymmetry is broken by a gauge singlet field in the hidden sector [3],[5]. Next, we shall assume that the Kähler potential is generation-blind at the GUT scale so that FCNC will be suppressed at low energies, and that the superpotential is restricted to renormalizable interactions (although an understanding of the full quark and lepton mass hierarchy may require inclusion of higher dimensioned operators). With the above assumptions, one can generate a low energy effective theory below the GUT scale, where the gauge group is $SU(3)_C\times SU(2)_L\times U(1)_Y$ and the superheavy sector is integrated out. The effective potential of this theory is given by $V_{eff} = V_{SS}+V_{SB}$ where V_{SS} is the supersymmetry-invariant part and V_{SB} is the symmetry-breaking part. In supergravity a very simple form emerges for V_{SB}:

$$V_{SB} = \sum_i m_0^2 z_i z_i^+ + \left(A_0 W^{(3)} + B_0 W^{(2)} + M_{H_3}^{-1} W^{(4)}\right) , \tag{1.2}$$

where $W^{(2)}$ is the quadratic, $W^{(3)}$ is the cubic and $W^{(4)}$ the quartic part of the effective superpotential below the GUT scale. After SUSY breaking, Eq. (1.1) gives a gaugino mass term of the form $m_{\frac{1}{2}}\bar{\lambda}^\alpha\lambda^\alpha$, so that together with Eq. (1.2) one has a total of four soft SUSY-breaking parameters

$$m_0,\ m_{\frac{1}{2}},\ A_0,\ B_0\ . \tag{1.3}$$

A phenomenologically interesting GUT group in general could be an SU(5), SO(10), E(6), etc. We shall assume that the specific model we are considering is or contains an SU(5). For the spectrum, we shall assume three generations of quarks and leptons in $5+\overline{10}$ representations, and two plets of Higgs (H_1, H_2) in $(5,\bar{5})$ representations. At low energy, the effective superpotential will then have a quadratic term of the form $W^{(2)} = \mu_0 H_1 H_2$, where H_1, H_2 now refer to the Higgs doublets in the respective multiplets.

The outline of the rest of the paper is as follows: in Section 2, we discuss the analysis using radiative electroweak symmetry breaking under a number of constraints which are physically desirable. In Section 3, we discuss the question whether Super Kamiokande and ICARUS can test SUGRA GUTs. In Section 4, we discuss $b \to s\gamma$ decay in SU(5) SUGRA GUTs. Conclusions are given in Section 5.

II. Analysis using Radiative Electroweak Symmetry Breaking

As is well known an attractive feature of supergravity grand unification is that it can generate its own electroweak symmetry breaking via renormalization group effects [9]. Under the assumption of charge and colour conservation, the potential that governs symmetry breaking in supergravity models is given by $V = V_0 + \Delta V_1$, where V_0 is the renormalization group improved semi-classical tree potential

$$\begin{aligned} V_0(Q) &= m_1^2(t)|H_1|^2 + m_2^2(t)|H_2|^2 - m_3^2(t)(H_1H_2 + \text{h.c.}) \\ &+ \frac{1}{2}(g_2^2 + g_Y^2)(|H_1|^2 - |H_2|^2)^2 \end{aligned} \tag{2.1}$$

and ΔV_1 gives the one-loop correction [10]–[12]

$$\Delta V_1 = \frac{1}{64\pi^2}\sum_i(-1)^{2j_i+1}n_i\left[M_i^4\left(\log\frac{M_i^2}{Q^2} - \frac{3}{2}\right)\right] . \tag{2.2}$$

The analysis of electroweak symmetry breaking is carried out by what has now become a standard procedure. One evolves the renormalization group equations on gauge and Yukawa couplings and on soft SUSY-breaking terms. In addition to the charge and colour conservation, one imposes CDF and LEP lower bounds and a naturalness upper limit of 1 TeV on m_0 and m_{gluino}.

Additionally, in supergravity GUTs one imposes the proton lifetime lower limits from Kamiokande/IMB experiments. The radiative breaking analysis [13] determines the parameter μ_0 by fixing M_Z, and the parameter B_0 can be traded in favour of $\tan\beta$. Thus the theory may be described by the four parameters

$$m_0\ , \quad m_{\frac{1}{2}}\ , \quad A_0\ , \quad \tan\beta\ . \tag{2.3}$$

There are 32 supersymmetric particles in the theory, whose masses can be computed in terms of the four parameters of Eq. (2.3). Thus there are 28 predictions relating the SUSY masses in supergravity unification [13]–[24].

Some of the main results of the analysis described above are now discussed:

(1) Scaling laws [13]–[15]: For a large part of the parameter space one finds $\mu \gg M_Z$, and scaling laws emerge. Specifically for the neutralino and chargino masses one finds the relation

$$2m_{\tilde{Z}_1} \simeq m_{\tilde{W}_1} \simeq m_{\tilde{Z}_2} \tag{2.4a}$$

$$m_{\tilde{Z}_3} \simeq m_{\tilde{Z}_4} \simeq m_{\tilde{W}_2} \simeq |\mu| \tag{2.4b}$$

$$m_{\tilde{W}_1} \simeq \frac{1}{4} m_{\tilde{g}} (\mu > 0), \quad m_{\tilde{W}_1} \simeq \frac{1}{3} m_{\tilde{g}} (\mu < 0) \ . \tag{2.4c}$$

Similarly the four Higgs bosons, except the lightest Higgs (h^0), are seen to have essentially degenerate masses:

$$m_{H^+} \simeq m_{H^0} \simeq m_A \ . \tag{2.5}$$

(2) Limits on the light Higgs and the top [13],[14]: The analysis gives upper limits that are

$$m_{h^0} \lesssim 110\text{--}120 \text{ GeV} \tag{2.6a}$$

$$m_t \lesssim 180\text{--}190 \text{ GeV} \ . \tag{2.6b}$$

(3) Other spectra: Radiative electroweak symmetry breaking and proton stability coonstrain other SUSY mass spectra as well. The proton stability constraint can be easily understood in a qualitative fashion in the limit when m_0 is large. This limit gives, for the dressing loop function B [see Eq. (3.2)], the result

$$B \simeq -2 \left(\frac{\alpha_2}{\alpha_3} \sin 2\beta \right) \left(m_{\tilde{g}} / m_{\tilde{q}}^2 \right) \tag{2.7}$$

and the approximate relation $m_{\tilde{q}}^2 = m_0^2 + 0.65 m_{\tilde{g}}^2$. Thus a small gluino mass, a large squark mass and a small $\tan\beta$ are favoured by proton stability. Typically this implies that the first two generations fo squarks will be essentially degenerate in mass and heavier than the gluino. Similarly, masses of the three generations of sleptons will be essentially degenerate and large, but lighter than the first two generations of squarks. Masses of the third generation of squarks is more complicated, because a heavy top mass can lead to a very small mass for the lighter of the two stop masses. In fact the condition that the stop masses not turn tachyonic acts as a constraint on the parameter space of the model. Proton stability constraints also give a lower limit on the mass of the A-Higgs boson, which is significantly larger than what is obtained from electroweak symmetry breaking alone. In this context we recall that there exists a hole in the parameter space of the MSSM which extends roughly from 100 to 200 GeV (with $\tan\beta$ in the range $5 \lesssim \tan\beta \lesssim 20$) which cannot be explored by LEP2 and LHC experiments [25]. The constraints of radiative electroweak symmetry breaking with proton stability already give a lower bound on the A-Higgs mass that closes this gap.

III. Can the SUGRA GUT be Tested at Super Kamiokande and ICARUS?

There already exist stringent limits on the proton lifetime, and these limits are expected to improve significantly in the new generation of proton stability experiments, i.e. Super Kamiokande and ICARUS. One may ask if the expected increase in the sensitivity of these proton decay experiments will be able to test in a conclusive fashion at least a class of SUGRA GUT models. To quantify the discussion, we shall assume that the GUT group G we are dealing with is or contains an SU(5). We assume further that at the GUT scale, the GUT group G breaks into the Standard Model gauge group $SU(3)_C \times SU(2)_L \times U(1)_Y$. We also assume the existence of just two doublets of Higgs, which are embedded in $5+\bar{5}$ of SU(5). Finally, we assume that there are no discrete symmetries in the model which forbid proton decay. We shall focus here on the dominant decay mode of $p \rightarrow \bar{\nu} K^+$. One can easily show that there is a model-independent amplitude for the above mode which goes via the exchange of the colour triplet Higgsino field $\tilde{H}_3$, and one may write the decay width as

$$\Gamma(p \rightarrow \bar{\nu} K^+) = \text{Const} \left(\frac{\beta_p}{m_{H_3}} \right)^2 |B|^2 . \tag{3.1}$$

In Eq. (3.1), the Const is a product of phase-space and chiral Lagrangian factors [26], β_p is the three-quark matrix element of the proton, whose value is known from lattice gauge theory, M_{H_3} is the Higgsino triplet mass and B is the dressing loop function defined in Refs. [13]–[14]. Using the current experimental lower limit on this mode, which is [27] $\tau(p \rightarrow \bar{\nu} K^+) > 1.0 \times 10^{32}$ yr, one finds on using Eq. (3.1) an upper limit of

$$B \lesssim 100(M_{H_3}/M_G)\ \text{GeV}^{-1} , \tag{3.2}$$

where we expect $M_{H_3}/M_G \lesssim 10$. In the future, Super Kamiokande and ICARUS are expected to reach lower limits of

$$\text{Super Kamiokande [28]}\ \tau(p \rightarrow \bar{\nu} K^+) > 2 \times 10^{33}\text{yr} \tag{3.3a}$$

$$\text{ICARUS [29]}\ \tau(p \rightarrow \bar{\nu} K^+) > 5 \times 10^{33}\text{yr} \tag{3.3b}$$

In view of the increased sensitivity expected in these experiments, one may ask if Eq. (3.3) will exhaust the full parameter space of the SU(5)-type SUGRA GUT. A detailed analysis of this question shows [30] that Eq. (3.3) would not itself be able to exhaust the full parameter space of the SU(5)-type GUT. However, it was found that Super Kamiokande and ICARUS experiments, along with maximum achievable superparticle mass limits at LEP2 and at the Tevatron can exhaust the full parameter space [30]. Specifically one finds the following results [30]:

(a) If $\tau(p \rightarrow \bar{\nu} K^+) > 1.5 \times 10^{33}$yr, then either $m_{h^0} \lesssim 95$ GeV or $m_{\tilde{W}_1} < 100$ GeV. Thus either h^0 or $\tilde{W}_1$ (and possibly both) will be observable at LEP2, provided LEP2 can reach its optimum energy and luminosity.

(b) Either the $\bar{\nu} K^+$ mode should be seen at the Super Kamiokande and ICARUS experiments, or the $\tilde{W}_1$ should be seen at LEP2.

The result of case (b) above is exhibited in Fig. 1, where the maximum value of $\tau(p \rightarrow \bar{\nu} K)$ is given when $m_{\tilde{W}_1} > 100$ GeV, for the case $\mu > 0$ and $m_t = 150$ GeV. The maximum lifetime at a given m_0 is obtained by exhausting the allowed domain in the rest of the parameter space, i.e. $m_{\tilde{g}}, A_t, \tan\beta$. Figure 1 shows that with $M_{H_3}/M_G < 10$, ICARUS will exhaust the entire parameter space of the SU(5)-type SUGRA GUT under the restriction that $m_{\tilde{W}_1} > 100$ GeV. Thus the conclusion of (b) above follows. The above analysis shows that the SU(5)-type SUGRA GUT can be tested by using a combination of accelerator and non-accelerator experiments.

IV. $b \to s\gamma$ DECAY IN SUGRA GUT

Recently CLEO [31] has obtained a new upper bound on the inclusive process $b \to s\gamma$ with a value $BR(b \to s\gamma) < 5.4 \times 10^{-4}$ [at 95% CL]. They also observe a non-vanishing result for the exclusive mode $B \to K^*\gamma$ with a branching ratio of 5×10^{-5}. Assuming that the $K^*\gamma$ contributes $\approx 15\%$ to the inclusive process, as is indicated by lattice gauge calculations [32], one obtains also a lower limit $BR(b \to s\gamma) > 1.5 \times 10^{-5}$. The branching ratio measurements on $b \to s\gamma$ are expected to improve in the future, both with analysis of additional data at CLEO and from data that would emerge from B-factories, where one expects luminosities that would be an order of magnitude larger than at current machines. The Standard Model result would then be put to a severe test, and any deviation from it would signal the onset of new physics beyond the SM.

In the SM, $b \to s\gamma$ proceeds via a penguin, which involves exchange of a W boson and gives $BR(b \to s\gamma) \simeq 3.5 \times 10^{-4}$. This result is ambiguous up to few per cent since the BR is a sensitive function of the inputs such as quark masses and α_3, which currently have some inherent experimental errors. Additionally, the current analyses of the BR are done only to leading order QCD corrections, and there may be important beyond the leading-order corrections which have not yet been fully computed [33]. There are additional penguins in supergravity which contribute to this branching ratio. These involve the exchange of charged Higgses, charginos, gluinos and neutralinos [34]–[36]. We summarize here results of a recent analysis of these contributions within the framework of electroweak symmetry [37],[38] and SUGRA GUT constraints including the constraint of proton stability [37].

In the SUGRA GUT analysis here we compute the contributions from W, charged Higgses and charginos and neglect small contributions from the neutralino and gluino exchanges. To leading order QCD corrections, one has [39],[34],[35]

$$\frac{BR(b \to s\gamma)}{BR(b \to ce\nu)} = \frac{6\alpha}{\pi} \frac{\left[\eta^{\frac{16}{32}} A_\gamma + \frac{8}{3}\left(\eta^{\frac{14}{23}} - \eta^{\frac{16}{32}} A_g\right) + C\right]^2}{P\left(\frac{m_c}{m_b}\right)\left[1 - \frac{2}{3\pi}\alpha_s(m_b) f\left(\frac{m_c}{m_b}\right)\right]} \tag{4.1}$$

where $\eta = \alpha_s(m_Z)/\alpha_s(m_b), f(m_c/m_b) = 0.241$, C is an operator mixing coefficient and P is a phase space factor. For C we use the recent analysis of Ref. [40].

For the case of the SUGRA SU(5) GUT one finds [37] that there is a significant region of the parameter space where the branching ratio $BR(b \to s\gamma)$ lies within the current experimental bound.

Results for the case $\mu > 0$ and $m_t = 150$ GeV are exhibited in Fig. 2. The vertical line at $B = 1000$ GeV^{-1} gives the current value of the Kamiokande experimental bound with $m_{H_3}/M_G = 10$. The region to the right of this vertical line is thus forbidden by p-stability while the region to the left is allowed. We see that the branching ratio can be either larger and smaller than the SM value in the allowed domain, with significant deviations from the SM value occurring in both directions. The analysis of the $\mu < 0$ case, although quantitatively different, is similar to the $\mu > 0$ case. Again one finds here a significant region of the parameter space where the model gives a $b \to s\gamma$ branching ratio within the current experimental bounds.

The $b \to s\gamma$ branching ratio is also a sensitive function of the hidden sector parameter A_t, and of $\tan\beta$. In SUGRA SU(5) GUT $\tan\beta$ is limited by p-stability in such a way that $\tan\beta < 7 - 8$. If the constraint of p-stability is eliminated, $\tan\beta$ becomes unrestricted and much larger variations in the $b \to s\gamma$ branching ratio can be generated [38].

b–τ unification: Recent analyses on b–τ masses within SUSY SU(5)-type models point to rather stringent constraints on $\tan\beta$. Specifically it is found that the constraint of the equality of the b–τ Yukawa couplings at the GUT scale implies that the top mass lies close to its fixed

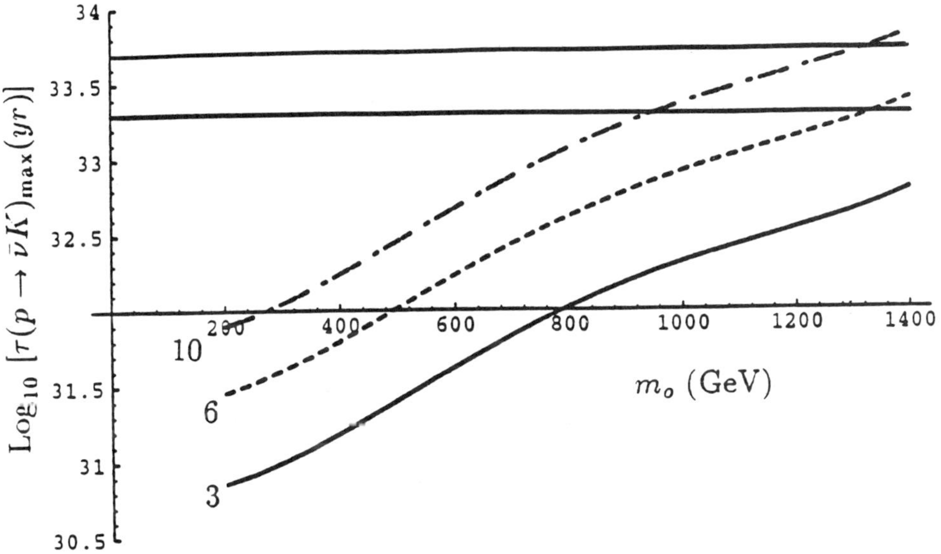

Fig. 1. Maximum value of $\tau(p \to \bar{\nu}K^+)$ when $m_{\tilde{W}_1} > 100$ GeV as a function of m_0, for the case $m_t = 150$ GeV and $\mu > 0$ from Ref. [30]. The solid, dashed and dot-dashed lines correspond to $m_{H_3}/M_G = 3, 6$ and 10. The lowest horizontal line is the current experimental limit. For the two horizontal lines above, the lower and higher lines are the upper bound for Super Kamiokande and ICARUS, i.e. the experiments are sensitive to lifetimes below these lines.

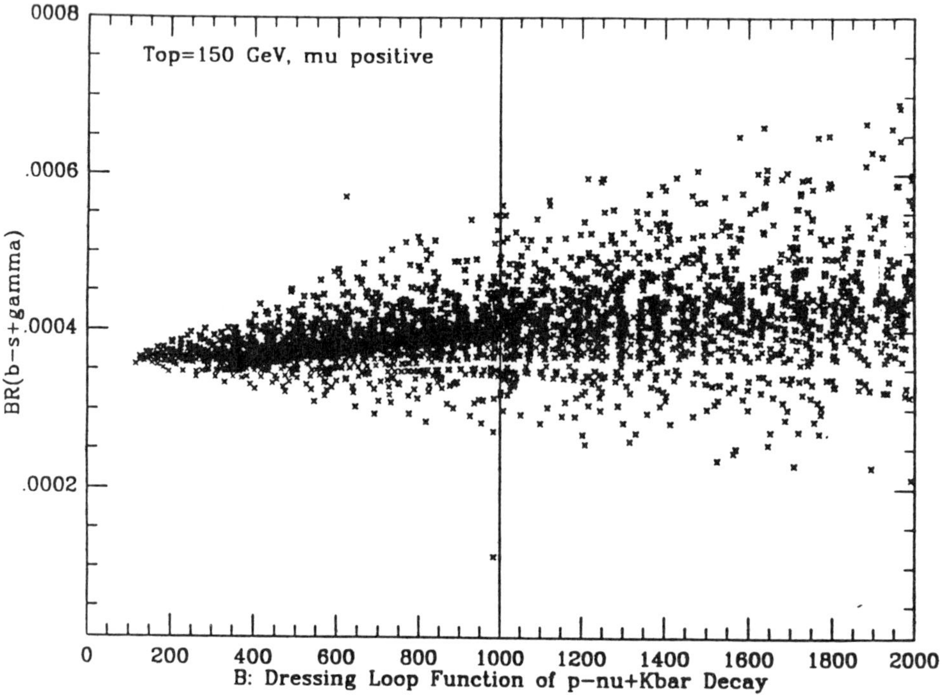

Fig. 2. $BR(b \to s\gamma)$ in the SU(5) supergravity GUT for $\mu > 0$ and $m_t = 150$ GeV.

point value [41]. The analyses reveal two branches in $\tan\beta$ for fixed m_t. For the range of the top mass in the LEP favoured region of 130–170 GeV, $\tan\beta$ is found to be either small (< 2) or very large (> 50). These constraints on $\tan\beta$ are rather severe, and it is reasonable to ask how rigid they are. Before a fixed conclusion can be drawn, it is necessary to carry out a re-evaluation of the inputs used in the analyses. These include the b and τ masses, α_3, as well as specific assumptions made in the renormalization group analyses, such as the treatment of the low energy and GUT thresholds. Aside from the above there is the philosophical issue of whether it is reasonable to impose a strict b–τ unification since s–μ and d–e unification are not nearly as good. This latter feature points to the possibility of higher-dimensional operators or Planck slop terms at the GUT scale. The existence of such operators appears reasonable in view of the fact that we are dealing with an effective theory, even at the GUT scale, and there may be quantum gravity corrections induced at this scale due to new physics at the Planck scale. Inclusion of such Planck slop terms via higher-dimensional operators indicates a loosening of the stringent constraints discussed above [42].

V. CONCLUSIONS

Supergravity SU(5) GUT is a leading contender for the unification of electroweak and strong interactions. The model is presently consistent with all known experiment, and makes predictions that are accessible at current accelerator and non-accelerator experiments, and at machines that are expected to go on line in the near future. Specifically in this review, it was shown that although Super Kamiokande and ICARUS by themselves cannot exhaust the full parameter space of SU(5)-type supergravity GUTs, they can do so when combined with the maximum the maximum achievable superparticle mass limits at LEP2 and the Tevatron. An analysis of the $b \to s\gamma$ decay in the SU(5) supergravity GUT was also given. It is found that there is a significant region of the parameter space where the $BR(b \to s\gamma)$ predicted by the theory lies within the current experimental bounds. It is pointed out that improved experimental limits may be able to put more stringent constraints on the parameters of supergravity GUTs.

ACKNOWLEDGEMENTS

This research was supported in part by NSF Grant Nos. PHY-19306906 and PHY-916593.

References

[1] M. Davier, Proc. Lepton-Photon High Energy Phys. Conference, Geneva, 1991, eds. S. Hegarty et al. (World Scientific, Singapore, 1991);
H. Bethke, Proc. XXVI Conference on High Energy Physics, Dallas, 1992, ed. J. Sanford, AIP Proc. No. 272 (1993).

[2] P. Langacker, Proc. PASCOS90-Symposium, eds. P. Nath and S. Reucroft (World Scientific, Singapore, 1990);
J. Ellis, S. Kelley and D.V. Nanopoulos, *Phys. Lett.* **249B** (1990) 441; **B260** (1991) 131;
U. Amaldi, W. de Boer and H. Fürstenau, *Phys. Lett.* **260B** (1991) 447;
F. Anselmo, L. Cifarelli, A. Peterman and A. Zichichi, *Nuovo Cim.* **104A** (1991) 1817; **115A** (1992) 581;
M. Carena, S. Pokorski and C.E.M. Wagner, Munich preprint MPI-Ph/93-10.

[3] A.H. Chamseddine, R. Arnowitt and P. Nath, *Phys. Rev. Lett.* **29** (1982) 970.

[4] P. Nath, R. Arnowitt and A.H. Chamseddine, "Applied $N = 1$ Supergravity" (World Scientific, Singapore, 1984).

[5] R. Barbieri, S. Ferrara and C.A. Savoy, *Phys. Lett.* **B119** (1983) 343.

[6] L. Hall, J. Lykken and S. Weinberg, *Phys. Rev.* **D22** (1983) 2359.

[7] P. Nath, R. Arnowitt and A.H. Chamseddine, *Nucl. Phys.* **B227** (1983) 121;
S. Soni and A. Weldon, *Phys. Lett.* **B216** (1983) 215.

[8] S. Dimopoulos and H. Georgi, *Nucl. Phys.* **B1983** (1981) 150;
N. Sakai, *Z. Phys.* **C11** (1981) 153.

[9] K. Inoue et al., *Prog. Theor. Phys.* **68** (1982) 927;
L. Ibañez and G.G. Ross, *Phys. Lett.* **B110** (1982) 227;
L. Alvarez-Gaumé, J. Polchinski and M.B. Wise, *Nucl. Phys.* **B250** (1983) 495;
J. Ellis, J. Hagelin, D.V. Nanopoulos and K. Tamvakis, *Phys. Lett.* **125B** (1983) 275;
L.E. Ibañez, C. Lopez and C. Muñoz, *Nucl. Phys.* **B256** (1985) 218.

[10] S. Coleman and E. Weinberg, *Phys. Rev.* **D7** (1973) 1888;
S. Weinberg, *Phys. Rev.* **D7** (1973) 2887.

[11] G. Gamberini, G. Ridolfi and F. Zwirner, *Nucl. Phys.* **B331** (1990) 331.

[12] R. Arnowitt and P. Nath, *Phys. Rev.* **D46** (1992) 725.

[13] R. Arnowitt and P. Nath, *Phys. Rev. Lett.* **69** (1992) 725.

[14] P. Nath and R. Arnowitt, *Phys. Lett.* **B289** (1992) 368.

[15] S. Kelley, J. Lopez, D.V. Nanopoulos, H. Pois and K. Yuan, *Phys. Lett.* **B272** (1991) 423.

[16] G.G. Ross and R.G. Roberts, *Nucl. Phys.* **B377** (1992) 971.

[17] M. Drees and M.M. Nojiri, *Nucl. Phys.* **B369** (1992) 54.

[18] K. Inoue et al., *Phys. Rev.* **D45** (1992) 387.

[19] S.P. Martin and P. Ramond, preprint NUB-3067-93 TH/SSCL-439 (1993).

[20] M. Olechowski and S. Pokorski, *Nucl. Phys.* **B404** (1993) 590.

[21] D.J. Castaño, E.J. Piard and P. Ramond, Univ. Florida preprint UFIFT-HEP-93-18 (1993).

[22] V. Barger, M.S. Berger and P. Ohman, Univ. Wisconsin preprint MAD/PH/801 (1993).

[23] G.L. Kane, C. Kolda, L. Roszkowski and J.D. Wells, Univ. Michigan preprint UM-TH-93-24 (1993).

[24] W. de Boer, R. Ehret and D. Kazakov, Karlsruhe preprint IEKP-KA/93/13 (1993).

[25] Z. Kunszt and F. Zwirner, *Nucl. Phys.* **B385** (1992) 3;
H. Baer, C. Kao and X. Tata, *Phys. Lett.* **B303** (1993) 289.

[26] R. Arnowitt, A.H. Chamseddine and P. Nath, *Phys. Lett.* **156B** (1985) 215;
P. Nath, R. Arnowitt and A.H. Chamseddine, *Phys. Rev.* **32D** (1985) 2348;
J. Hisano, H. Murayama and T. Yanagida, *Nucl. Phys.* **B402** (1993) 46;
J. Ellis, D.V. Nanopoulos and S. Rudaz, *Nucl. Phys.* **B202** (1982) 43 and references quoted therein.

[27] R. Becker-Szendy et al., *Phys. Rev.* **D47** (1993) 4028.

[28] Y. Totsuka, Proc. XXIV Conf. on High Energy Physics, Munich, 1988, eds. R. Kotthaus and J.H. Kühn (Springer Verlag, Berlin Heidelberg, 1989).

[29] ICARUS Detector Group, Int. Symposium on Neutrino Astrophysics, Takayama, 1992.

[30] R. Arnowitt and P. Nath, preprint CTP-TAMU-32/93/NUB-TH-3066/93/SSCL-440 (1993).

[31] E. Thorndike et al., *Bull. Am. Phys. Soc.* **38** (1993) 922;
R. Ammar et al., *Phys. Rev. Lett.* **71** (1993) 674.

[32] C. Bernard, P. Hsieh and A. Soni, Washington Univ. preprint HEP/93/35 (1993).

[33] M. Misiak, *Nucl. Phys.* **B393** (1993) 23.

[34] S. Bertolini, F. Borzumati, A. Masiero and G. Ridolfi, *Nucl. Phys.* **B353** (1991) 591.

[35] R. Barbieri and G. Giudice, *Phys. Lett.* **B309** (1993) 86;
J.L. Hewett, *Phys. Rev. Lett.* **70** (1993) 1045;
V. Barger, M.S. Berger and R.J.N. Phillips, *Phys. Rev. Lett.* **70** (1993) 1369;
M.A. Diaz, *Phys. Lett.* **B304** (1993) 278;
N. Oshima, *Nucl. Phys.* **B404** (1993) 20;
R. Garisto and J.N. Ng, *Phys. Lett.* **B315** (1993) 372.

[36] J.L. Lopez, D.V. Nanopoulos and G.T. Park, *Phys. Rev.* **D48** (1994) R974;
F.M. Borzumati, preprint DESY 93-90 (1993).

[37] P. Nath and R. Arnowitt, in preparation.

[38] J. Wu, R. Arnowitt and P. Nath, in preparation.

[39] B. Grinstein, R. Springer and M. Wise, *Nucl. Phys.* **B339** (1990) 269;
M. Misiak, *Phys. Lett.* **B269** (1991) 161.

[40] P. Cho and M. Misiak, Caltech preprint CALT-68-1893 (1968).

[41] V. Barger, M.S. Berger and P. Ohmann, *Phys. Rev.* **D47** (1993) 1094;
P. Langacker and N. Polonski, Univ. Pennsylvania preprint UPR-0556-T (1993);
W. Bardeen, M. Carena, S. Pokorski and C.E.M. Wagner, Munich preprint MPI-Ph/93-58 (1993).

[42] T. Dasgupta, Private Communication.

DYNAMICAL PROBLEMS OF BARYOGENESIS

John M. Cornwall

Department of Physics
University of California
Los Angeles, CA 90024-1547

ABSTRACT

Above a certain temperature T_c any non-Abelian gauge theory has a strongly-coupled sector, consisting of the three-dimensional magnetic gauge fields. We argue that in this sector the electroweak part of the standard model develops an entropy-dominated condensates of Z_2 strings, and that associated with those strings is a strongly-fluctuating Chern-Simons condensate driven by the strings' linking, twisting, and writhing. (For $SU(N)$ gauge groups with $N \geq 3$ there are links with trivalent vertices; these vertices also give rise to Chern-Simons fluctuations.) Similar phenomena happen for QCD; we discuss the effects of these, as well as QCD sphalerons, on baryogenesis. We outline an ongoing program, including study of the reaction of pre-existing Chern-Simons condensates on the strings and sphalerons, and the problem of finding the non-perturbative electroweak free energy near the phase transition.

INTRODUCTION

Understanding baryogenesis in the early universe is a long way off: one needs to identify the fundamental CP-violation and baryogenesis mechanisms, as well as out-of-equilibrium dynamics (e.g., bubbles near a first-order transition). Here we will discuss only some dynamical issues, really more related to baryoexodus than baryogenesis, having to do with the non-perturbative and strongly-coupled three-dimensional sectors of both QCD and electroweak SU(2) at temperatures T above their phase transition temperatures T_c so that the Higgs expectation value $V(T)$ vanishes. These sectors (which we call magnetic, since only the spatial components of the gauge potentials will be considered) are strongly-coupled because of their severe infrared divergences. Non-perturbative effects generated here infect other sectors ($N \neq 0$ Matsubara modes, Higgs bosons, fermions, electric gauge sector) but do not,

Unified Symmetry: In the Small and in the Large,
Edited by B.N. Kursunoglu *et al.*, Plenum Press, New York, 1995

except as mentioned below, receive strong feedback from them, so it is reasonable to begin by considering only magnetic effects.

The static (and unstable) sphaleron[1] is an example of a non-perturbative effect potentially of great importance in baryogenesis via the coupling of the divergence of the B and L currents to the topological anomaly interpolated by the sphaleron. Naively, one might expect the electroweak (EW) sphaleron to play a significant role only for $T < T_c$, since $V(T) \neq 0$ seems to be required for its existence, and a popular view[2,3] is that for temperatures such that $V(T) = 0$ there is no sphaleron and no barrier to topological charge-changing transitions.

The major theme of our work is that there are sphalerons at $T > T_c$, both QCD sphalerons[4] and EW sphalerons;[5] that these sphalerons may constitute a formidable barrier to topological charge change; and that there are other EW and QCD configurations, notably vortices, which really do promote rapid B+L washout for $T > T_c$ no matter what sphalerons in isolation do. (Of course, QCD vortices or sphalerons do not directly affect B+L, but they do affect baryonic helicities and thereby certain mechanisms[6] for B+L violation.)

So the usual picture of unsuppressed B+L-violating mechanisms at high T is correct in its broadest terms, but the details of the dynamics of these mechanisms are still obscure. We will begin by discussing symmetric-phase (high-T) sphalerons and argue that even though they exist in principle, they may not play an important dynamical role, because they probably lack the configurational entropy which would cause them to condense at a significant density.

We then argue[7] that there must be an entropy-driven condensate in the $d = 3$ gauge sector, on the basis of an infinite sequence of zero-momentum sum rules for the condensate operator G_{ij}^2 which show that the thermal-vacuum free energy is negative,[7,8] as expected when the entropy dominates the internal energy. These sum rules also show that $\langle G_{ij}^2 \rangle > 0$ and that there is a mass gap. The sum rules and consequent effective potential for G_{ij}^2 are qualitatively similar to those found, if only at the one-loop level, for $d = 4$ gauge theories[9]; in $d = 3$, they are exact.

These sum rules are only consistent if a dynamical gauge boson mass is generated; for EW $SU(2)$, $M_W \sim \alpha_W T$ [10,11] (and, of course, for QCD gluons $M_G \sim \alpha_S T$). Such a W-boson mass means that there are symmetric-phase sphalerons,[4,12] but numerical evidence suggests that if only sphalerons were present, there would be no entropy domination and the free energy would be positive, contrary to the sum rules.

What then is the condensate? We argue that it consists of closed strings, with both internal energy and entropy proportional to length, and thickness $\sim M^{-1}$, where M is either M_W or M_G. Entropy domination means that the strings like to be very long, and the condensate is rather like a polymer condensate. Because the gauge theories are in symmetric phases, the strings carry magnetic flux quantized in the center Z_N (for $SU(N)$). There are several variants on this basic closed-string condensate, some of which we will discuss later (but not the obvious possibility of an open string joining a monopole-antimonopole pair).

We remark parenthetically that this string condensate can contribute (whether in a GUT or the standard model) to cosmological Maxwell magnetic fields, serving as a seed for dynamo amplification to observed magnetic field strengths. This possibility will be discussed elsewhere.

The interesting thing about these strings is that two apparently disparate phenomena are related by the same topological concept: Both the space-like area law (i.e., $d = 3$ string tension) and the Chern-Simons density are expressed in terms of the canonical Gauss linking integral, averaged in different ways over the distribution of strings. (It was argued years ago[13] that confinement and the area law in $d = 4$ QCD were expressed via the $d = 4$ version of the Gauss integral as an average linking of closed two-surfaces with a Wilson loop.) Localized Chern-Simons density is produced by the mutual linking of separate strings, as well as by self-linking on one string (twist and writhe). The finite string thickness provides an automatic regularization of potential self-linking singularities.

The interplay between classical knot theory and the Chern-Simons term is familiar through Witten's[14] work, but with a different emphasis from what we explore here. Witten's action is pure Chern-Simons, while the action we use has no Chern-Simons term. The relation between (Abelian) Chern-Simons terms and linking goes back a long way in physics (e.g., Ref. 15); indeed, the mathematical investigation of knots was inspired by Lord Kelvin's work on atoms as vortices.

$SU(N)$ gauge theories offer another source of Chern-Simons fluctuations; for $N > 2$, these are associated with three strings meeting at a point, subject to a fusion rule on their fluxes. Closed strings with trivalent vertices afford vistas in knot theory apparently not fully explored yet, which may realize the full mathematical richness inherent in Witten's[14] work.

After describing these $d = 3$ gauge theory features in the next sections, we close with some remarks on how this sector of the high-T standard model can be joined with fermions, Higgs particles, $N \neq 0$ Matsubara modes, and the like, to produce quantitative dynamics of B+L violation. At present, we can go no further than the standard lore that rates scale like $(\alpha T)^4$, but there is hope for doing better.

In what follows we need not consider the fermions explicitly, since $\dot{B}$ and $\dot{L}$ are given in terms of the rate of change of the Chern-Simons term W.

SYMMETRIC-PHASE SPHALERONS

The conventional sphaleron[1] is a semiclassical object built from EW gauge fields and the Higgs field; if the Higgs mass is not too large, the sphaleron mass M_{SP} is about $3M_W/\alpha_W$. To the extent that M_W is driven solely by a Higgs VEV, the sphaleron mass disappears above the phase transition. However, even a gauge theory in its symmetric phase has dynamical gauge-boson mass generation, whether in $d = 3$ [10,11] or $d = 4$,[11] which gives rise to a **quantum** sphaleron, that is, one whose mass is determined by non-perturbative quantum phenomena such as a gauge-field condensate. This mass scales like αT ($\alpha = \alpha_W, \alpha_S$). Detailed (but approximate) investigations[9,11,12] show that there is a characteristic dimensionless number b_3, where

$$b_3 = \frac{15N}{32\pi} \quad (\text{for } SU(N)) \tag{1}$$

such that the dynamical gauge-boson mass M is

$$M = \xi b_3 g^2 T \tag{2}$$

with $\xi = O(1)$. Various authors have estimated ξ, both on the lattice and analytically, finding ξ in the range 0.4 to 2, roughly.

The next step is to estimate the sphaleron mass M_{SP}. The original calculation[4,16] gave

$$M_{\mathrm{SP}} = 5.3\,\frac{M}{\alpha} \tag{3}$$

which was essentially confirmed by Yaffe,[17] who found the more accurate coefficient 5.41 in (3). It seems reasonable to use these values in the one-loop calculation[18] of the sphaleron rate, which gives

$$\Gamma_{\mathrm{SP}} \sim 10^{-2}(\alpha T)^4 \left(\frac{M_{\mathrm{SP}}}{T}\right)^7 e^{-M_{\mathrm{SP}}/T}\,. \tag{4}$$

With $M_{\mathrm{SP}}/T = 20\xi$ for $SU(2)$ and 30ξ for $SU(3)$, rather small values for ξ would be needed to make Γ_{SP} of order unity times $(\alpha T)^4$ (it maximizes at $M_{SP}/T = 7$). Values of ξ in the range 0.2-0.4 are not excluded, of course, and furthermore there is doubt about the coefficient in (3), for the following reason. The formal calculations leading to (3) are equivalent both to taking the Higgs mass to infinity in the conventional sphaleron, and to assuming that the gauge-boson mass M is constant at all distance scales. In fact, M decreases at large momentum, as Lavelle[19] has shown; for $d = 3$, M depends on momentum as

$$M^2(k) \underset{k\to\infty}{\to} \left(\frac{29}{30}\right)\frac{g^2\langle G_{ij}^2\rangle}{k^2}\left(\frac{N}{N^2-1}\right) \tag{5}$$

vanishing like k^{-2} (modulo logarithms); just as a quark constituent mass does. Since there is really only one mass scale (see (2)) in $d = 3$ gauge theory, it may be more reasonable to suppose that the large-momentum vanishing of M is effectively the same as saying that the coefficient of M/α in (3) is more like that of the original sphaleron, that is, about 3. Then $\xi \simeq 0.5 - 1$ gives results for Γ_{SP} near the maximum possible, at least for $SU(2)$.

Still, it is far from clear that there is enough entropy (essentially $6\ln(M_{\mathrm{SP}}/T)$) to overcome the dimensionless free energy per sphaleron, M_{SP}/T and to contribute significantly to the condensate. So in the next section we will look further afield for the ingredients of the condensate. But before doing so we take the opportunity to mention the special role that QCD sphalerons or strings would have on mechanisms for B+L violation. They are coupled, through the QCD anomaly, to the divergence of the $U_{A(1)}$ current, which measures the difference in helicities of quarks. The fluctuating QCD condensate flips helicities at a rate $\sim (\alpha_S T)^4$, much faster than typical EW rates $\sim (\alpha_W T)^4$. As a result, any B+L-violation mechanism operating on EW time scales that requires quark helicity to be preserved on this scale (e.g., the charge-transport mechanism of Cohen, Kaplan, and Nelson[6]) will be seriously affected.

During the preparation of this material for the conference, we received a preprint from Giudice and Shaposhnikov[20] which goes into much more detail on this subject.

EFFECTIVE ACTION FOR $d = 3$ GAUGE THEORY

In $d = 4$, there are well known[21,22] sum rules for the zero-momentum matrix elements of the condensate operator $G^2_{\mu\nu}$, as well as[22] for even powers of the topological charge operator $G_{\mu\nu}\,\tilde{G}_{\mu\nu}$, both valid only at the one-loop level. Corresponding sum rules (and an effective action) can be derived[7] for $d = 3$ gauge theory (with no matter fields) and they are exact. Define

$$\theta \equiv \frac{1}{4T}\Sigma(G^a_{ij})^2\,; \tag{6}$$

then the effective potential $\Gamma(\theta)$ is given by

$$\Gamma(\theta) \equiv \beta F = \int d^3x \left(\theta - \frac{4}{3}\,\theta^{3/4}\langle\theta\rangle^{1/4}\right) \tag{7}$$

where F is the Helmholtz free energy in the thermal interpretation of $d = 3$ gauge theory and $\beta = 1/T$. Observe that Γ has a minimum at $\theta = \langle\theta\rangle$, and the minimum value is negative, provided that (as we will show) $\langle\theta\rangle > 0$:

$$\Gamma(\langle\theta\rangle) = -\frac{1}{3}\langle\theta\rangle\,. \tag{8}$$

Clearly, the first term in (7) is the usual internal energy, and the second term represents the entropy of a condensate. Aside from the free-energy result (8) (also derived by Shaposhnikov[8]), there are an infinite number of sum rules following from (7):

$$\int d^3x_1 \cdots d^3x_{N-1}\langle\theta(x_1)\cdots\theta(x_{N-1})\theta(0)\rangle_{CONN} = \frac{(N+2)!}{3!}\langle\theta\rangle\,. \tag{9}$$

The $N = 2$ sum rule has a positive LHS, establishing that $\langle\theta\rangle > 0$.

These results follow from the fact that $d = 3$ pure gauge theory has only one mass scale, given by the square of the three-dimensional coupling constant

$$g_3^2 \equiv g^2 T\,. \tag{10}$$

The sum rules are derived by writing the relevant part of the partition function as

$$Z = Tr\; e^{-\beta H} = \int (d\bar{A})\exp\left[-\frac{1}{4g_3^2}\int d^3x \bar{G}^2_{ij}\right] \tag{11}$$

where $\bar{G}_{ij} = g\,G_{ij}$, and then repeatedly differentiating $\ln Z$ by $(g_3)^3\,\partial/\partial g_3$.

While exact, this approach gives no hint as to the numerical value of $\langle\theta\rangle$, except that it is proportional to $(g^2T)^3$. Refs. 7, 9 develop various non-perturbative approximations to determine $\langle\theta\rangle$ which, however, may not be very accurate; Ref. 8 also discusses some quite different estimates for $\langle\theta\rangle$. Later we will remark further on the significance of the negative free energy in the EW phase transition below; for now our only concern is that it tells us of the existence of an entropy-dominated condensate.

THE LINKING, TWISTING, AND WRITHING STRING CONDENSATE

In $d = 4$, it was argued long ago[11,13] that there is for QCD an entropy-driven condensate of closed two-surfaces, self-consistently connected with dynamical gluon mass generation. For this case it is essential that the coupling be strong, or entropy cannot dominate. But in $d = 3$ there is no small dimensionless parameter (g^2 always appears multiplied by T), so entropy can dominate even if $g^2 \ll 1$, and we have seen above that it does indeed dominate.

Given a dynamically-generated gauge-boson mass, the natural condensate candidate is what the closed two-surfaces became in $d = 3$, namely, closed strings. These are described by an effective action [4,11,13] which has a gauge-invariant mass term in addition to the usual G_{ij}^2 term. Well above T_c the Higgs field couples only weakly and has no VEV; upon omitting it both EW $SU(2)$ and QCD have the same features in the symmetric phase. The strings are described by a closed path $\vec{z}(\ell)$, and the holonomy group of Wilson loops is just the center Z_N for $SU(N)$:

$$U = P \exp\left[ig \oint_\Gamma d\vec{x} \cdot \vec{A} \right] \in Z_N \tag{12}$$

where Γ is a contour linking with the string, described by the vector potential $\vec{A}$; the contour dimensions are large compared to M^{-1}. The vector potential for a single string is a solution to the equations of motion for the effective action, including the mass term, and is given by

$$g\vec{A}(\vec{x}) = 2\pi Q \vec{\nabla} \times \oint d\vec{z} \left[\nabla_M(\vec{x} - \vec{z}) - \nabla_0(\vec{x} - \vec{z}) \right] . \tag{13}$$

Here

$$\Delta_M(\vec{R}) = -\frac{1}{4\pi R} e^{-MR} \tag{14}$$

is the $d = 3$ Green's function for mass M; Δ_0 is that for $M = 0$; and Q is an element of the Lie algebra of $SU(N)$ such that $\exp[2\pi i Q] \in Z_N$, as in (12).

We now show that both the string tension and the Chern-Simons condensate are governed by the same topological quantity, namely the Gauss linking integral $L(\Gamma_1, \Gamma_2)$, defined by

$$L(\Gamma_1, \Gamma_2) = \frac{1}{4\pi} \oint_{\Gamma_1} \oint_{\Gamma_2} d\vec{z}_1 \times d\vec{z}_2 \cdot \frac{(\vec{z}_1 - \vec{z}_2)}{|\vec{z}_1 - \vec{z}_2|^3} . \tag{15}$$

L takes on positive or negative integral values, counting the signed linkages of Γ_1 and Γ_2, and is zero for unlinked contours. Let us calculate the Wilson-loop expectation value for a large flat loop Γ_W all of whose length scales $\hat{\ell}$ obey $M\hat{\ell} \gg 1$. In this case, we may drop the Δ_M term in (13), and find

$$\begin{aligned} WL(\Gamma_W) &\equiv \langle Tr\, P \exp i \oint_{\Gamma_W} g\, d\vec{x} \cdot \vec{A} \rangle \\ &= \langle Tr \exp 2\pi i Q L(\Gamma_W, \Gamma) \rangle \end{aligned} \tag{16}$$

where the expectation value is an average over the strings Γ of the condensate. For $SU(2)$, we choose $Q = \pm\frac{1}{2}\tau_3$, and (16) is just $\langle e^{i\pi L} \rangle$, where $L \in Z$ (0 if there is no linkage). The loop is linked with $N \gg 1$ strings, each contributing ± 1 to L with equal probability. Then the probability distribution $P(L; A)$ for L in the Wilson loop, which has minimal area A, is a Poisson distribution:

$$P(L; A) = (4\pi\rho A)^{-1/2} e^{-L^2/2\rho A} , \tag{17}$$

where the number N of linked strings has been replaced by ρA. Here ρ is the density per unit area of strings, a fundamental quantity of the theory which remains to be calculated; it scales like $(\alpha_W T)^2$. It is now elementary to find

$$WL(\Gamma_W) = \langle e^{i\pi L} \rangle = e^{-\pi^2 \rho A/2} \tag{18}$$

establishing an area law with string tension (in the fundamental representation of EW $SU(2)$)

$$K_F = \pi^2 \rho/2 \,. \tag{19}$$

Other approximate calculations, which we will not report here, relate K_F or ρ to $\langle \theta \rangle$ and M^2.

Now consider the Chern-Simons term W:

$$W = \frac{g^2}{8\pi^2} \int d^3x \epsilon^{ijk} \left(A_i^a \partial_j A_k^a + \frac{2g}{3} \epsilon_{abc} A_i^a A_j^b A_k^c \right) . \tag{20}$$

For strings with Abelian holonomy, the cubic term does not contribute. First, calculate W for two strings, by using (13) in (20):

$$\begin{aligned} W(\Gamma_1, \Gamma_2) =& Tr\, Q_1 Q_2 \oint_{\Gamma_1} \oint_{\Gamma_2} d\vec{z}_1 \times d\vec{z}_2 \cdot \frac{\vec{R}}{4\pi R^3} F(MR) \\ \vec{R} =& \vec{z}_1 - \vec{z}_2 \end{aligned} \tag{21}$$

Here we have saved the Δ_M term in (13), which gives rise to $F(MR)$:

$$F(MR) = \frac{1}{2} \int_0^{MR} du\, u^2 e^{-u} \,. \tag{22}$$

For long, well-separated strings MR is always large, and since $F(\infty) = 1$, W_{12} is proportional to the link integral $L(\Gamma_1, \Gamma_2)$. By calculating $Tr\, Q_1 Q_2$ in $SU(N)$, one finds that W_{12} is fractional, and $NW_{12} \equiv O(\text{mod}\, 1)$. This would suggest, via the connection between the gauge-theory functional Schrödinger equation for the vacuum functional and $d = 3$ gauge theory, that the $d = 4$ vacuum energy at $T = 0$ has periodicity $2\pi N$ in the vacuum angle. However, self-linking effects do not contribute to W as simple fractions. We will not discuss the dependence of vacuum energy on the vacuum angle any further here.

A single string can contribute to W through the self-linking effects of twist and writhe; for string Γ the contribution to W is $\frac{1}{2}W(\Gamma, \Gamma)$ (see (21)). The contributions to the classic link integral (15) from $\vec{R} = 0$ are singular and require regularization, a well-known problem in knot theory.[23] The usual solution is to adopt some sort of ribbon framing by displacing Γ infinitesimally perpendicular to itself and calculating the mutual linkage of the two curves. The result is that the linking number L is the sum of two terms, neither uniquely defined. If the ribbon is created by displacement along the Frénet-Serret binormal, then

$$L = T_W + W_R \tag{23}$$

where the twist is the integral

$$T_W = \frac{1}{2\pi} \int \hat{e}_3 \cdot d\hat{e}_2 \tag{24}$$

in terms of the Frénet-Serret binormal unit vector $\hat{e}_3$ and curvature vector $\hat{e}_2$; this is well-defined as long as the curvature does not vanish. The writhe is an integral over a unit two-sphere mapped out by the chord vectors $\vec{z}(\ell_1)-\vec{z}(\ell_2)$ as ℓ_1, ℓ_2 independently trace out the string; its exact definition will not be needed here.[23,24] A framing is not unique, and T_W, W_R can be separately changed by integers, with their sum remaining constant.

In our case, physics rather than mathematics regularizes the self-linking, via the function $F(MR)$ in (21) which vanishes like $(MR)^3$ for small MR. For a string whose length scales (including curvature, etc.) $\hat{\ell}$ are large, with $M\hat{\ell} \gg 1$, we can split (21) into terms where $MR \leq 1$ and $MR > 1$. Simply by expanding $\vec{z}(\ell_2)$ in a Taylor series about ℓ_1, where ℓ_1 is the argument of the other $\vec{z}$, one finds that the $MR \leq 1$ contribution is the Frénet-Serret twist, and the remainder is the writhe. However, for strings whose length scales are such that $M\hat{\ell}$ is not large, there is no reason to expect any sort of quantization of W, at least not locally (in the vicinity of the string). Of course, there may well be good reasons to insist that W is an integer globally, which can easily be done.[25]

There is yet one more source of W, discussed below, but like linking, twisting, and writhing, there are $O(1)$ fluctuations of W everywhere, simply because there is a string condensate. All of these W sources contribute both to the strings' internal energy and to the entropy, and when included in the overall energy-entropy budget (including unlinked, untwisted strings) one will have a way of relating Chern-Simons fluctuations to the condensate and string tension. We expect to find sum rules of the following sort, akin to those[22] for $G_{\mu\nu}\tilde{G}_{\mu\nu}$ in $d = 4$: Write the Chern-Simons term W of (20) as the integral of a density $\mathcal{W}(\vec{x})$. Then one can derive a sum rule like

$$\int d^3x \langle \mathcal{W}(\vec{x})\,\mathcal{W}(0)\rangle = \xi\langle\theta\rangle,\ \ \xi \approx 1\,. \tag{25}$$

A standard technique in polymer physics for dealing with closed-string condensates is to write the string partition function as that of a scalar field-theory with negative squared mass (see, e.g., Ref. 26). We will report elsewhere on calculations of this sort, which lead to estimates of ξ in (25) (and also the density ρ in the string tension; see (19)).

ANOTHER SOURCE OF W AND THE CORRESPONDING KNOT THEORY

It was pointed out long ago[27] that the Klinkhamer-Manton sphaleron can be thought of as a bead on a string which interpolates, in $SU(2)$, between two segments of a Z_2 string on which the EW magnetic field points in opposite directions. Of course, this sphaleron has $W = 1/2$. For $SU(N)$ with $N > 2$, from 3 to N strings can meet in a point, and that point has localized Chern-Simons density. Each string has flux in Z_N, labeled by an integer N_i, $0 \leq |N_i| \leq N$, and the fusion rule for joining is $\Sigma N_i \equiv 0 \pmod N$. The only fundamental vertex is trivalent, and has $N_1 = N_2 = 1$, $N_3 = N - 2$; all other trivalent vertices can be formed by adding simple strings. We have constructed a gauge-field **ansatz** for this fundamental vertex, involving three $SU(2)$ generators and the Z_N generator Q (see (13)). The Chern-Simons number from such configurations is expected to be quantized in units of 1/N.

So for $SU(N)$ with $N \geq 3$ new string links can be formed, where the links have trivalent vertices and can link, twist, and writhe. To the author's knowledge, the knot theory of such links is as yet relatively unexplored, and might be an interesting subject for mathematicians as well as physicists.

FOR THE FUTURE

A number of projects are underway, which we can but list here:

1. EW Free Energy. A number of authors[28] have tried to calculate the EW free energy near $T = T_c$ perturbatively, but problems arise because the IR divergences of the $d = 3$ magnetic sector are not cured in perturbation theory. Moreover, there are condensate effects which actually drive the phase transition more first-order.[8] The gauge-invariant techniques of Ref. 7 are being applied to this problem.

2. Dynamics. B+L violation involves calculating (out of equilibrium) rates, which requires consideration of time dependence and coupling of the $d = 3$ magnetic sector to all other sectors. The mere existence of Chern-Simons fluctuations in the static $d = 3$ magnetic sector in itself proves nothing; it has to be shown that the full dynamics allows for a long-time net change in the Chern-Simons number, or equivalently that non-zero instanton numbers are possible. In the case of the $d = 3$ string condensate, this happens when reconnection allows for changes in the linking, twisting, and writhing of the string condensate. In other words, one is not so much interested in $W(t)$, the Chern-Simons number at time t, but its rate of change $\dot{W}(t)$. Perhaps the simplest connection one can make for rates of change is to consider the **canonical** commutator for $W(t)$:

$$\langle [\dot{W}(0), W(0)] \rangle = \frac{iT}{2\pi^2} \alpha_W^2 \int d^3x \langle \theta \rangle \, . \tag{26}$$

By comparison with (25), one deduces a rate scale $\alpha_W^2 T$, but this is wrong; the actual rate scale is $\alpha_W T$. It is true that $\alpha_W^2 T$ is a natural scale of energy differences for the sphaleron; its non-relativistic kinetic energy $p^2/2M_{SP}$ is of this order, with $p \sim \alpha_W T$, $M_{SP} \sim T$. But to get the right answer $\alpha_W T$ requires coupling of the $d = 3$ magnetic sector to other sectors, as we will discuss elsewhere.

3. Back Reaction. In the presence of a fermionic source for B+L violation other than those considered here, the partition function may require a chemical potential for W, that is, an explicit Chern-Simons term in the action. Sphalerons and strings will be thereby modified. A preliminary investigation was done in Ref. 12, and the author and B. Yan are now extending this earlier work.

4. Generalizations. It is interesting to study the Chern-Simons terms and generalized knot theory of Section V associated with the most general symmetry-breaking pattern of a gauge theory with Higgs particles, rather than just to stick to the symmetric phase, as we have here, with its Z_N strings.

Added Note: I learned at this Conference of the interesting work of Vachaspati, Barriola, and Perivolaropoulos[29] which is also concerned with the connection between Chern-Simons number and linking, etc., of electroweak strings. However, their emphasis is on a different category of strings, generated by Higgs effects below the phase transition.

ACKNOWLEDGEMENTS

I thank Drs. Peter Van Driel and Daniel Cangemi for interesting discussions. This work was support in part by the National Science Foundation under Grant PHY-9218990.

REFERENCES

1. The gauge and Higgs field configuration making up a sphaleron was first constructed by Dashen, Hasslacher, and Neveu (Phys. Rev. D**10**, 4138 (1974)), but not interpreted topologically. This was rediscovered and shown to interpolate topological charge by Manton and Klinkhamer (N. Manton, Phys. Rev. D**28**, 2019 (1983); F. Klinkhamer and N. Manton, *ibid.* **30**, 2212 (1984)).
2. V. A. Kuzmin, V. A. Rubakov, and M. E. Shaposhnikov, Phys. Lett. B**155**, 36 (1985).
3. P. Arnold and L. McLerran, Phys. Rev. D**36**, 581 (1987); *ibid.* D**37**, 1020 (1988).
4. J. M. Cornwall, in *Deeper Pathways in High-Energy Physics*, ed. A. Perlmutter and L. F. Scott (Plenum Press, New York, 1977), p. 683. The QCD sphaleron was presented as a glueball, with no interpretation of its connection to the axial anomaly. Eq. (19) of this reference is a misprint.
5. J. M. Cornwall, Phys. Rev. D**40**, 4130 (1989).
6. A. Cohen, D. Kaplan, and A. Nelson, Nucl. Phys. B**373**, 453 (1992); Preprint UCSD-PTH-93-02 (1993).
7. J. M. Cornwall, Nucl. Phys. B —— (1994).
8. M. B. Shaposhnikov, Phys. Lett. B**316**, 112 (1993).
9. For a discussion of how $d = 4$ and $d = 3$ gauge theories resemble each other dynamically, see J. M. Cornwall, Physica A**158**, 97 (1989).
10. A. Linde, Phys. Lett. B**96**, 289 (1980).
11. J. M. Cornwall, Phys. Rev. D**26**, 1453 (1982);
    J. M. Cornwall, W.-S. Hou, and J. E. King, Phys. Lett. B**153**, 173 (1985).
12. J. M. Cornwall, Phys. Rev. D**40**, 4130 (1989).
13. J. M. Cornwall, Nucl. Phys. B**157**, 392 (1979).
14. E. Witten, Comm. Math. Phys. B**121**, 351 (1989).
15. H. K. Moffatt, J. Fluid Mech. **35**, 117 (1989).
16. The calculation of Ref. 4 was felt by the author not to be highly accurate, so he requested someone to recalculate it, and the new answer was claimed to have a coefficient 3.1 in place of 5.3. This (erroneous) value was quoted in Ref. 12 and elsewhere, unfortunately.
17. L. Yaffe, Phys. Rev. D**40**, 3463 (1989).
18. First paper of Ref. 3. See also O. Philipsen, Phys. Lett. B**304**, 134 (1993), which has further references.
19. M. Lavelle, Phys. Rev. D**44**, R26 (1991).
20. G. F. Guidice and M. E. Shaposhnikov, CERN Preprint TH 7080/93 (November 1993, unpublished).
21. V. A. Novikov, M. A. Shifman, A. I. Vainshtein, and V. I. Zakharov, Nucl. Phys. B**191**, 301 (1981).

22. J. M. Cornwall and A. Soni, Phys. Rev. D**29**, 1424 (1984).
23. J. White, Am. J. Math. **91**, 693 (1969).
24. L. H. Kauffmann, *Knots and Physics* (World Scientific, Singapore, 1991), p. 488.
25. Compare to the situation with fractional instantons (J. M. Cornwall and G. Tiktopoulos, Phys. Lett. B**181**, 353 (1986)). Two instantons, each carrying fractional charge but with integral total charge, must be joined by a sphaleron-like world line which is not detectable from the sphere at infinity, but an isolated fractional instanton would have the sphaleron branch line running all the way to infinity.
26. S. F. Edwards, J. Phys. A**1**, 15 (1968).
27. J. M. Cornwall and G. Tiktopoulos, Ref. 25;
M. Hindmarsh and T. W. B. Kibble, Phys. Rev. Lett. **55**, 2398 (1985).
28. J. R. Espinosa, M. Quiros, and F. Zwirner, Phys. Lett. B**314**, 206 (1993); W. Buchmüller, Z. Fodor, T. Helbig, and D. Walliser, DESY Preprint 93-021 (1993).
29. See these Proceedings, and T. Vachaspati and G. B. Field, preprint.

INDEX